Problem-Solving and Selected Topics in Number Theory

T0192193

Michael Th. Rassias

Problem-Solving
and Selected Topics
in Number Theory

In the Spirit of the Mathematical Olympiads

Foreword by Preda Mihăilescu

 Springer

Michael Th. Rassias
Department of Pure Mathematics
and Mathematical Statistics
University of Cambridge
Cambridge CB3 0WB, UK
mthrassias@yahoo.com

ISBN 978-1-4899-8194-3   ISBN 978-1-4419-0495-9 (eBook)
DOI 10.1007/978-1-4419-0495-9
Springer New York Dordrecht Heidelberg London

Mathematics Subject Classification (2010): 11-XX, 00A07

Printed on acid-free paper

Springer is part of Springer Science+Business Media (www.springer.com)

To my father Themistocles

Contents

Foreword

The International Mathematics Olympiad (IMO), in the last two decades, has become an international institution with an impact in most countries throughout the world, fostering young mathematical talent and promoting a certain approach to complex, yet basic, mathematics. It lays the ground for an open, unspecialized understanding of the field to those dedicated to this ancient art.

The tradition of mathematical competitions is sometimes traced back to national contests which were organized in some countries of central Europe already at the beginning of the last century. It is very likely that a slight variation of the understanding of mathematical competition would reveal even more remote ancestors of the present IMO. It is, however, a fact that the present tradition was born after World War II in a divided Europe when the first IMO took place in Bucharest in 1959 among the countries of the Eastern Block. As an urban legend would have it, it came about when a high school mathematics teacher from a small Romanian town began to pursue his vision for an organized event that would help improve the teaching of mathematics.

Since the early beginnings, mathematical competitions of the international olympiad type have established their own style of problems, which do not require wide mathematical background and are easy to state. These problems are nevertheless difficult to solve and require imagination plus a high degree of original thinking. The Olympiads have reached full maturity and worldwide status in the last two decades. There are presently over 100 participating countries.

Accordingly, quite a few collections of Olympiad problems have been published by various major publishing houses. These collections include problems from past olympic competitions or from among problems proposed by various participating countries. Through their variety and required detail of solution, the problems offer valuable training for young students and a captivating source of challenges for the mathematically interested adult.

In the so-called Hall of Fame of the IMO, which includes numerous presently famous mathematicians and several Fields medalists, one finds a list of the participants and results of former mathematical olympiads (see [HF]). We find in the list of the participants for Greece, in the year 2003, the name of Michael Th. Rassias. At the age of 15 at that time, he won a silver medal and achieved the highest score on the Greek team. He was the first Greek of such a young age in over a decade, to receive a silver medal. He is the author of the present book: one more book of Olympiad Problems among other similar beautiful books.

Every single collection adds its own accent and focus. The one at hand has a few particular characteristics which make it unique among similar problem books. While most of these books have been written by experienced mathematicians after several decades of practicing their skills as a profession, Michael wrote this present book during his undergraduate years in the Department of Electrical and Computer Engineering of the National Technical University of Athens. It is composed of some number theory fundamentals and also includes some problems that he undertook while training for the olympiads. He focused on problems of number theory, which was the field of mathematics that began to capture his passion. It appears like a confession of a young mathematician to students of his age, revealing to them some of his preferred topics in number theory based on solutions of some particular problems—most of which also appear in this collection. Michael does not limit himself to just those particular problems. He also deals with topics in classical number theory and provides extensive proofs of the results, which read like "all the details a beginner would have liked to find in a book" but are often omitted.

In this spirit, the book treats Legendre symbols and quadratic reciprocity, the Bertrand Postulate, the Riemann ζ-function, the Prime Number Theorem, arithmetic functions, diophantine equations, and more. It offers pleasant reading for young people who are interested in mathematics. They will be guided to an easy comprehension of some of the jewels of number theory. The problems will offer them the possibility to sharpen their skills and to apply the theory.

After an introduction of the principles, including Euclid's proof of the infinity of the set of prime numbers, follows a presentation of the extended Euclidean algorithm in a simple matricial form known as the Blankinship method. Unique factorization in the integers is presented in full detail, giving thus the basics necessary for the proof of the same fact in principal ideal domains. The next chapter deals with rational and irrational numbers and supplies elegant comprehensive proofs of the irrationality of e and π, which are a first taste of Rassias's way of breaking down proofs in explicit, extended steps.

The chapter on arithmetic functions presents, along with the definition of the Möbius μ and Euler ϕ functions, the various sums of divisors

$$\sigma_a(n) = \sum_{d|n} d^a,$$

as well as nice proofs and applications that involve the Möbius inversion formula. We find a historical note on Möbius, which is the first of a sequence of such notes by which the author adds a temporal and historical frame to the mathematical material.

The third chapter is devoted to algebraic aspects, perfect numbers, Mersenne and Fermat numbers, and an introduction to some open questions related to these. The fourth deals with congruences, the Chinese Remainder Theorem, and some results on the rings $\mathbb{Z}/(n \cdot \mathbb{Z})$ in terms of congruences. These results open the door to a large number of problems contained in the second part of the book.

Chapter 5 treats the symbols of Legendre and Jacobi and gives Gauss's first geometric proof of the law of quadratic reciprocity. The algorithm of Solovay and Strassen—which was the seminal work leading to a probabilistic perspective of fundamental notions of number theory, such as primality—is described as an application of the Jacobi symbol. The next chapters are analytic, introducing the ζ and Dirichlet series. They lead to a proof of the Prime Number Theorem, which is completed in the ninth chapter. The tenth and eleventh chapters are, in fact, not only a smooth transition to the problem part of the book, containing already numerous examples of solved problems, they also, at the same time, lead up to some theorems. In the last two subsections of the appendix, Michael discusses special cases of Fermat's Last Theorem and Catalan's conjecture.

I could close this introduction with the presentation of *my favorite problem*, but instead I shall present and briefly discuss another short problem which is included in the present book. It is a *conjecture* that Michael Rassias conceived of at the age of 14 and tested intensively on the computer before realizing its intimate connection with other deep conjectures of analytic number theory. These conjectures are still today considered as intractable.

Rassias Conjecture. For any prime p with $p > 2$ there are two primes p_1, p_2, with $p_1 < p_2$ such that

$$p = \frac{p_1 + p_2 + 1}{p_1}. \tag{1}$$

The conjecture was verified empirically on a computer and was published along with a series of problems from international Olympiads (see [A]). The purpose of this short note is to put this conjecture in its mathematical context and relate it to further known conjectures.

At first glance, the expression (1) is utterly surprising and it could stand for some unknown category of problems concerning representation of primes. Let us, though, develop the fraction in (1):

$$(p - 1)p_1 = p_2 + 1.$$

Since p is an odd prime, we obtain the following slightly more general conjecture: *For all $a \in \mathbb{N}$ there are two primes p, q such that*

$$2ap = q + 1. \tag{2}$$

Of course, if (2) admits a solution for any $a \in \mathbb{N}$, then a fortiori (1) admits a solution. Thus, the Rassias conjecture is true. The new question has the particularity that it only asks to prove the existence of a single solution. We note, however, that this question is related to some famous problems, in which one asks more generally to show that there is an infinity of primes verifying certain conditions.

For instance, the question if there is an infinity of Sophie Germain primes p, i.e., primes such that $2p + 1$ is also a prime, has a similar structure. While in the version (2) of the Rassias conjecture, we have a free parameter a and search for a pair (p, q), in the Sophie Germain problem we may consider p itself as a parameter subject to the constraint that $2p + 1$ is prime, too. The fact that there is an infinity of Sophie Germain primes is an accepted conjecture, and one expects the density of such primes to be $O(x/\ln^2(x))$ [Du]. We obtain from this the modified Rassias conjecture by introducing a constant a as factor of 2 and replacing $+1$ by -1. Thus $q = 2p + 1$ becomes $q = 2ap - 1$, which is (2). Since a is a parameter, in this case we do not know whether there are single solutions for each a. When a is fixed, this may of course be verified on a computer or symbolically.

A further related problem is the one of Cunningham chains. Given two coprime integers m, n, a Cunningham chain is a sequence $p_1, p_2, \ldots, p_k$ of primes such that $p_{i+1} = mp_i + n$ for $i > 1$. There are competitions for finding the longest Cunningham chains, but we find no relevant conjectures related to either length or frequencies of such chains. In relation to (2), one would rather consider the Cunningham chains of fixed length 2 with $m = 2a$ and $n = -1$. So the question (2) reduces to the statement: *there are Cunningham chains of length two with parameters $2a, -1$, for any $a \in \mathbb{N}$.*

By usual heuristic arguments, one should expect that (2) has an infinity of solutions for every fixed a. The solutions are determined by one of p or q via (2). Therefore, we may define

$$S_x = \{p < ax \; : \; p \text{ is prime and verifies (2)}\}$$

and the counting function $\pi_r(x) = |S_x|$. There are $O(\ln(x))$ primes $p < x$, and $2ap - 1$ is an odd integer belonging to the class -1 modulo $2a$. Assuming that the primes are equidistributed in the residue classes modulo $2a$, we obtain the expected estimate:

$$\pi_r(x) \sim x/\ln^2(x) \tag{3}$$

for the density of solutions to the extended conjecture (2) of Rassias.

Probably the most general conjecture on distribution of prime constellations is Schinzel's *Conjecture H*:

Conjecture H. Consider s polynomials $f_i(x) \in \mathbb{Z}[X], i = 1, 2, \ldots, s$ with positive leading coefficients and such that the product $F(X) = \prod_{i=1}^s f_i(x)$ is not divisible, as a polynomial, by any integer different from ± 1. Then there is at least one integer x for which all the polynomials $f_i(x)$ take prime values.

Of course, the Rassias conjecture follows for $s = 2$ with $f_1(x) = x$ and $f_2(x) = 2ax - 1$. Let us finally consider the initial problem. Can one prove that (2) has at least one solution in primes p, q, for arbitrary a? In [SW], Schinzel and Sierpiński show that Conjecture H can be stated for one value of x or for infinitely many values of x, since the two statements are equivalent. Therefore, solving the conjecture of Rassias is as difficult as showing that there are infinitely many prime pairs verifying (2). Of course, this does not exclude the possibility that the conjecture could be proved easier for certain particular families of values of the parameter a.

The book is self-contained and rigorously presented. Various aspects of it should be of interest to graduate and undergraduate students in number theory, high school students and the teachers who train them for the Putnam Mathematics Competition and Mathematical Olympiads as well as, naturally, to scholars who enjoy learning more about number theory.

Bibliography

[A] T Andreescu and D. Andrica, *Number Theory*, Birkhäuser, Boston, (2009), p. 12.

[Du] H Dubner, *Large Sophie-Germain primes*, Math. Comp., 65(1996), pp. 393–396.

[HF] *http://www.imo-official.org/country_hall.aspx?code=HEL*

[R] Michael Th. Rassias, *Open Problem No. 1825*, Octogon Mathematical Magazine, 13(2005), p. 885. See also Problem 25, Newsletter of the European Mathematical Society, 65(2007), p. 47.

[SW] A Schinzel and W. Sierpiński, *Sur certaines hypothèses concernant les nombres premiers*, Acta Arith., 4(1958), pp. 185–208.

Preda Mihăilescu
Mathematics Institute
University of Göttingen
Germany

Acknowledgments

I wish to express my gratitude to Professors A. Papaioannou and V. Papanicolaou for their invaluable assistance and inspirational guidance, both during my studies at the National Technical University of Athens and the preparation of this book.

I feel deeply honored that I had the opportunity to communicate with Professor Preda Mihăilescu, who has been my mentor in Mathematics since my high school years and has written the Foreword of the book.

I would like to thank Professors M. Filaseta, S. Konyagin, V. Papanicolaou and J. Sarantopoulos for their very helpful comments concerning the step-by-step analysis of Newman's proof of the Prime Number Theorem. Professor P. Pardalos has my special appreciation for his valuable advice and encouragement. I would like to offer my sincere thanks to Professors K. Drakakis, J. Kioustelidis, V. Protassov and J. Sandor for reading the manuscript and providing valuable suggestions and comments which have helped to improve the presentation of the book.

This book is essentially based on my undergraduate thesis on computational number theory, which I wrote under the supervision of Professors A. Papaioannou, V. Papanicolaou and C. Papaodysseus at the National Technical University of Athens. I have added a large number of problems with their solutions and some supplementary number theory on special topics.

I would like to express my thanks to my teachers for their generous advice and encouragement during my training for the Mathematical Olympiads and throughout my studies.

Finally, it is my pleasure to acknowledge the superb assistance provided by the staff of Springer for the publication of the book.

Michael Th. Rassias

1

Introduction

God created the natural numbers. The rest is the work of man.

Leopold Kronecker (1823–1891)

Number Theory is one of the most ancient and active branches of pure mathematics. It is mainly concerned with the properties of integers and rational numbers. In recent decades, number theoretic methods are also being used in several areas of applied mathematics, such as cryptography and coding theory.

In this section, we shall present some basic definitions, such as the definition of a prime number, composite number, rational number, etc. In addition, we shall present some basic theorems.

1.1 Basic notions

Definition 1.1.1. *An integer p greater than 1 is called a **prime number**, if and only if it has no positive divisors other than 1 and itself.*

Hence, for example, the integers 2, 3, 13, 17 are *prime numbers*, but 4, 8, 12, 15, 18, 21 are not.

The natural number 1 is not considered to be a prime number.

Definition 1.1.2. *All integers greater than one which are not prime numbers are called **composite numbers**.*

Definition 1.1.3. *Two integers a and b are called **relatively prime** or **coprime** if and only if there does not exist another integer c greater than 1, which can divide both a and b.*

For example, the integers 12 and 17 are *relatively prime*.

M.Th. Rassias, *Problem-Solving and Selected Topics in Number Theory: In the Spirit of the Mathematical Olympiads*, DOI 10.1007/978-1-4419-0495-9_1, © Springer Science+Business Media, LLC 2011

Prime numbers are, in a sense, the building blocks with which one can construct all integers. At the end of this chapter we are going to prove the *Fundamental Theorem of Arithmetic* according to which every natural number greater than one can be represented as the product of powers of prime numbers in a unique way.

This theorem was used by the ancient Greek mathematician Euclid, in order to prove that prime numbers are infinitely many.

We shall now present the proof of the fact that the number of primes is infinite. The following proof is due to Euclid and is considered to be one of the most elementary and elegant proofs of this theorem.

Lemma 1.1.4. *The least nontrivial divisor of every positive integer greater than 1 is a prime number.*

Proof. Let $n \in \mathbb{N}$, with $n > 1$ and d_0 be the least nontrivial divisor of n. Let us also suppose that d_0 is a composite positive integer. Then, since d_0 is composite, it must have a divisor m, with $1 < m < d_0$. But, in that case, m would also divide n and therefore d_0 would not be the least nontrivial divisor of n. That contradicts our hypothesis and hence completes the proof of the lemma. □

Theorem 1.1.5 (Euclid). *The number of primes is infinite.*

Proof. Let us suppose that the number of primes is finite and let p be the greatest prime number. We consider the integer

$$Q = p! + 1.$$

Therefore, if Q is a prime number it must be greater than p. But, this contradicts the property of p being the greatest prime number. On the other hand, if Q is not a prime number, then by the previous lemma it follows that it will certainly have prime divisors. However, if Q is divided by any prime number less than or equal to p, it leaves remainder 1. Thus, every prime divisor of Q is necessarily greater than p, which again contradicts the property of p.

So, the hypothesis that the number of primes is finite, leads to a contradiction. Hence, the number of primes is infinite. □

We shall now proceed to the proof of a theorem which is known as **Bezout's Lemma** or the **extended Euclidean algorithm.**

Theorem 1.1.6. *Let $a, b \in \mathbb{Z}$, where at least one of these integers is different than zero. If d is the greatest positive integer with the property $d \mid a$ and $d \mid b$, then there exist $x, y \in \mathbb{Z}$ such that $d = ax + by$.*

Proof. Let us consider the nonempty set

$$A = \{ax + by \mid a, b, x, y \in \mathbb{Z}, \text{ with } ax + by > 0\}.$$

We shall prove that the integer d is the least element in A.

Let d' be the least element in A. Then, there exist integers q, r, such that

$$a = d'q + r, \ 0 \le r < d.$$

We are going to prove that $d' \mid a$. In other words, we will show that $r = 0$.
Let $r \ne 0$, then

$$r = a - d'q = a - (ax_1 + by_1)q,$$

for some integers x_1, y_1.
Therefore,

$$r = a(1 - x_1 q) + b(-y_1 q).$$

But, by the assumption we know that $r \ne 0$. Hence, it is evident that $r > 0$ and $r = ax_2 + by_2$, with $x_2 = 1 - x_1$, $y_2 = -y_1 q \in \mathbb{Z}$. However, this is impossible due to the assumption that d' is the least element in A. Thus, $r = 0$, which means that $d' \mid a$. Similarly, we can prove that $d' \mid b$.

So, d' is a common divisor of a and b. We shall now prove that d' is the greatest positive integer with that property.

Let m be a common divisor of a and b. Then $m \mid ax + ay$ and thus $m \mid d'$, from which it follows that $m \le d'$. Consequently, we obtain that

$$d' = d = ax + by, \text{ for } x, y \in \mathbb{Z}. \qquad \square$$

Remark 1.1.7. The positive integer d with the property stated in the above theorem is unique. This happens because if there were two positive integers with that property, then it should hold $d_1 \le d_2$ and $d_2 \le d_1$. Thus, $d_1 = d_2$.

As a consequence of the above theorem we obtain the following corollary.

Corollary 1.1.8. *For every integer e with $e \mid a$ and $e \mid b$, it follows that $e \mid d$.*

Definition 1.1.9. *Let $a, b \in \mathbb{Z}$, where at least one of these integers is nonzero. An integer $d > 0$ is called the **greatest common divisor of a and b** (and we write $d=\gcd(a, b)$) if and only if $d \mid a$ and $d \mid b$ and for every other positive integer e for which $e \mid a$ and $e \mid b$ it follows that $e \mid d$.*[1]

Theorem 1.1.10. *Let $d = \gcd(a_1, a_2, \ldots, a_n)$, where $a_1, a_2, \ldots, a_n \in \mathbb{Z}$. Then*

$$\gcd\left(\frac{a_1}{d}, \frac{a_2}{d}, \ldots, \frac{a_n}{d}\right) = 1.$$

Proof. It is evident that $d \mid a_1, d \mid a_2, \ldots, d \mid a_n$. Hence,

$$a_1 = k_1 d, a_2 = k_2 d, \ldots, a_n = k_n d, \qquad (1)$$

---

[1] Similarly one can define the greatest common divisor of n integers, where at least one of them is different than zero.

where $k_i \in \mathbb{Z}$ for $i = 1, 2, \ldots, n$. Let

$$\left(\frac{a_1}{d}, \frac{a_2}{d}, \ldots, \frac{a_n}{d}\right) = d' > 1.$$

Then, similarly we obtain

$$d' \mid \frac{a_1}{d}, d' \mid \frac{a_2}{d}, \ldots, d' \mid \frac{a_n}{d}.$$

Consequently, there exist integers $k'_1, k'_2, \ldots, k'_n$, for which

$$\frac{a_1}{d} = k'_1 d', \frac{a_2}{d} = k'_2 d', \ldots, \frac{a_n}{d} = k'_n d'. \tag{2}$$

Therefore, by (1) and (2) we get

$$a_1 = k'_1 d' d, a_2 = k'_2 d' d, \ldots, a_n = k'_n d' d.$$

Thus,

$$d'd \mid a_1, d'd \mid a_2, \ldots, d'd \mid a_n.$$

Hence, $dd' \mid d$, which is impossible since $d' > 1$. Therefore, $d' = 1$. $\qquad \square$

Theorem 1.1.11. *Let $a, b, c \in \mathbb{Z}$ and $a \mid bc$. If $\gcd(a, b) = 1$, then $a \mid c$.*

Proof. If $\gcd(a, b) = 1$, then

$$1 = ax + by, \quad \text{where } x, y \in \mathbb{Z}.$$

Therefore,

$$c = acx + bcy.$$

But, since $a \mid acx$ and $a \mid bcy$, it yields $a \mid c$. $\qquad \square$

1.2 Basic methods to compute the greatest common divisor

Let $a, b \in \mathbb{Z}$. One way to compute the greatest common divisor of a and b is to find the least element in the set

$$A = \{ax + by \mid a, b, x, y \in \mathbb{Z}, \text{ with } ax + by > 0\}.$$

However, there is a much more effective method to compute $\gcd(a, b)$ and is known as the *Euclidean algorithm*.

1.2.1 The Euclidean algorithm

In case we want to compute the $\gcd(a, b)$, without loss of generality we can suppose that $b \leq a$. Then $\gcd(a, b) = \gcd(b, r)$, with r being the remainder when a is divided by b.

This happens because $a = bq + r$ or $r = a - bq$, for some integer q and therefore $\gcd(a, b) \mid r$. In addition, $\gcd(a, b) \mid b$. Thus, by the definition of the greatest common divisor, we obtain

$$\gcd(a, b) \mid \gcd(b, r). \tag{1}$$

Similarly, since $a = bq + r$, we get $\gcd(b, r) \mid b$ and $\gcd(b, r) \mid a$. Hence,

$$\gcd(b, r) \mid \gcd(a, b). \tag{2}$$

By (1) and (2) it is evident that $\gcd(a, b) = \gcd(b, r)$.

If $b = a$, then $\gcd(a, b) = \gcd(a, 0) = \gcd(b, 0) = a = b$ and the algorithm terminates. However, generally we have

$$\gcd(a, b) = \gcd(b, r_1) = \gcd(r_1, r_2) = \cdots = \gcd(r_{n-1}, r_n) = \gcd(r_n, 0) = r_n,$$

with

$$a = bq_1 + r_1, \text{ because } b \leq a$$
$$b = r_1 q_2 + r_2, \text{ because } 0 \leq r_1 < b$$
$$r_1 = r_2 q_3 + r_3, \text{ because } 0 \leq r_2 < r_1$$
$$\vdots$$
$$r_{n-2} = r_{n-1} q_n + r_n, \text{ because } 0 \leq r_{n-1} < r_n$$
$$r_{n-1} = r_n q_{n+1} + 0, \text{ because } 0 \leq r_n < r_{n-1}.$$

Therefore, r_n is the greatest common divisor of a and b.

1.2.2 Blankinship's method

Blankinship's method is a very practical way to compute the greatest common divisor of two integers a and b. Without loss of generality, let us suppose that $a > b > 0$. Then, the idea of this method is the following. Set

$$A = \begin{pmatrix} a & 1 & 0 \\ b & 0 & 1 \end{pmatrix}.$$

By doing row operations we try to transform A, so that it takes the form

$$\begin{pmatrix} d & x & y \\ 0 & x' & y' \end{pmatrix} \text{ or } \begin{pmatrix} 0 & x' & y' \\ d & x & y \end{pmatrix}$$

In other words, we try to transform A so that it has a zero element in its first column. Then, $\gcd(a,b) = d$, as it appears in the first column of the transformed matrix. More specifically, we have

$$d = ax + by.$$

Hence, by the above argument, it follows that by Blankinship's method, not only can we compute $\gcd(a,b)$, but also the coefficients x, y, which appear in Bezout's Lemma.

EXAMPLE 1.2.1. Compute the greatest common divisor of the integers 414 and 621.

Consider the matrix

$$A = \begin{pmatrix} 621 & 1 & 0 \\ 414 & 0 & 1 \end{pmatrix}.$$

Let R_i be the ith row of A. We substitute R_1 with $R_1 - R_2$ and we get

$$A \sim \begin{pmatrix} 207 & 1 & -1 \\ 414 & 0 & 1 \end{pmatrix}.$$

Furthermore, if we substitute R_2 with $R_2 - 2R_1$, we get

$$A \sim \begin{pmatrix} 207 & 1 & -1 \\ 0 & -2 & 3 \end{pmatrix}.$$

Hence, we succeeded to transform A so that it has a zero element in its first column. Therefore, we obtain that $\gcd(414, 621) = 207$ and more specifically

$$207 = 621 \cdot 1 + 414 \cdot (-1). \qquad \square$$

1.3 The fundamental theorem of arithmetic

Theorem 1.3.1 (Euclid's First Theorem). *Let p be a prime number and $a, b \in \mathbb{Z}$. If $p \mid ab$, then*

$$p \mid a \quad or \quad p \mid b.$$

Proof. Let us suppose that p does not divide a. Then, it is evident that $\gcd(a,p) = 1$ and by Bezout's Lemma we have $1 = ax + py$ and thus $b = abx + pby$, where $x, y \in \mathbb{Z}$. But, $p \mid abx$ and $p \mid pby$. Therefore, $p \mid b$.

Similarly, if p does not divide b, we can prove that $p \mid a$. Hence, $p \mid a$ or $p \mid b$. $\qquad \square$

Theorem 1.3.2 (The Fundamental Theorem of Arithmetic). *Every positive integer greater than 1 can be represented as the product of powers of prime numbers in a unique way.*

Proof.

Step 1. We shall prove that every positive integer $n > 1$ can be represented as the product of prime numbers.

If d is a divisor of n, then $1 < d \leq n$. Of course, if n is a prime number, then $n = d$ and the theorem holds true. On the other hand, if n is a composite integer, then it obviously has a least divisor $d_0 > 1$. But, by the above lemma we know that the least nontrivial divisor of every integer is always a prime number. Hence, d_0 is a prime number and there exists a positive integer n_1 for which it holds

$$n = d_0 n_1.$$

Similarly, the positive integer n_1 has a least nontrivial divisor d_2 which must be prime. Therefore, there exists another positive integer n_2, for which

$$n = d_1 d_2 n_2.$$

If we continue the same process, it is evident that n can be represented as the product of prime numbers. Furthermore, because of the fact that some prime numbers may appear more than once in this product, we can represent n as the product of powers of distinct primes. Namely,

$$n = p_1^{a_1} p_2^{a_2} \cdots p_k^{a_k}, \quad \text{where} \quad k \in \mathbb{N}.$$

The above form of representation of a positive integer n is called the **canonical** or **standard form** of n.

Step 2. We shall now prove that **the canonical form is unique**.

Let us suppose that the positive integer n can be represented as the product of powers of prime numbers in two different ways. Namely,

$$n = p_1^{a_1} p_2^{a_2} \cdots p_k^{a_k} = q_1^{b_1} q_2^{b_2} \cdots q_\lambda^{b_\lambda}, \quad \text{where} \quad k, \lambda \in \mathbb{N}.$$

Then, by Euclid's first theorem, we obtain

$$p_i \mid p_1^{a_1} p_2^{a_2} \cdots p_k^{a_k}$$

which yields

$$p_i \mid q_1^{b_1} q_2^{b_2} \cdots q_\lambda^{b_\lambda}$$

and therefore, for every i, with $1 < i < k$, there exists a j, with $1 < j < \lambda$ for which $p_i \mid q_j$ and thus $p_i = q_j$. Thus, it is clear that $k = \lambda$ and the sets $\{p_1, p_2, \ldots, p_k\}$, $\{q_1, q_2, \ldots, q_\lambda\}$ are identical.

Hence, it suffices to prove that

$$a_i = b_i,$$

for every i, where $i = 1, 2, \ldots, k = \lambda$. Let us suppose that $a_i > b_i$. Then, we have

$$p_1^{a_1} p_2^{a_2} \cdots p_i^{a_i} \cdots p_k^{a_k} = q_1^{b_1} q_2^{b_2} \cdots q_i^{b_i} \cdots q_k^{b_k}$$
$$= p_1^{b_1} p_2^{b_2} \cdots p_i^{b_i} \cdots p_k^{b_k}.$$

Thus,

$$p_1^{a_1} p_2^{a_2} \cdots p_i^{a_i - b_i} \cdots p_k^{a_k} = p_1^{b_1} p_2^{b_2} \cdots p_{i-1}^{b_{i-1}} p_{i+1}^{b_{i+1}} \cdots p_k^{b_k}. \tag{1}$$

But, $a_i - b_i \geq 1$ and therefore by (1) we obtain

$$p_i \mid p_1^{a_1} p_2^{a_2} \cdots p_k^{a_k} \quad \text{and} \quad p_i \nmid p_1^{b_1} p_2^{b_2} \cdots p_k^{b_k},$$

which is a contradiction.

Similarly, we are led to a contradiction in the case $a_i < b_i$. Therefore, it is evident that $a_i = b_i$ must hold true for every $i = 1, 2, \ldots, k$. This completes the proof. $\qquad \square$

Definition 1.3.3. *A positive integer n is said to be **squarefree**, if and only if it cannot be divided by the square of any prime number.*

Lemma 1.3.4. *Every positive integer n can be represented in a unique way as the product $a^2 b$ of two integers a, b, where b is a squarefree integer.*

Proof. Since for $n = 1$ the lemma obviously holds true, we suppose that $n > 1$. By the Fundamental Theorem of Arithmetic we know that every positive integer greater than 1 can be represented as the product of powers of prime numbers in a unique way. Therefore, we have

$$n = p_1^{a_1} p_2^{a_2} \cdots p_k^{a_k}, \quad \text{where} \quad n \in \mathbb{N}.$$

Let us consider the set $A = \{a_1, a_2, \ldots, a_k\}$. If m_i are the even integers in A and h_i are the odd integers in A, then clearly we have

$$n = (p_{i_1}^{m_1} p_{i_2}^{m_2} \cdots p_{i_\lambda}^{m_\lambda})(p_{j_1}^{h_1} p_{j_2}^{h_2} \cdots p_{j_\mu}^{h_\mu})$$
$$= (p_{i_1}^{2e_1} p_{i_2}^{2e_2} \cdots p_{i_\lambda}^{2e_\lambda})(p_{j_1}^{2f_1+1} p_{j_2}^{2f_2+1} \cdots p_{j_\mu}^{2f_\mu+1})$$
$$= (p_{i_1}^{e_1} p_{i_2}^{e_2} \cdots p_{i_\lambda}^{e_\lambda} \cdot p_{j_1}^{f_1} p_{j_2}^{f_2} \cdots p_{j_\mu}^{f_\mu})^2 \cdot (p_{j_1} \cdots p_{j_\mu})$$
$$= a^2 \cdot b.$$

The integers p_i, q_i are unique and thus the integers m_i, h_j are unique. Hence, the integers a and b are unique. $\qquad \square$

1.4 Rational and irrational numbers

Definition 1.4.1. *Any number that can be expressed as the quotient p/q of two integers p and q, where $q \neq 0$, is called a rational number.*

The set of rational numbers (usually denoted by $\mathbb{Q}$) is a countable set. An interesting property of this set is that between any two members of it, say a and b, it is always possible to find another rational number, e.g., $(a + b)/2$.

In addition, another interesting property is that the decimal expansion of any rational number either has finitely many digits or can be formed by a certain sequence of digits which is repeated periodically.

The notion of rational numbers appeared in mathematics relatively early, since it is known that they were examined by the ancient Egyptians. It is worth mentioning that for a long period of time mathematicians believed that every number was rational. However, the existence of irrational numbers (i.e., real numbers which are not rational) was proved by the ancient Greeks. More specifically, a proof of the fact that $\sqrt{2}$ is an irrational number appears in the 10th book of Euclid's *Elements*.

But, we must mention that because of the fact that real numbers are uncountable and rational numbers countable, it follows that *almost all* real numbers are irrational.

We shall now present some basic theorems concerning irrational numbers.

Theorem 1.4.2. *If p is a prime number, then $\sqrt{p}$ is an irrational number.*

Proof. Let us suppose that $\sqrt{p}$ is a rational number. Then, there exist two relatively prime integers a, b, such that

$$\sqrt{p} = \frac{a}{b}.$$

Thus, we have

$$a^2 = p\,b^2. \tag{1}$$

However, by Euclid's first theorem (see 1.3.1), it follows that $p \mid a$. Hence, there exists an integer k, such that

$$a = kp.$$

Therefore, by (1) we obtain that

$$k^2 p = b^2.$$

But, by the above relation it follows similarly that $p \mid b$. Thus, the prime number p divides the integers a and b simultaneously, which is impossible since $\gcd(a, b) = 1$.

Therefore, the assumption that $\sqrt{p}$ is a rational number leads to a contradiction and hence $\sqrt{p}$ is irrational. $\qquad\square$

Corollary 1.4.3. *By the above theorem it follows that $\sqrt{2}$ is an irrational number, since 2 is a prime number.*

Theorem 1.4.4. *The number e is irrational.*

Let us suppose that e is a rational number. Then there exist two relatively prime integers p, q, such that $e = p/q$. It is a standard fact that

$$e = 1 + \frac{1}{1!} + \frac{1}{2!} + \cdots + \frac{1}{n!} + \cdots .$$

Thus, it is evident that

$$0 < e - \left(1 + \frac{1}{1!} + \frac{1}{2!} + \cdots + \frac{1}{n!}\right) = \frac{1}{(n+1)!} + \frac{1}{(n+2)!} + \cdots$$

$$= \frac{1}{n!}\left(\frac{1}{n+1} + \frac{1}{(n+1)(n+2)} + \cdots\right)$$

$$< \frac{1}{n!}\left(\frac{1}{n+1} + \frac{1}{(n+1)^2} + \frac{1}{(n+1)^3} + \cdots\right) = \frac{1}{n!} \cdot \frac{1}{n}.$$

Therefore, we have

$$0 < e - \left(1 + \frac{1}{1!} + \frac{1}{2!} + \cdots + \frac{1}{n!}\right) < \frac{1}{n!} \cdot \frac{1}{n}, \tag{1}$$

for every natural number n.

Hence, (1) will also hold true for every natural number $n \geq q$. Consequently, for $n \geq q$ we obtain

$$0 < \frac{p}{q}n! - \left(1 + \frac{1}{1!} + \frac{1}{2!} + \cdots + \frac{1}{n!}\right)n! < \frac{1}{n} < 1$$

and

$$\frac{p}{q}n! - \left(1 + \frac{1}{1!} + \frac{1}{2!} + \cdots + \frac{1}{n!}\right)n! \in \mathbb{N},$$

which is impossible, since every natural number is greater than or equal to 1.
□

We shall now present a proof of the fact that π^2 is an irrational number, due to Ivan Niven. But first we will prove a useful lemma.

Lemma 1.4.5. *Let*
$$f(x) = \frac{x^n(1-x)^n}{n!},$$
where $n \in \mathbb{N}$. For every natural number n, the following statements hold true:
(a) The function $f(x)$ is a polynomial of the form

$$f(x) = \frac{1}{n!}(a_n x^n + a_{n+1} x^{n+1} + \cdots + a_{2n} x^{2n}),$$

where $a_n, a_{n+1}, \ldots, a_{2n} \in \mathbb{Z}$.

(b) For $x \in (0,1)$, it holds

$$0 < f(x) < \frac{1}{n!}.$$

(c) For every integer $m \geq 0$, the derivatives

$$\frac{d^m f(0)}{dx^m}, \quad \frac{d^m f(1)}{dx^m}$$

are integers.

Proof. (a) It is a standard fact that

$$(1-x)^n = 1 - \binom{n}{1}x + \binom{n}{2}x^2 - \cdots + (-1)^n x^n.$$

Thus, for $f(x)$ we get

$$f(x) = \frac{x^n(1-x)^n}{n!} = \frac{1}{n!}\left(x^n - \binom{n}{1}x^{n+1} + \binom{n}{2}x^{n+2} - \cdots + (-1)^n x^{2n} \right).$$

Hence, the first statement is obviously true.
(b) Since $0 < x < 1$, it is clear that

$$0 < x^n < 1$$

and

$$0 < (1-x)^n < 1,$$

for every $n \in \mathbb{N}$. Therefore, we obtain

$$0 < x^n(1-x)^n < 1$$

and thus

$$0 < \frac{x^n(1-x)^n}{n!} < \frac{1}{n!}.$$

This completes the proof of the second statement.
(c) By the definition of the function $f(x)$, it follows that

$$f(x) = f(1-x).$$

But, by the above formula, it is clear that

$$\frac{d^m f(x)}{dx^m} = (-1)^m \frac{d^m f(1-x)}{dx^m}.$$

Therefore, for $x = 1$, we obtain

$$\frac{d^m f(1)}{dx^m} = (-1)^m \frac{d^m f(0)}{dx^m}. \tag{1}$$

Hence, it suffices to prove that either $d^m f(1)/dx^m$ or $d^m f(0)/dx^m$ is an integer.

However, for $m < n$ or $m > 2n$, it is evident that

$$\frac{d^m f(0)}{dx^m} = 0$$

and for $n \leq m \leq 2n$, it follows that

$$\frac{d^m f(0)}{dx^m} = \frac{m\,!a_m}{n\,!} \in \mathbb{Z}.$$

Thus, by (1) it follows that

$$\frac{d^m f(1)}{dx^m}$$

is also an integer.

This completes the proof of the third statement.   □

Theorem 1.4.6. *The number π^2 is irrational.*

Proof (Ivan Niven, 1947). Let us assume that the number π^2 is rational. In that case, there exist two positive integers p, q, such that $\pi^2 = p/q$. Consider the function

$$F(x) = q^n(\pi^{2n} f(x) - \pi^{2n-2} f^{(2)}(x) + \pi^{2n-4} f^{(4)}(x) - \cdots + (-1)^n f^{(2n)}(x)),\,^2$$

where $f(x)$ is as in Lemma 1.4.5. Then

$$\frac{d}{dx}(F^{(1)}(x)\sin(\pi x) - \pi F(x)\cos(\pi x)) = (F^{(2)}(x) + \pi^2 F(x))\sin(\pi x). \quad (1)$$

But

$$\pi^2 F(x) = q^n(\pi^{2n+2} f(x) - \pi^{2n} f^{(2)}(x) + \pi^{2n-2} f^{(4)}(x) - \cdots$$
$$+ (-1)^n \pi^2 f^{(2n)}(x)). \quad (2)$$

In addition, we have

$$F^{(2)}(x) = q^n(\pi^{2n} f^{(2)}(x) - \pi^{2n-2} f^{(4)}(x) + \pi^{2n-4} f^{(6)}(x) - \cdots$$
$$+ (-1)^n f^{(2n+2)}(x)). \quad (3)$$

Hence, by (2) and (3) we get

$$F^{(2)}(x) + \pi^2 F(x) = q^n \pi^{2n+2} f(x) + (-1)^n \pi^2 f^{(2n+2)}(x).$$

---

^{2} $f^{(k)}(x)$ denotes the kth derivative of the function $f(x)$.

However, since

$$f(x) = \frac{x^n(1-x)^n}{n!}$$

it is clear that $f^{(2n+2)}(x) = 0$. Thus, we obtain

$$F^{(2)}(x) + \pi^2 F(x) = q^n \pi^{2n+2} f(x). \tag{4}$$

By (1) and (4), it follows that

$$\frac{d}{dx}(F^{(1)}(x)\sin(\pi x) - \pi F(x)\cos(\pi x)) = q^n \pi^{2n+2} f(x)\sin(\pi x)$$

$$= q^n (\pi^2)^{n+1} f(x)\sin(\pi x)$$

$$= q^n \left(\frac{p}{q}\right)^{n+1} f(x)\sin(\pi x)$$

$$= \frac{p}{q} p^n f(x)\sin(\pi x)$$

$$= \pi^2 p^n f(x)\sin(\pi x).$$

Therefore, it is evident that

$$[F^{(1)}(x)\sin(\pi x) - \pi F(x)\cos(\pi x)]_0^1 = \pi^2 p^n \int_0^1 f(x)\sin(\pi x)dx$$

$$\Leftrightarrow \ \pi^2 p^n \int_0^1 f(x)\sin(\pi x)dx = \pi(F(0) + F(1)). \tag{5}$$

However, in the previous lemma, we proved that

$$\frac{d^m f(0)}{dx^m} \quad \text{and} \quad \frac{d^m f(1)}{dx^m}$$

are integers. Thus, it is clear that $F(0)$ and $F(1)$ are also integers. Hence, by (5) it follows that

$$I = \pi p^n \int_0^1 f(x)\sin(\pi x)dx$$

is an integer.

Furthermore, in the previous lemma we proved that for $n \in \mathbb{N}$ and $x \in (0,1)$ it holds

$$0 < f(x) < \frac{1}{n!}.$$

Thus, we get

$$0 < \pi p^n \int_0^1 f(x)\sin(\pi x)dx < \frac{\pi p^n}{n!},$$

for every $n \in \mathbb{N}$.

Moreover, since

$$\lim_{n \to +\infty} \frac{\pi p^n}{n!} = 0$$

it is evident that there exists a positive integer N_0, such that for every $n \geq N_0$ we have

$$\frac{\pi p^n}{n!} < 1.$$

Hence, for $n \geq N_0$ we get

$$0 < \pi p^n \int_0^1 f(x) \sin(\pi x) dx < 1,$$

which is impossible, since I is an integer. Therefore, the assumption that π^2 is a rational number leads to a contradiction.

Hence, the number π^2 is irrational. $\square$

Corollary 1.4.7. *The number π is irrational.*

Proof. Let us suppose that the number π is rational. Then there exist two positive integers p and q, such that $\pi = p/q$. However, in that case

$$\pi^2 = \frac{p^2}{q^2} = \frac{a}{b},$$

where a, $b \in \mathbb{Z}$. That is a contradiction, since π^2 is an irrational number. $\square$

Open Problem. It has not been proved yet whether the numbers

$$\pi + e, \quad \pi^e$$

are irrational or not.

Note. It has been proved that the numbers

$$e^\pi, \quad e^\pi + \pi$$

are irrational.

2

Arithmetic functions

The pleasure we obtain from music comes from counting,
but counting unconsciously.
Music is nothing but unconscious arithmetic.
Gottfried Wilhelm Leibniz (1646–1716)

In this chapter we shall define the arithmetic functions Möbius $\mu(n)$, Euler $\phi(n)$, the functions $\tau(n)$ and $\sigma_a(n)$ and, in addition, we shall prove some of their most basic properties and several formulas which are related to them. However, we shall first define some introductory notions.

2.1 Basic definitions

Definition 2.1.1. *An **arithmetic function** is a function $f : \mathbb{N} \to \mathbb{C}$ with domain of definition the set of natural numbers $\mathbb{N}$ and range a subset of the set of complex numbers $\mathbb{C}$.*

Definition 2.1.2. *A function f is called an **additive function** if and only if*

$$f(mn) = f(m) + f(n), \qquad (1)$$

*for every pair of coprime integers m, n. In case (1) is satisfied for every pair of integers m, n, which are not necessarily coprime, then the function f is called **completely additive**.*

Definition 2.1.3. *A function f is called a **multiplicative function** if and only if*

$$f(1) = 1 \quad and \quad f(mn) = f(m)f(n), \qquad (2)$$

*for every pair of coprime integers m, n. In case (2) is satisfied for every pair of integers m, n, which are not necessarily coprime, then the function f is called **completely multiplicative**.*

M.Th. Rassias, *Problem-Solving and Selected Topics in Number Theory: In the Spirit of the Mathematical Olympiads*, DOI 10.1007/978-1-4419-0495-9_2,
© Springer Science+Business Media, LLC 2011

2.2 The Möbius function

Definition 2.2.1. *The Möbius function* $\mu(n)$ *is defined as follows:*

$$\mu(n) = \begin{cases} 1, & \text{if } n = 1 \\ (-1)^k, & \text{if } n = p_1 p_2 \ldots p_k \text{ where } p_1, p_2, \ldots, p_k \text{ are } k \text{ distinct primes} \\ 0, & \text{in every other case.} \end{cases}$$

For example, we have

$$\mu(2) = -1, \ \mu(3) = -1, \ \mu(4) = 0, \ \mu(5) = -1, \ \mu(6) = 1$$

Remark 2.2.2. The Möbius function is a *multiplicative function*, since

$$\mu(1) = 1 \ \text{ and } \ \mu(mn) = \mu(m)\mu(n),$$

for every pair of coprime integers m, n.

However, it is not a *completely multiplicative* function because, for example, $\mu(4) = 0$ and $\mu(2)\mu(2) = (-1)(-1) = 1$.

Theorem 2.2.3.

$$\sum_{d|n} \mu(d) = \begin{cases} 1, & \text{if } n = 1 \\ 0, & \text{if } n > 1, \end{cases}$$

where the sum extends over all positive divisors of the positive integer n.

Proof.

- If $n = 1$, then the theorem obviously holds true, since by the definition of the Möbius function we know that $\mu(1) = 1$.
- If $n > 1$, we can write

$$n = p_1^{a_1} p_2^{a_2} \cdots p_k^{a_k},$$

where $p_1, p_2, \ldots, p_k$ are distinct prime numbers.
Therefore,

$$\sum_{d|n} \mu(d) = \mu(1) + \sum_{1 \le i \le k} \mu(p_i) + \sum_{\substack{i \ne j \\ 1 \le i,j \le k}} \mu(p_i p_j) + \cdots + \mu(p_1 p_2 \cdots p_k), \quad (1)$$

where generally the sum

$$\sum_{i_1 \ne i_2 \ne \cdots \ne i_\lambda} \mu(p_{i_1} p_{i_2} \cdots p_{i_\lambda})$$

extends over all possible products of λ distinct prime numbers. (By the definition of $\mu(m)$, we know that if in the canonical form of m some prime number appears multiple times, then $\mu(m) = 0$.) Hence, by (1) and the binomial identity, we obtain

$$\sum_{d|n} \mu(d) = 1 + \binom{k}{1}(-1) + \binom{k}{2}(-1)^2 + \cdots + \binom{k}{k}(-1)^k$$

$$= (1 - 1)^k = 0.$$

Therefore,

$$\sum_{d|n} \mu(d) = 0, \;\; if \;\; n > 1. \qquad \square$$

Theorem 2.2.4 (The Möbius Inversion Formula). *Let* $n \in \mathbb{N}$. *If*

$$g(n) = \sum_{d|n} f(d),$$

then

$$f(n) = \sum_{d|n} \mu\left(\frac{n}{d}\right) g(d).$$

The inverse also holds.

Proof.

- Generally, for every arithmetic function $m(n)$, it holds

$$\sum_{d|n} m(d) = \sum_{d|n} m\left(\frac{n}{d}\right),$$

since $n/d = d'$ and d' is also a divisor of n. Therefore, it is evident that

$$\sum_{d|n} \mu\left(\frac{n}{d}\right) g(d) = \sum_{d|n} \mu(d) g\left(\frac{n}{d}\right). \qquad (1)$$

But

$$\sum_{d|n} \mu(d) g\left(\frac{n}{d}\right) = \sum_{d|n} \left(\mu(d) \cdot \sum_{\lambda | \frac{n}{d}} f(\lambda) \right). \qquad (2)$$

At this point, we are going to express (2) in an equivalent form, where there will be just one sum at the left-hand side. In order to do so, we must find a common condition for the sums $\sum_{d|n}$ and $\sum_{\lambda | \frac{n}{d}}$. The desired condition is $\lambda d|n$.

Hence, we get

$$\sum_{d|n} \mu(d) g\left(\frac{n}{d}\right) = \sum_{\lambda d|n} \mu(d) f(\lambda).$$

Similarly,

$$\sum_{\lambda|n}\left(f(\lambda)\cdot\sum_{d|\frac{n}{\lambda}}\mu(d)\right)=\sum_{\lambda d|n}\mu(d)f(\lambda).$$

Thus,

$$\sum_{d|n}\mu(d)g\left(\frac{n}{d}\right)=\sum_{\lambda|n}\left(f(\lambda)\cdot\sum_{d|\frac{n}{\lambda}}\mu(d)\right). \qquad (3)$$

However, by the previous theorem

$$\sum_{d|\frac{n}{\lambda}}\mu(d)=1 \text{ if and only if } \frac{n}{\lambda}=1,$$

and in every other case the sum is equal to zero. Thus, for $n=\lambda$ we obtain

$$\sum_{\lambda|n}\left(f(\lambda)\cdot\sum_{d|\frac{n}{\lambda}}\mu(d)\right)=f(n). \qquad (4)$$

Therefore, by (1), (3) and (4) it follows that if

$$g(n)=\sum_{d|n}f(d),$$

then

$$f(n)=\sum_{d|n}\mu\left(\frac{n}{d}\right)g(d).$$

- Conversely, we shall prove that if

$$f(n)=\sum_{d|n}\mu\left(\frac{n}{d}\right)g(d),$$

then

$$g(n)=\sum_{d|n}f(d).$$

We have

$$\sum_{d|n}f(d)=\sum_{d|n}f\left(\frac{n}{d}\right)$$

$$=\sum_{d|n}\sum_{\lambda|\frac{n}{d}}\mu\left(\frac{n}{\lambda d}\right)g(\lambda)$$

$$=\sum_{d\lambda|n}\mu\left(\frac{n}{\lambda d}\right)g(\lambda)$$

$$=\sum_{\lambda|n}g(\lambda)\sum_{d|\frac{n}{\lambda}}\mu\left(\frac{n}{\lambda d}\right).$$

The sum

$$\sum_{d \mid \frac{n}{\lambda}} \mu \left(\frac{n}{\lambda d} \right) = 1$$

if and only if $n = \lambda$ and in every other case it is equal to zero. Hence, for $n = \lambda$ we obtain

$$\sum_{d \mid n} f(d) = g(n). \qquad \qquad \square$$

Historical Remark. August Ferdinand Möbius, born on the 17th of November 1790 in Schulpforta, was a German mathematician and theoretical astronomer. He was first introduced to mathematical notions by his father and later on by his uncle. During his school years (1803–1809), August showed a special skill in mathematics. In 1809, however, he started law studies at the University of Leipzig. Not long after that, he decided to quit these studies and concentrate in mathematics, physics and astronomy. August studied astronomy and mathematics under the guidance of Gauss and Pfaff, respectively, while at the University of Göttingen. In 1814, he obtained his doctorate from the University of Leipzig, where he also became a professor.

Möbius's main work in astronomy was his book entitled *Die Elemente den Mechanik des Himmels* (1843) which focused on celestial mechanics. Furthermore, in mathematics, he focused on projective geometry, statics and number theory. More specifically, in number theory, the *Möbius function* $\mu(n)$ and the *Möbius inversion formula* are named after him.

The most famous of Möbius's discoveries was the *Möbius strip* which is a nonorientable two-dimensional surface.

Möbius is also famous for the *five-color problem* which he presented in 1840. The problem's description was to find the least number of colors required to draw the regions of a map in such a way so that no two adjacent regions have the same color (this problem is known today as the *four-color theorem*, as it has been proved that the least number of colors required is four). A. F. Möbius died in Leipzig on the 26th of September, 1868.

Problem 2.2.5. Let f be a multiplicative function and

$$n = p_1^{a_1} p_2^{a_2} \cdots p_k^{a_k}, \quad where \quad k \in \mathbb{N},$$

be the canonical form of the positive integer n.

Prove that

$$\sum_{d \mid n} \mu(d) f(d) = \prod_{i=1}^{k} (1 - f(p_i)).$$

Proof. The nonzero terms of the sum

$$\sum_{d \mid n} \mu(d) f(d)$$

correspond to divisors d, for which

$$d = p_1^{q_1} p_2^{q_2} \cdots p_k^{q_k}, \quad \text{where} \quad q_i = 0 \text{ or } 1 \text{ and } 1 \leq i \leq k. \tag{1}$$

Therefore,

$$\sum_{d|n} \mu(d)f(d) = \sum_{q_i=0 \text{ or } 1} (-1)^k f(p_1^{q_1} p_2^{q_2} \cdots p_k^{q_k}), \tag{2}$$

where the sum at the right-hand side of (2) extends over all divisors d obeying the property (1). However, if we carry over the operations in the product

$$(1 - f(p_1))(1 - f(p_2)) \cdots (1 - f(p_k)),$$

we get a sum of the form

$$\sum_{q_i=0 \text{ or } 1} (-1)^k f(p_1^{q_1}) f(p_2^{q_2}) \cdots f(p_k^{q_k}) = \sum_{q_i=0 \text{ or } 1} (-1)^k f(p_1^{q_1} p_2^{q_2} \cdots p_k^{q_k}).$$

Hence, by (2) it is evident that

$$\prod_{i=1}^{k} (1 - f(p_i)) = \sum_{q_i=0 \text{ or } 1} (-1)^k f(p_1^{q_1} p_2^{q_2} \cdots p_k^{q_k})$$

$$= \sum_{d|n} \mu(d)f(d). \qquad \square$$

Remark 2.2.6. In the special case when $f(d) = 1$ for every divisor d of n, it follows that

$$\sum_{d|n} \mu(d)f(d) = \sum_{d|n} \mu(d) = \prod_{i=1}^{k} (1 - 1) = \left\lfloor \frac{1}{n} \right\rfloor, \quad [1]$$

which is exactly Theorem 2.2.3.

2.3 The Euler function

Definition 2.3.1. *The Euler function $\phi(n)$ is defined as the number of positive integers which are less than or equal to n and at the same time relatively prime to n. Equivalently, the Euler function $\phi(n)$ can be defined by the formula*

$$\phi(n) = \sum_{m=1}^{n} \left\lfloor \frac{1}{\gcd(n, m)} \right\rfloor.$$

---

[1] $\lfloor r \rfloor$ denotes the *integer part* (also called *integral part*) of a real number r.

For example, we have

$$\phi(1) = 1, \; \phi(2) = 1, \; \phi(3) = 2, \; \phi(6) = 2, \; \phi(9) = 6.$$

Before we proceed on proving theorems concerning the Euler function $\phi(n)$, we shall present two of its most basic properties.

Proposition 2.3.2. *For every prime number p, it holds*

$$\phi(p^k) = p^k - p^{k-1}.$$

Proof. The only positive integers which are less than or equal to p^k and at the same time not relatively prime to p^k are the integers

$$p, \; 2p, \; 3p, \ldots, \; p^{k-1}p.$$

Thus, the number of these integers is p^{k-1} and therefore the number of positive integers which are less than or equal to p^k and at the same time relatively prime to p^k are

$$p^k - p^{k-1}. \qquad \square$$

The Euler function $\phi(n)$ is a multiplicative function, since

$$\phi(1) = 1 \; \text{ and } \; \phi(mn) = \phi(m)\phi(n),$$

for every pair of coprime integers m, n.

We shall present the proof of the above fact at the end of this section.

Theorem 2.3.3. *For every positive integer n, it holds*

$$\phi(n) = \sum_{d|n} \mu(d)\frac{n}{d}.$$

Proof. In the previous section, we proved that the sum $\sum_{d|n} \mu(d)$ is equal to 1 if $n = 1$ and equal to 0 in any other case. Hence, equivalently we have

$$\sum_{d|n} \mu(d) = \left\lfloor \frac{1}{n} \right\rfloor.$$

Thus, we can write

$$\phi(n) = \sum_{m=1}^{n} \left\lfloor \frac{1}{\gcd(n,m)} \right\rfloor = \sum_{m=1}^{n} \sum_{d|\gcd(n,m)} \mu(d). \tag{1}$$

In the above sums it is evident that

$$1 \leq m \leq n, \; d|n,$$

and

$$d \mid m.$$

Therefore,

$$\sum_{m=1}^{n} \sum_{d \mid \gcd(n,m)} \mu(d) = \sum_{d \mid n} \sum_{\lambda=1}^{n/d} \mu(d) = \sum_{d \mid n} \frac{n}{d} \mu(d). \tag{2}$$

Thus, by (1) and (2) we finally get

$$\phi(n) = \sum_{d \mid n} \mu(d) \frac{n}{d}. \qquad \square$$

Theorem 2.3.4. *For every positive integer n it holds*

$$\sum_{d \mid n} \phi(d) = n.$$

Proof. It is clear that every positive integer k which is less than or equal to n has some divisibility relation with n. More specifically, either k and n are coprime or $\gcd(n,k) = d > 1$. Generally, if $\gcd(n,k) = d$, then

$$\left(\frac{n}{d}, \frac{k}{d} \right) = 1.$$

Hence, the number of positive integers for which $\gcd(n,k) = d$ is equal to $\phi(n/d)$. However, since the number of positive integers k with $k \leq n$ is clearly equal to n we obtain

$$\sum_{d \mid n} \phi \left(\frac{n}{d} \right) = n.$$

But, it is evident that

$$\sum_{d \mid n} \phi \left(\frac{n}{d} \right) = \sum_{d \mid n} \phi(d),$$

thus,

$$\sum_{d \mid n} \phi(d) = n. \qquad \square$$

Remark 2.3.5. Another proof of the above theorem can be given by the use of the Möbius Inversion Formula.

Theorem 2.3.6. *Let n be a positive integer and $p_1, p_2, \ldots, p_k$ be its prime divisors. Then*

$$\phi(n) = n \left(1 - \frac{1}{p_1} \right) \left(1 - \frac{1}{p_2} \right) \cdots \left(1 - \frac{1}{p_k} \right)$$

and therefore, for any pair of positive integers n_1, n_2 it holds

$$\phi(n_1 n_2) = \phi(n_1)\phi(n_2)\frac{d}{\phi(d)},$$

where $d = \gcd(n_1, n_2)$.

Proof. We can write

$$\left(1 - \frac{1}{p_1}\right)\left(1 - \frac{1}{p_2}\right)\cdots\left(1 - \frac{1}{p_k}\right) = 1 + \sum \frac{(-1)^\lambda}{p_{m_1}p_{m_2}\cdots p_{m_\lambda}},$$

where m_i are λ distinct integers in the set $\{1, 2, \ldots, k\}$ and hence the sum extends over all possible products of the prime divisors of n. However, by the definition of the Möbius function we know that

$$\mu(p_{m_1}p_{m_2}\cdots p_{m_\lambda}) = (-1)^\lambda,$$

where $\mu(1) = 1$ and $\mu(r) = 0$ if the positive integer r is divisible by the square of any of the prime numbers $p_1, p_2, \ldots, p_k$. Therefore, we get

$$\left(1 - \frac{1}{p_1}\right)\left(1 - \frac{1}{p_2}\right)\cdots\left(1 - \frac{1}{p_k}\right) = \sum_{d|n} \frac{\mu(d)}{d}$$

$$= \frac{\phi(n)}{n}.$$

Hence,

$$\phi(n) = n\left(1 - \frac{1}{p_1}\right)\left(1 - \frac{1}{p_2}\right)\cdots\left(1 - \frac{1}{p_k}\right).$$

We shall now prove that

$$\phi(n_1 n_2) = \phi(n_1)\phi(n_2)\frac{d}{\phi(d)}.$$

From the first part of the theorem, it follows

$$\phi(n_1 n_2) = (n_1 n_2) \prod_{p|n_1 n_2} \left(1 - \frac{1}{p}\right).$$

But, if $n_1 n_2 = p_1^{q_1} p_2^{q_2} \ldots p_m^{q_m}$, then each of the prime numbers $p_1, p_2, \ldots, p_m$ appears exactly once in the product

$$\prod_{p|n_1 n_2} \left(1 - \frac{1}{p}\right).$$

More specifically, distinct primes p appear in distinct factors

$$1 - \frac{1}{p}.$$

On the other hand, in the product

$$\prod_{p|n_1} \left(1 - \frac{1}{p}\right) \cdot \prod_{p|n_2} \left(1 - \frac{1}{p}\right),$$

the prime numbers from the set $\{p_1, p_2, \ldots, p_m\}$ which divide both n_1 and n_2, appear twice.

Hence, according to the above arguments it is evident that

$$\prod_{p|n_1 n_2} \left(1 - \frac{1}{p}\right) = \frac{\prod_{p|n_1} \left(1 - \frac{1}{p}\right) \cdot \prod_{p|n_2} \left(1 - \frac{1}{p}\right)}{\prod_{\substack{p|n_1 \\ p|n_2}} \left(1 - \frac{1}{p}\right)}.$$

Thus, we have

$$\phi(n_1 n_2) = \frac{n_1 \prod_{p|n_1} \left(1 - \frac{1}{p}\right) \cdot n_2 \prod_{p|n_2} \left(1 - \frac{1}{p}\right)}{\prod_{\substack{p|n_1 \\ p|n_2}} \left(1 - \frac{1}{p}\right)}$$

$$= \frac{\phi(n_1)\phi(n_2)}{\prod_{p|d} \left(1 - \frac{1}{p}\right)}$$

$$= \frac{\phi(n_1)\phi(n_2)}{\frac{\phi(d)}{d}}$$

$$= \phi(n_1)\phi(n_2)\frac{d}{\phi(d)}.$$

Therefore,

$$\phi(n_1 n_2) = \phi(n_1)\phi(n_2)\frac{d}{\phi(d)}. \qquad \square$$

Note. For further reading concerning the Möbius and Euler functions the reader is referred to [38].

2.4 The τ-function

Definition 2.4.1. *The function $\tau(n)$ is defined as the number of positive divisors of a positive integer n, including 1 and n. Equivalently, the function $\tau(n)$ can be defined by the formula*

$$\tau(n) = \sum_{\substack{d|n \\ d \geq 1}} 1.$$

Remark 2.4.2. The function $\tau(n)$ is a *multiplicative function*, since

$$\tau(1) = 1 \quad and \quad \tau(mn) = \tau(m)\tau(n),$$

for every pair of coprime integers m, n.

This property is very useful for the computation of the number of divisors of large integers.

Theorem 2.4.3. *Let $n = p_1^{a_1} p_2^{a_2} \cdots p_k^{a_k}$ be the canonical form of the positive integer n. Then it holds*

$$\tau(n) = (a_1 + 1)(a_2 + 1) \cdots (a_k + 1).$$

Proof. We shall follow the Mathematical Induction Principle.

For $k = 1$ we have

$$\tau(n) = \tau(p_1^{a_1}).$$

Since the divisors of n, where $n = p_1^{a_1}$, are the positive integers $1, p_1, p_1^2, \ldots, p_1^{a_1}$, it is evident that

$$\tau(n) = a_1 + 1.$$

Let $m = p_1^{a_1} p_2^{a_2} \cdots p_{k-1}^{a_{k-1}}$ and assume that

$$\tau(m) = (a_1 + 1)(a_2 + 1) \cdots (a_{k-1} + 1). \tag{1}$$

In order to determine the divisors of the positive integer n, where

$$n = p_1^{a_1} p_2^{a_2} \cdots p_k^{a_k},$$

it suffices to multiply each divisor of m by the powers of the prime p_k (i.e., $p_k^0, p_k^1, p_k^2, \ldots, p_k^{a_k}$).

Therefore, if d_n and d_m denote the positive divisors of n and m, respectively, then

$$\tau(n) = \sum_{d_n|n} 1 = \sum_{d_m|m} 1 + \sum_{d_m p_k|n} 1 + \sum_{d_m p_k^2|n} 1 + \cdots + \sum_{d_m p_k^{a_k}|n} 1$$

and since the number of divisors d_m is $\tau(m)$, we obtain

$$\tau(n) = \tau(m) + \tau(m) + \tau(m) + \cdots + \tau(m) = \tau(m)(a_k + 1). \tag{2}$$

Hence, by (1) and (2) we obtain

$$\tau(n) = (a_1 + 1)(a_2 + 1) \cdots (a_k + 1),$$

which is the desired result. $\qquad\square$

Remark 2.4.4. Generally, for every positive integer n with

$$n = p_1^{a_1} p_2^{a_2} \cdots p_k^{a_k},$$

it holds

$$\tau(n) = \tau(p_1^{a_1})\tau(p_2^{a_2}) \cdots \tau(p_k^{a_k}).$$

EXAMPLES 2.4.5.

$$\tau(126) = \tau(2 \cdot 3^2 \cdot 7) = (1+1)(2+1)(1+1) = 12$$

$$\tau(168) = \tau(2^3 \cdot 3 \cdot 7) = (3+1)(1+1)(1+1) = 16$$

$$\tau(560) = \tau(2^4 \cdot 5 \cdot 7) = (4+1)(1+1)(1+1) = 20$$

$$\tau(1,376,375) = \tau(5^3 \cdot 7 \cdot 11^2 \cdot 13) = (3+1)(1+1)(2+1)(1+1) = 48.$$

2.5 The generalized σ-function

Definition 2.5.1. *The function $\sigma_a(n)$ is defined as the sum of the a-th powers of the positive divisors of a positive integer n, including 1 and n, where a can be any complex number. Equivalently, the function $\sigma(n)$ can be defined by the formula*

$$\sigma_a(n) = \sum_{\substack{d|n \\ d \geq 1}} d^a,$$

where the sum extends over all positive divisors of n.

Remark 2.5.2. For $k = 0$ we obtain

$$\sigma_0(n) = \tau(n).$$

Remark 2.5.3. The function $\sigma_a(n)$ is a *multiplicative function*, since

$$\sigma_a(1) = 1 \quad \text{and} \quad \sigma_a(mn) = \sigma_a(m)\sigma_a(n),$$

for every pair of coprime integers m, n.

Theorem 2.5.4. *Let $n = p_1^{a_1} p_2^{a_2} \cdots p_k^{a_k}$ be the canonical form of the positive integer n. Then*

$$\sigma_1(n) = \frac{p_1^{a_1+1} - 1}{p_1 - 1} \cdot \frac{p_2^{a_2+1} - 1}{p_2 - 1} \cdots \frac{p_k^{a_k+1} - 1}{p_k - 1}.$$

Proof. We shall follow the Mathematical Induction Principle.
For $k = 1$ we obtain

$$\sigma_1(n) = \sigma_1(p_1^{a_1}).$$

But, since the divisors of n, where $n = p_1^{a_1}$, are the integers $1, p_1, p_1^2, \ldots, p_1^{a_1}$, it is evident that

$$\sigma_1(n) = 1 + p_1 + p_1^2 + \cdots + p_1^{a_1} = \frac{p_1^{a_1+1} - 1}{p_1 - 1}.$$

Now let $m = p_1^{a_1} p_2^{a_2} \cdots p_{k-1}^{a_{k-1}}$ and assume that

$$\sigma_1(m) = \frac{p_1^{a_1+1} - 1}{p_1 - 1} \cdot \frac{p_2^{a_2+1} - 1}{p_2 - 1} \cdots \frac{p_{k-1}^{a_{k-1}+1} - 1}{p_{k-1} - 1}. \tag{1}$$

Similarly to the proof of Theorem 2.2.3 let d_n and d_m denote the positive divisors of n and m, respectively, where $n = p_1^{a_1} p_2^{a_2} \cdots p_k^{a_k}$. Then we have

$$\sigma_1(n) = \sum_{d_n | n} d_n$$

$$= \sum_{d_m | m} d_m + \sum_{d_m | m} d_m p_k + \sum_{d_m | m} d_m p_k^2 + \cdots + \sum_{d_m | m} d_m p_k^{a_k}$$

$$= 1 \cdot \sum_{d_m | m} d_m + p_k \sum_{d_m | m} d_m + p_k^2 \sum_{d_m | m} d_m + \cdots + p_k^{a_k} \sum_{d_m | m} d_m$$

$$= (1 + p_k + p_k^2 + \cdots + p_k^{a_k}) \sum_{d_m | m} d_m.$$

Therefore, by the above result and relation (1), we obtain

$$\sigma_1(n) = \frac{p_1^{a_1+1} - 1}{p_1 - 1} \cdot \frac{p_2^{a_2+1} - 1}{p_2 - 1} \cdots \frac{p_k^{a_k+1} - 1}{p_k - 1}. \qquad \square$$

Remark 2.5.5. For the function $\sigma_a(n)$, it holds

$$\sigma_a(p_1^{a_1} p_2^{a_2} \cdots p_k^{a_k}) = \sigma_a(p_1^{a_1}) \sigma_a(p_2^{a_2}) \cdots \sigma_a(p_k^{a_k}).$$

EXAMPLES 2.5.6.

$$\sigma(126) = \sigma(2 \cdot 3^2 \cdot 7) = \frac{2^2 - 1}{2 - 1} \frac{3^3 - 1}{3 - 1} \frac{7^2 - 1}{7 - 1} = 312$$

$$\sigma(168) = \sigma(2^3 \cdot 3 \cdot 7) = \frac{2^4 - 1}{2 - 1} \frac{3^2 - 1}{3 - 1} \frac{7^2 - 1}{7 - 1} = 480$$

$$\sigma(560) = \sigma(2^4 \cdot 5 \cdot 7) = \frac{2^5 - 1}{2 - 1} \frac{5^2 - 1}{5 - 1} \frac{7^2 - 1}{7 - 1} = 1488.$$

Application. We shall use Theorem 2.5.4 in order to prove that *the number of primes is infinite.*

Proof (George Miliakos). Let us suppose that the number of primes is finite. If $m = p_1^{q_1} p_2^{q_2} \cdots p_k^{q_k}$ and $n = m! = p_1^{a_1} p_2^{a_2} \cdots p_k^{a_k}$ are the canonical forms of m and n, respectively, then it is evident that $a_1 \geq q_1, a_2 \geq q_2, \ldots, a_k \geq q_k$. Therefore, by Theorem 2.5.4, we obtain

$$\frac{\sigma(n)}{n} = \frac{p_1 - 1/p_1^{a_1}}{p_1 - 1} \cdot \frac{p_2 - 1/p_2^{a_2}}{p_2 - 1} \cdots \frac{p_k - 1/p_k^{a_k}}{p_k - 1}. \tag{1}$$

But, for $q_1, q_2, \ldots, q_k \to \infty$ it follows that $a_1, a_2, \ldots, a_k \to \infty$. Thus, it follows that $n \to \infty$.

Therefore, by (1) we get

$$\lim_{n \to \infty} \frac{\sigma(n)}{n} = \frac{p_1}{p_1 - 1} \cdot \frac{p_2}{p_2 - 1} \cdots \frac{p_k}{p_k - 1}. \tag{2}$$

However, it is clear that

$$\frac{n}{1} + \frac{n}{2} + \frac{n}{3} + \cdots + \frac{n}{m} \leq \sigma(n)$$

and hence

$$1 + \frac{1}{2} + \frac{1}{3} + \cdots + \frac{1}{m} \leq \frac{\sigma(n)}{n}. \tag{3}$$

But, it is a standard fact in mathematical analysis that $\sum_{n=1}^{\infty} 1/n = \infty$. Consequently, for $m \to \infty$, by (2) and (3) we obtain

$$\frac{p_1}{p_1 - 1} \cdot \frac{p_2}{p_2 - 1} \cdots \frac{p_k}{p_k - 1} = \infty,$$

which is obviously a contradiction, since we have assumed that the number of primes is finite. Hence, the number of primes must be infinite. □

3

Perfect numbers, Fermat numbers

Perfect numbers like perfect men are very rare.
René Descartes (1596–1650)

In this chapter we shall define *perfect numbers* and *Fermat numbers* and we are going to provide proofs of some of their most basic properties and theorems which are related to them. Furthermore, some related open problems will be presented.

3.1 Perfect numbers

Definition 3.1.1. *A positive integer n is said to be a **perfect number** if and only if it is equal to the sum of its positive divisors without counting n in the summation. Symbolically, n is a perfect number if and only if*

$$\sigma_1(n) = 2n.$$

For example, 6 is the first perfect number, with $6 = 1 + 2 + 3$ and 28 is the second perfect number, with $28 = 1 + 2 + 4 + 7 + 14$.

Research related to perfect numbers has its roots in ancient times and particularly in ancient Greece. Euclid in his *Elements*[1] presented one of the most important theorems regarding perfect numbers.

---

[1] Euclid's Elements comprise thirteen volumes that Euclid himself composed in Alexandria in about 300 BC. More specifically, the first four volumes deal with *figures*, such as triangles, circles and quadrilaterals. The fifth and sixth volumes study topics such as *similar figures*. The next three volumes deal with a primary form of elementary number theory and the rest study topics related to geometry. It is believed that the *Elements* founded logic and modern science. In addition, it is the oldest and best established surviving ancient-Greek work and has been suggested to be the second work published after the Bible.

M.Th. Rassias, *Problem-Solving and Selected Topics in Number Theory: In the Spirit of the Mathematical Olympiads*, DOI 10.1007/978-1-4419-0495-9_3, © Springer Science+Business Media, LLC 2011

Theorem 3.1.2 (Euclid). *For every positive integer n for which $2^n - 1$ is a prime number, it holds that $2^{n-1}(2^n - 1)$ is a perfect number.*

Proof. It suffices to prove that

$$\sigma_1(2^{n-1}(2^n - 1)) = 2^n(2^n - 1).$$

Hence, in order to do so, we shall determine the divisors of the positive integer $2^{n-1} \cdot p$, where $p = 2^n - 1$. It is clear though that the integers

$$1, 2, 2^2, \ldots, 2^{n-1}, p, 2p, 2^2 p, \ldots, 2^{n-1} p$$

are the desired divisors. Hence, we get

$$
\begin{aligned}
\sigma_1(2^{n-1} p) &= 1 + 2 + 2^2 + \cdots + 2^{n-1} + p + 2p + 2^2 p + \cdots + 2^{n-1} p \\
&= (p+1)(1 + 2 + 2^2 + \cdots + 2^{n-1}) \\
&= (p+1)(2^n - 1) \\
&= 2^n(2^n - 1).
\end{aligned}
$$

Therefore,

$$\sigma_1(2^{n-1} p) = 2^n(2^n - 1),$$

which is the desired result. $\square$

Theorem 3.1.3 (Euler). *Every even perfect number can be represented in the form $2^{n-1}(2^n - 1)$, where n is a positive integer and $2^n - 1$ is a prime number.*

Proof. Let k be a perfect number and $n - 1$ be the greatest power of 2 which divides k. Then, for some positive integer m it holds

$$
\begin{aligned}
2k = \sigma_1(k) &= \sigma_1(2^{n-1} m) \\
&= \sigma_1(2^{n-1})\sigma_1(m),
\end{aligned}
$$

since 2^{n-1} and m are relatively prime integers. However, by Theorem 2.5.4 we know that if p is a prime number, then

$$\sigma_1(p^k) = \frac{p^{k+1} - 1}{p - 1}.$$

Therefore,

$$2k = (2^n - 1)\sigma_1(m)$$

or

$$2^n m = (2^n - 1)\sigma_1(m) \tag{1}$$

or

$$\frac{m}{\sigma_1(m)} = \frac{2^n - 1}{2^n}.$$

Clearly the fraction $(2^n - 1)/2^n$ is irreducible, since $2^n - 1$ and 2^n are relatively prime integers. Thus, it is evident that

$$m = c(2^n - 1) \quad \text{and} \quad \sigma_1(m) = c2^n,$$

for some positive integer c. We are now going to prove that c can only be equal to 1.

Let us suppose that $c \neq 1$. In that case,

$$\sigma_1(m) \geq m + c + 1,$$

since m has at least m, c, 1 as its divisors. Hence,

$$\sigma_1(m) \geq c(2^n - 1) + c + 1 = 2^n c + 1 > \sigma_1(m),$$

since $\sigma_1(m) = 2^n c$, which is a contradiction.

Therefore, we obtain that $c = 1$ and consequently we get $m = 2^n - 1$ and $k = 2^{n-1}(2^n - 1)$ since $k = 2^{n-1}m$. The only question which remains unanswered is whether $2^n - 1$ is a prime number. However, by (1) we have

$$(2^n - 1)\sigma_1(2^n - 1) = 2^n(2^n - 1)$$

or

$$\sigma_1(2^n - 1) = 2^n = (2^n - 1) + 1.$$

Thus, the only divisors of $2^n - 1$ are the number itself and 1. Thus, clearly $2^n - 1$ is a prime number. □

3.1.1 Related open problems

(i) We observe that Euler's theorem strictly refers to even perfect numbers. Thus, naturally, the following question arises:

Are there any odd perfect numbers?

The above question is one of the oldest open problems in number theory and most probably one of the oldest in the history of mathematics.

In 1993, R. P. Brent, G. L. Cohen and H. J. J. te Riele in their joint paper [14] proved that if there exist odd perfect numbers n, then it must hold

$$n > 10^{300}.$$

(ii) Another well-known open problem related to perfect numbers is the following:

Are there infinitely many even perfect numbers?

Remark 3.1.4. In the previous two theorems we examined the case when the integer $2^n - 1$ was a prime number. Hence, it is worth mentioning that the integers of that form are called **Mersenne numbers**, after the mathematician Marin Mersenne who first investigated their properties.

Marin Mersenne (1588–1648) maintained correspondence with Pierre de Fermat (1601–1665) and hence Fermat also investigated the prime numbers of the form $2^n - 1$. His research in this topic led him to the discovery of the theorem which is known as *Fermat's Little Theorem*. In 1644, Mersenne formulated the conjecture that the integer

$$M_p = 2^p - 1$$

is a prime number for $p = 2, 3, 5, 7, 13, 17, 19, 31, 67, 127, 257$ and composite for the prime numbers p for which $p < 257$. But, Pervusin and Seelhoff in the years 1883 and 1886, respectively, independently proved Mersenne's conjecture to be false by giving a counterexample. They proved that for $p = 61$ the integer $2^p - 1$ is a prime number.

3.2 Fermat numbers

Definition 3.2.1. *The integers F_n of the form*

$$F_n = 2^{2^n} + 1,$$

where $n \in \mathbb{N} \cup \{0\}$, are called **Fermat numbers**.

For example, $F_0 = 3$, $F_1 = 5$, $F_2 = 17$, $F_3 = 257$, $F_4 = 65537$.
Fermat formulated the conjecture[2] that every number of the form

$$2^{2^n} + 1$$

is prime. However, in 1732 Leonhard Euler (1707–1783) proved that

$$F_5 = 2^{2^5} + 1$$

is a composite integer and therefore disproved Fermat's conjecture. We will present Euler's proof at the end of this chapter.

3.2.1 Some basic properties

Corollary 3.2.2. *For all Fermat numbers F_m, where $m \in \mathbb{N}$, it holds*

$$F_m - 2 = F_0 F_1 \cdots F_{m-1}.$$

---

[2] More specifically, the conjecture first appeared in a letter addressed by Fermat to Mersenne on December 25, 1640.

Proof. We shall follow the Mathematical Induction Principle.

For $m = 1$ we get

$$F_1 - 2 = F_0,$$

which is obviously true.

Let us suppose that for some integer k with $k > 1$ it holds

$$F_k - 2 = F_0 F_1 \cdots F_{k-1}.$$

It suffices to prove that

$$F_{k+1} - 2 = F_0 F_1 \cdots F_k.$$

But,

$$(F_0 F_1 \cdots F_{k-1}) F_k = (F_k - 2) F_k = (2^{2^k} - 1)(2^{2^k} + 1)$$

or

$$F_0 F_1 \cdots F_k = (2^{2^k})^2 - 1 = 2^{2^k} \cdot 2^{2^k} - 1 = 2^{2 \cdot 2^k} - 1 = 2^{2^{k+1}} - 1$$

or

$$F_0 F_1 \cdots F_k = F_{k+1} - 2. \qquad \square$$

Corollary 3.2.3. *For all Fermat numbers F_n, with $n \in \mathbb{N}$, it holds*

$$F_n \mid 2^{F_n} - 2.$$

Proof. By Corollary 3.2.2 we know that

$$F_m - 2 = F_0 F_1 \cdots F_{m-1}.$$

Let $n \in \mathbb{N}$ with $n < m$. Then, it is evident that F_n is one of the integers

$$F_0, \ F_1, \ldots, F_{m-1}.$$

By the Mathematical Induction Principle we can easily prove that $n + 1 \leq 2^n$, for every positive integer n. Therefore, we can set m to be 2^n. In that case, we obtain

$$F_{2^n} - 2 = F_0 F_1 \cdots F_{2^n - 1}.$$

Hence,

$$F_n \mid F_{2^n} - 2.$$

However,

$$F_{2^n} = 2^{2^{2^n}} + 1 = 2^{F_n - 1} + 1.$$

Thus, it follows that

$$F_n \mid (2^{F_n - 1} - 1).$$

Therefore, obviously

$$F_n \mid 2(2^{F_n - 1} - 1)$$

or

$$F_n \mid (2^{F_n} - 2). \qquad \square$$

Theorem 3.2.4. *Fermat numbers are coprime.*

Proof. Let us suppose that Fermat numbers are not coprime. Then, there exists a prime number p, for which

$$p \mid F_m \quad \text{and} \quad p \mid F_n,$$

for some positive integers m, n.

Without loss of generality we may suppose that $n < m$. By Corollary 3.2.2 we obtain

$$F_n \mid F_m - 2$$

and thus

$$p \mid F_m - 2.$$

But, because of the fact that p also divides F_m, we get $p \mid F_m - (F_m - 2)$ and consequently $p = 2$. But, that is a contradiction since Fermat numbers are odd integers and therefore are not divisible by 2. □

G. Pólya (1887–1985) used Corollary 3.2.2 in order to give a new proof of the fact that the number of primes is infinite. His proof is presented below.

Theorem 3.2.5 (Pólya's Proof). *The number of primes is infinite.*

Proof. The idea of G. Pólya was to determine an infinite sequence of coprime integers. Let (a_k) be such a sequence. In that case, if a prime number p_k divides the kth term of the sequence (a_k), then it cannot divide any other term of that sequence. Hence, the terms $a_{k+1}, a_{k+2}, \ldots$ are divisible by distinct prime numbers. But, the terms of the sequence are infinite and therefore the number of primes must be infinite, too.

The sequence which Pólya used to demonstrate his argument was the sequence of Fermat numbers. □

We have previously mentioned that in 1732 Euler proved that

$$F_5 = 2^{2^5} + 1$$

is a composite integer. It is truly remarkable that while Euler composed that proof he was completely blind. His proof is presented below.

Euler observed that F_5 is divisible by 641 and this is true because

$$641 = 5^4 + 2^4 = 5 \cdot 5^3 + 16 = 5 \cdot 125 + 15 + 1$$

$$= 5(125 + 3) + 1 = 5 \cdot 128 + 1 = 5 \cdot 2^7 + 1.$$

Hence,

$$641 = 5^4 + 2^4 = 5 \cdot 2^7 + 1. \tag{1}$$

But,

$$(5^4 + 2^4)(2^7)^4 = 5^4(2^7)^4 + 2^4(2^7)^4 = (5 \cdot 2^7)^4 + (2 \cdot 2^7)^4$$
$$= 5^4 \cdot 2^{28} + 2^{32},$$

therefore

$$641 \mid 5^4 \cdot 2^{28} + 2^{32}. \tag{2}$$

Furthermore, we have

$$5^4 \cdot 2^{28} - 1 = 5^2 \cdot 2^{14} \cdot 5^2 \cdot 2^{14} - 1$$
$$= (5^2 \cdot 2^{14})^2 - 1$$
$$= (5^2 \cdot 2^{14} - 1)(5^2 \cdot 2^{14} + 1)$$
$$= (5 \cdot 2^7 - 1)(5 \cdot 2^7 + 1)(5^2 \cdot 2^{14} + 1).$$

Therefore, by (1), we obtain that $641 \mid 5^4 \cdot 2^{28} - 1$ and thus by (2) we get

$$641 \mid (5^4 \cdot 2^{28} + 2^{32}) - (5^4 \cdot 2^{28} - 1)$$

or

$$641 \mid 2^{32} + 1 = 2^{2^5} + 1 = F_5.$$

Hence, F_5 is not a prime number. More specifically, it holds

$$F_5 = 641 \cdot 6700417. \qquad \square$$

It is worth mentioning that 148 years after the presentation of Euler's proof, E. Landau proved that F_6 is not a prime number. A lot of other Fermat numbers have been proven to be composite since then.

4

Congruences

Miracles are not to be multiplied beyond necessity.
Gottfried Wilhelm Leibniz (1646–1716)

4.1 Basic theorems

Definition 4.1.1. *Two integers a and b are said to be congruent modulo m, where m is a nonzero integer, if and only if m divides the difference $a - b$. In that case we write*

$$a \equiv b \pmod{m}.$$

On the other hand, if the difference $a - b$ is not divisible by m, then we say that a is not congruent to b modulo m and we write

$$a \not\equiv b \pmod{m}.$$

Theorem 4.1.2 (Fermat's Little Theorem). *Let p be a prime number and a be an integer for which the $\gcd(a, p) = 1$. Then it holds*

$$a^{p-1} \equiv 1 \pmod{p}.$$

Proof. Firstly, we shall prove that

$$a^p \equiv a \pmod{p}$$

for every integer value of a. In order to do so, we will distinguish two cases.

Case 1. At first, we assume that a is a positive integer. If $a = 1$, then the congruence $a^p \equiv a \pmod{p}$ obviously holds for every prime number p. Let us now assume that $a^p \equiv a \pmod{p}$ is true. We shall prove that

$$(a + 1)^p \equiv a + 1 \pmod{p}.$$

M.Th. Rassias, *Problem-Solving and Selected Topics in Number Theory: In the Spirit of the Mathematical Olympiads*, DOI 10.1007/978-1-4419-0495-9_4,
© Springer Science+Business Media, LLC 2011

By the binomial identity we know that

$$(a+1)^p = a^p + \binom{p}{1}a^{p-1} + \cdots + \binom{p}{p-1}a + 1.$$

But, since p divides each of the integers

$$\binom{p}{1}, \binom{p}{2}, \ldots, \binom{p}{p-1}$$

it is clear that

$$(a+1)^p \equiv a^p + 1 \,(\text{mod } p). \tag{1}$$

But, we have made the hypothesis that $a^p \equiv a \,(\text{mod } p)$. Therefore, (1) takes the form

$$(a+1)^p \equiv a + 1 \,(\text{mod } p).$$

Hence, by the Mathematical Induction Principle we have proved that

$$a^p \equiv a \,(\text{mod } p),$$

for every positive integer a.

Case 2. We shall prove that $a^p \equiv a \,(\text{mod } p)$ for $a \le 0$.

If $a = 0$, then $a^p \equiv a \,(\text{mod } p)$ is obviously true. If $a < 0$, then for $p = 2$ we have

$$a^2 = (-a)^2 \equiv (-a) \,(\text{mod } 2),$$

since $-a$ is a positive integer. Therefore, $2 \mid a^2 + a$ and thus $2 \mid a^2 + a - 2a$ or $2 \mid a^2 - a$ which is equivalent to $a^2 \equiv a \,(\text{mod } 2)$.

If $p \ne 2$, in which case p is an odd integer, we get

$$a^p = -(-a)^p \equiv -(-a) \,(\text{mod } p),$$

since $-a$ is a positive integer. Therefore,

$$a^p \equiv a \,(\text{mod } p),$$

for every integer a.

By the above relation, we get

$$p \mid a(a^{p-1} - 1).$$

Hence, by Euclid's First Theorem, the prime number p must divide either a or $a^{p-1}-1$. But, since $\gcd(a, p) = 1$, it is evident that $p \mid a^{p-1}-1$, or equivalently

$$a^{p-1} \equiv 1 \,(\text{mod } p). \qquad \square$$

Theorem 4.1.3 (Fermat–Euler Theorem). *For every pair of coprime integers a, m it holds*

$$a^{\phi(m)} \equiv 1 \,(\text{mod } m)$$

(where $\phi(m)$ is the Euler function).

Proof. We shall first prove the theorem in the special case when m is a perfect power of a prime number p. Thus, let $m = p^k$ for some positive integer k. Then, for $k = 1$, by Fermat's Little Theorem we get

$$a^{\phi(m)} = a^{\phi(p)} = a^{p-1} \equiv 1 \,(\mathrm{mod}\, p).$$

We now assume that

$$a^{\phi(p^k)} \equiv 1 \,(\mathrm{mod}\, p^k)$$

and we are going to prove that

$$a^{\phi(p^{k+1})} \equiv 1 \,(\mathrm{mod}\, p^{k+1}).$$

We have

$$a^{\phi(p^k)} - 1 = cp^k,$$

for some integer c. Thus,

$$a^{p^k - p^{k-1}} = 1 + cp^k$$

or

$$a^{p^{k+1} - p^k} = (1 + cp^k)^p$$

or

$$a^{\phi(p^{k+1})} = (1 + cp^k)^p. \tag{1}$$

But, by the binomial identity we have

$$(1 + cp^k)^p = 1 + \binom{p}{1} cp^k + \cdots + \binom{p}{p-1}(cp^k)^{p-1} + (cp^k)^p,$$

and since p divides each of the integers

$$\binom{p}{1}, \binom{p}{2}, \ldots, \binom{p}{p-1},$$

it is evident that there exists an integer c', for which

$$(1 + cp^k)^p = 1 + c'p^{k+1}.$$

Therefore, by (1) we get

$$a^{\phi(p^{k+1})} = 1 + c'p^{k+1},$$

and equivalently

$$a^{\phi(p^{k+1})} \equiv 1 \,(\mathrm{mod}\, p^{k+1}).$$

Hence, by the Mathematical Induction Principle we deduce that

$$a^{\phi(p^k)} \equiv 1 \,(\mathrm{mod}\, p^k),$$

for every positive integer k.

If m is not a perfect power of a prime number, by the Fundamental Theorem of Arithmetic we can express m in the form

$$m = p_1^{k_1} p_2^{k_2} \cdots p_n^{k_n},$$

where $n \geq 2$ and $p_1, p_2, \ldots, p_n$ are the prime divisors of m.

Furthermore, we have proved that the Euler ϕ-function is multiplicative for every pair of coprime integers. Thus, we can write

$$(((a^{\phi(p_1^{k_1})})^{\phi(p_2^{k_2})})\cdots)^{\phi(p_n^{k_n})} \equiv (((1^{\phi(p_1^{k_1})})^{\phi(p_2^{k_2})})\cdots)^{\phi(p_n^{k_n})} \,(\mathrm{mod}\, p_1^{k_1})$$

or

$$a^{\phi(p_1^{k_1} p_2^{k_2} \cdots p_n^{k_n})} \equiv 1 \,(\mathrm{mod}\, p_1^{k_1})$$

or

$$a^{\phi(m)} \equiv 1 \,(\mathrm{mod}\, p_1^{k_1}).$$

Similarly, we can prove that

$$a^{\phi(m)} \equiv 1 \,(\mathrm{mod}\, p_2^{k_2}), \ldots, a^{\phi(m)} \equiv 1 \,(\mathrm{mod}\, p_n^{k_n}).$$

But, generally it is true that if $\alpha \equiv \beta \,(\mathrm{mod}\, \gamma_1)$ and $\alpha \equiv \beta \,(\mathrm{mod}\, \gamma_2)$, with $\gcd(\gamma_1, \gamma_2) = 1$, then $\alpha \equiv \beta \,(\mathrm{mod}\, \gamma_1 \gamma_2)$. Therefore, since

$$\gcd(p_1^{k_1}, p_2^{k_2}, \ldots, p_n^{k_n}) = 1,$$

we obtain

$$a^{\phi(m)} \equiv 1 \,(\mathrm{mod}\, m). \qquad \square$$

The above theorem is a generalization of Fermat's Little Theorem and was first proved by Leonhard Euler in 1758.

Theorem 4.1.4. *Let a, b, $c \in \mathbb{Z}$, where at least one of a, b is nonzero. If $d = \gcd(a, b)$ and $d \mid c$, then the diophantine equation*

$$ax + by = c$$

has infinitely many solutions of the form

$$x = x_0 + \frac{b}{d}n, \quad y = y_0 - \frac{a}{d}n,$$

where n is a positive integer and (x_0, y_0) is a solution of the equation.

In case $d \nmid c$, the diophantine equation

$$ax + by = c$$

has no solutions.

Proof.

Case 1. If $d \mid c$, then there exists an integer k for which $c = kd$. But, because of the fact that d is the greatest common divisor of a and b, by Bezout's Lemma we know that there exist integers k_1, k_2 such that

$$d = k_1 a + k_2 b$$

and thus

$$c = kk_1 a + kk_2 b.$$

Hence, there is at least one pair of integers $x_0 = kk_1$, $y_0 = kk_2$ which is a solution of the diophantine equation. In order to prove that there exist infinitely many solutions and specifically of the form

$$x = x_0 + \frac{b}{d} n, \quad y = y_0 - \frac{a}{d} n,$$

we set (x, y) to be an arbitrary solution of the diophantine equation. Then, we have

$$ax + by = c$$

and

$$ax_0 + by_0 = c.$$

Thus,

$$a(x - x_0) + b(y - y_0) = 0$$

or

$$a(x - x_0) = b(y_0 - y)$$

or

$$\frac{a}{d}(x - x_0) = \frac{b}{d}(y_0 - y). \tag{1}$$

Thus,

$$\frac{b}{d} \Big| \frac{a}{d}(x - x_0).$$

But

$$\gcd\left(\frac{a}{d}, \frac{b}{d}\right) = 1$$

and therefore

$$\frac{b}{d} \Big| (x - x_0).$$

Hence, there exists an integer n for which

$$x = x_0 + n\frac{b}{d}. \tag{2}$$

Let us suppose without loss of generality that $b \neq 0$. Then, by (1) and (2) we obtain

$$\frac{a}{d} n \frac{b}{d} = \frac{b}{d}(y_0 - y)$$

or
$$\frac{a}{d}n = y_0 - y$$

or
$$y = y_0 - \frac{a}{d}n.$$

For $a \neq 0$ the procedure is exactly the same and we deduce the same result. Hence, for a fixed integer n the pair (x, y), where

$$x = x_0 + \frac{b}{d}n, \quad y = y_0 - \frac{a}{d}n,$$

is a solution of the equation $ax + by = c$. But, if we consider an arbitrary integer t, for which

$$x = x_0 + \frac{b}{d}t, \quad y = y_0 - \frac{a}{d}t,$$

then we get

$$c = a\left(x_0 + \frac{b}{d}t\right) + b\left(y_0 - \frac{a}{d}t\right)$$

$$= ax_0 + \frac{ab}{d}t + by_0 - \frac{ba}{d}t$$

$$= ax_0 + by_0,$$

which holds true.

Therefore, the diophantine equation $ax + by = c$ has infinitely many solutions of the form

$$x = x_0 + \frac{b}{d}n, \quad y = y_0 - \frac{a}{d}n, \quad \text{for} \quad n \in \mathbb{N}$$

Case 2. Let us now suppose that $d \nmid c$. But, $d \mid a$ and $d \mid b$, thus

$$d \mid ax + by$$

and consequently $d \mid c$, which is a contradiction.

Thus, in this case, the diophantine equation $ax + by = c$ has no solutions.

$\square$

Theorem 4.1.5. *Let $a, b \in \mathbb{Z}$ and $m \in \mathbb{N}$. If $d = \gcd(a, m)$ and $d \mid b$, then the linear congruence*

$$ax \equiv b \,(\mathrm{mod}\, m)$$

has d, pairwise distinct, solutions modulo m.

If $d \nmid b$, then the linear congruence has no solutions.

Remark. Two solutions x_1 and x_2 are said to be distinct if and only if $x_1 \not\equiv x_2 \,(\mathrm{mod}\, m)$.

Proof.

Case 1. If $d \mid b$, then the linear congruence $ax \equiv b \, (\mathrm{mod}\, m)$ has a solution if the diophantine equation

$$ax - my = b \tag{1}$$

has a solution. But, (1) has infinitely many solutions with

$$x = x_0 - \frac{m}{d}n,$$

where (x_0, y_0) is a solution of (1).

We shall prove that from the infinitely many solutions of the linear congruence

$$ax \equiv b \, (\mathrm{mod}\, m),$$

exactly d are pairwise distinct.

We can observe that all integers

$$x_0, x_0 - \frac{m}{d}, \ x_0 - 2\frac{m}{d}, \ldots, x_0 - (d-1)\frac{m}{d}$$

are solutions of the linear congruence $ax \equiv b \, (\mathrm{mod}\, m)$. These solutions are pairwise distinct, because if there was a pair of these solutions for which

$$x_0 - n_1 \frac{m}{d} \equiv x_0 - n_2 \frac{m}{d} \, (\mathrm{mod}\, m),$$

where $n_1, n_2 \in \mathbb{N}$ with $1 \le n_1, n_2 \le d - 1$, then we would have

$$n_1 \frac{m}{d} \equiv n_2 \frac{m}{d} \, (\mathrm{mod}\, m)$$

or

$$m \left| (n_1 - n_2)\frac{m}{d} \right. \Rightarrow d \mid (n_1 - n_2),$$

which is a contradiction since $1 \le n_1, n_2 \le d - 1$. Therefore, the solutions

$$x_0, x_0 - \frac{m}{d}, \ x_0 - 2\frac{m}{d}, \ldots, x_0 - (d-1)\frac{m}{d}$$

are pairwise distinct. We shall now prove that there are no other solutions of the linear congruence $ax \equiv b \, (\mathrm{mod}\, m)$, such that all solutions remain pairwise distinct.

Let $k \in \mathbb{Z}$ be a solution of the linear congruence, different from the above. Then

$$ak \equiv b \, (\mathrm{mod}\, m),$$

while we know that $ax_0 \equiv b \, (\mathrm{mod}\, m)$ also holds. Therefore, we get

$$ak \equiv ax_0 \, (\mathrm{mod}\, m). \tag{2}$$

But, since $\gcd(a, m) = d$, we can write

$$a = \lambda_1 d, \quad m = \lambda_2 d,$$

where λ_1, λ_2 are relatively prime integers. Hence, by (2) we obtain

$$\lambda_1 dk \equiv \lambda_1 dx_0 \,(\text{mod } \lambda_2 d).$$

Thus,

$$\lambda_2 \mid \lambda_1(k - x_0).$$

But, since $\gcd(\lambda_1, \lambda_2) = 1$ it is evident that

$$\lambda_2 \mid (k - x_0).$$

Thus, there exists an integer ν for which

$$k = x_0 + \nu\lambda_2.$$

By the division algorithm we have

$$\nu = dq + r,$$

for some integers q, r with $0 \leq r < d$. Thus, we get

$$k = x_0 + d\lambda_2 q + \lambda_2 r$$

$$= x_0 + mq + \frac{m}{d}r$$

and therefore

$$mq = k - \left(x_0 + \frac{m}{d}r\right).$$

Hence, equivalently we can write

$$k \equiv x_0 + \frac{m}{d}r \,(\text{mod } m),$$

where $0 \leq r \leq d-1$. Thus, k is not considered to be a distinct solution, which is a contradiction. This completes the proof in the case that $d \mid b$.

Case 2. If $d \nmid b$, then the diophantine equation

$$ax - my = b$$

does not have any solutions in terms of x, y. Therefore, the linear congruence

$$ax \equiv b \,(\text{mod } m)$$

does not have any solutions.   □

Remark 4.1.6. In the special case when $\gcd(a, m) = 1$, the linear congruence $ax \equiv b \,(\text{mod } m)$ has a unique solution.

Theorem 4.1.7 (Lagrange Theorem). *Consider the polynomial*

$$f(x) = a_n x^n + a_{n-1} x^{n-1} + \cdots + a_1 x + a_0,$$

where $a_0, a_1, \ldots, a_n \in \mathbb{Z}$ and $a_n \neq 0$.

If p is a prime number and $a_n \not\equiv 0 \,(\mathrm{mod} p)$, then the polynomial congruence

$$f(x) \equiv 0 \,(\mathrm{mod}\, p)$$

has at most n solutions.

Proof. For $n = 1$ we have $f(x) = a_1 x + a_0$. But, by Remark 4.1.6 it is evident that the linear congruence

$$a_1 x + a_0 \equiv 0 \,(\mathrm{mod}\, p)$$

has a unique solution and thus

$$f(x) \equiv 0 \,(\mathrm{mod}\, p)$$

has exactly one solution. Therefore, in this case the theorem is proved.

We now assume that the theorem holds for polynomials up to $n - 1$ degree and the polynomial congruence

$$a_n x^n + a_{n-1} x^{n-1} + \cdots + a_1 x + a_0 \equiv 0 \,(\mathrm{mod}\, p)$$

has at least $n + 1$ solutions

$$x_0, x_1, \ldots, x_n.$$

In that case, we obtain

$$(a_n x_i^n + a_{n-1} x_i^{n-1} + \cdots + a_1 x_i + a_0) - (a_n x_0^n + a_{n-1} x_0^{n-1} + \cdots + a_1 x_0 + a_0)$$

$$= a_n(x_i^n - x_0^n) + a_{n-1}(x_i^{n-1} - x_0^{n-1}) + \cdots + a_1(x_i - x_0)$$

$$= (x_i - x_0)p(x_i), \quad i = 1, 2, \ldots, n,$$

where $p(x)$ is a polynomial of $n - 1$ degree with integer coefficients.

Because of the fact that

$$p \mid (a_n x_i^n + a_{n-1} x_i^{n-1} + \cdots + a_1 x_i + a_0)$$

and

$$p \mid (a_n x_0^n + a_{n-1} x_0^{n-1} + \cdots + a_1 x_0 + a_0)$$

it is evident that

$$p \mid (x_i - x_0)p(x_i), \quad i = 1, 2, \ldots, n.$$

The integers x_i, x_0 are distinct solutions and thus

$$x_i \not\equiv x_0 \,(\mathrm{mod}\,p).$$

Therefore,

$$p \mid p(x_i), \quad i = 1, 2, \ldots, n,$$

and consequently the polynomial congruence

$$p(x) \equiv 0 \,(\mathrm{mod}\,p)$$

has n solutions, which is impossible since the polynomial $p(x)$ is of $n-1$ degree and we have assumed that the theorem holds true for polynomials of degree up to $n - 1$.

Hence, by the Mathematical Induction Principle, it follows that the polynomial congruence

$$a_n x^n + a_{n-1} x^{n-1} + \cdots + a_1 x + a_0 \equiv 0 \,(\mathrm{mod}\,p),$$

where $a_n \not\equiv 0 \,(\mathrm{mod}\,p)$, has at most n solutions.   □

Theorem 4.1.8. *Consider the polynomial*

$$f(x) = a_n x^n + a_{n-1} x^{n-1} + \cdots + a_1 x + a_0,$$

where $a_0, a_1, \ldots, a_n \in \mathbb{Z}$ and $a_n \neq 0$. If p is a prime number and the polynomial congruence $f(x) \equiv 0 \,(\mathrm{mod}\,p)$ has more than n solutions, then p divides all the coefficients of the polynomial $f(x)$.

Proof. Since the polynomial congruence $f(x) \equiv 0 \,(\mathrm{mod}\,p)$ has more than n solutions, it follows that $p \mid a_n$. This happens because if $a_n \not\equiv 0 \,(\mathrm{mod}\,p)$, then by the Lagrange Theorem the congruence $f(x) \equiv 0 \,(\mathrm{mod}\,p)$ should have at most n solutions, which contradicts our hypothesis. Therefore, for each solution x_0 of $f(x) \equiv 0 \,(\mathrm{mod}\,p)$ we obtain

$$p \mid a_{n-1} x^{n-1} + \cdots + a_1 x + a_0.$$

Thus, the polynomial congruence

$$a_{n-1} x^{n-1} + \cdots + a_1 x + a_0 \equiv 0 \,(\mathrm{mod}\,p)$$

has more than n solutions. Hence, similarly it follows that

$$p \mid a_{n-1}.$$

According to the above arguments, it is evident that for every $\nu \leq n$, the polynomial congruence

$$a_\nu x^\nu + a_{\nu-1} x^{\nu-1} + \cdots + a_1 x + a_0 \equiv 0 \,(\mathrm{mod}\,p),$$

has more than ν solutions. Therefore, $p \mid a_\nu$, for every $\nu = 1, 2, \ldots, n$.   □

Theorem 4.1.9 (Chinese Remainder Theorem). *Let* $m_1, m_2, \ldots, m_k,$ $a_1, a_2, \ldots, a_k \in \mathbb{Z}$, *such that* $\gcd(m_i, m_j) = 1$, *for* $i \neq j$ *and* $\gcd(a_i, m_i) = 1,$ *for every* i, *where* $1 \leq i, j \leq k$. *If* $m = m_1 m_2 \cdots m_k$, *then the system of linear equations*

$$a_1 x \equiv b_1 \,(\mathrm{mod}\ m_1)$$

$$a_2 x \equiv b_2 \,(\mathrm{mod}\ m_2)$$

$$\vdots$$

$$a_k x \equiv b_k \,(\mathrm{mod}\ m_k)$$

has a unique solution modulo m.

Proof. At first, we shall prove that the system of linear congruences has a solution modulo m and afterwards we shall prove the uniqueness of that solution.

Set $r_i = m/m_i$. Then, it is obvious that $\gcd(r_i, m_i) = 1$ and thus, the linear congruence $r_i x \equiv 1 \,(\mathrm{mod}\ m_i)$ has a unique solution. If r_i' denotes that solution, we have

$$r_i r_i' \equiv 1 \,(\mathrm{mod}\ m_i), \quad \text{for} \quad i = 1, 2, \ldots, k.$$

Let x_i denote the unique solution of the linear congruence

$$a_i x \equiv b_i \,(\mathrm{mod}\ m_i).$$

We shall prove that the integer

$$x_0 = \sum_{i=1}^{k} x_i r_i r_i',$$

is a solution modulo m of the system.

Since x_i is the unique solution of $a_i x \equiv b_i \,(\mathrm{mod}\ m_i)$ we have

$$a_i x_i \equiv b_i \,(\mathrm{mod}\ m_i).$$

But, in addition we know that

$$r_i r_i' \equiv 1 \,(\mathrm{mod}\ m_i)$$

and therefore we obtain

$$a_i x_i r_i r_i' \equiv b_i \,(\mathrm{mod}\ m_i). \tag{1}$$

Furthermore, in case $i \neq j$ it is clear that $m_i \mid r_j$. Thus,

$$m_i \mid a_i x_j r_j r_j', \quad \text{for every} \quad j \neq i. \tag{2}$$

However, we have

$$a_i x_0 - b_i = a_i x_1 r_1 r_1' + a_i x_2 r_2 r_2' + \cdots + (a_i x_i r_i r_i' - b_i) + \cdots + a_i x_k r_k r_k'.$$

Hence, by (1) and (2) we obtain that

$$m_i \mid (a_i x_0 - b_i),$$

or

$$a_i x_0 \equiv b_i \, (\mathrm{mod}\ m_i).$$

Moreover, because of the fact that the integers $m_1, m_2, \ldots, m_k$ are coprime, it holds

$$a_i x_0 \equiv b_i \, (\mathrm{mod}\ m_1 m_2 \cdots m_k)$$

or

$$a_i x_0 \equiv b_i \, (\mathrm{mod}\ m).$$

Therefore, it suffices to prove that x_0 is the unique solution modulo m of the system. Let us assume that there exists another solution modulo m of the system and denote it by x_0'. Then

$$m \mid (a_i x_0' - b_i)$$

and thus

$$m_i \mid (a_i x_0' - b_i).$$

But, since $m_i \mid (a_i x_0 - b_i)$, it is evident that

$$m_i \mid a_i (x_0' - x_0), \quad \text{for every} \quad i = 1, 2, \ldots, k.$$

In addition, since $\gcd(a_i, m_i) = 1$, it yields

$$x_0' \equiv x_0 \, (\mathrm{mod}\ m_i).$$

But, since the integers $m_1, m_2, \ldots, m_k$ are coprime, it follows that

$$x_0' \equiv x_0 \, (\mathrm{mod}\ m).$$

Hence, the solutions x_0' and x_0 are not distinct and thus x_0 is the unique solution of the system of linear congruences. This completes the proof of the theorem. $\qquad\square$

Historical Remark. The Chinese Remainder Theorem is an ancient result which first appeared in the work of the Chinese mathematician Sun Tzu, entitled *Suanjing*, in about the 4th century AD. According to D. Wells (see [61]), Sun Tzu in his work mentions the following:

> There are certain things whose number is unknown.
> Repeatedly divided by 3, the remainder is 2,
> by 5 the remainder is 3,
> and by 7 the remainder is 2.
> What will be the number?

Notwithstanding the fact that the Chinese remainder theorem first appeared in the work of Sun Tzu, the complete theorem was presented for the first time in 1247, by the Chinese mathematician Qin Jiushao in his treatise entitled *Shùshū Jiǔzhāng*.

5

Quadratic residues

*Mathematics is concerned only with the enumeration
and comparison of relations.*
Carl Friedrich Gauss (1777–1855)

5.1 Introduction

Definition 5.1.1. *An integer a is called a **quadratic residue modulo** c, if $\gcd(a, c) = 1$ and the congruence $x^2 \equiv a \,(\mathrm{mod}\ c)$ has a solution. If the congruence does not have any solution, then a is called a **quadratic nonresidue modulo** c.*

For example,

$$3^2 \equiv 1 \,(\mathrm{mod}\ 4) \quad \text{and} \quad 6^2 \equiv 11 \,(\mathrm{mod}\ 5).$$

Therefore, 1 is a *quadratic residue modulo 4* and 11 is a *quadratic residue modulo 5*.

We shall now present some basic theorems concerning quadratic residues.

Theorem 5.1.2. *Let p be an odd prime number and a an integer for which $\gcd(a, p) = 1$. Then, the congruence*

$$x^2 \equiv a \,(\mathrm{mod}\ p), \tag{1}$$

will either have two distinct solutions[1] or no solutions at all.

---

[1] By distinct solutions we mean solutions which are not equivalent mod p.

M.Th. Rassias, *Problem-Solving and Selected Topics in Number Theory: In the Spirit of the Mathematical Olympiads*, DOI 10.1007/978-1-4419-0495-9_5,
© Springer Science+Business Media, LLC 2011

Proof. Let us assume that the congruence $x^2 \equiv a \,(\mathrm{mod}\,p)$ has a solution x_0. Then, we have

$$x_0^2 \equiv a \,(\mathrm{mod}\,p).$$

Thus, obviously, we also have

$$(-x_0)^2 \equiv a \,(\mathrm{mod}\,p).$$

Hence, if x_0 is a solution of (1), that yields that $-x_0$ is also a solution of (1). Moreover, these solutions are distinct, since

$$x_0 \not\equiv -x_0 \,(\mathrm{mod}\,p).$$

This happens because if $x_0 \equiv -x_0 \,(\mathrm{mod}\,p)$, then it follows that $p \mid x_0$, which is a contradiction since $p \mid (x_0^2 - a)$ and $\gcd(a, p) = 1$.

We shall now prove that there are no other distinct solutions of (1). This follows immediately by Theorem 4.1.8. However, here we will present a different proof.

Let x_0' be a solution of (1), different than x_0 and $-x_0$. Therefore, it is evident that

$$x_0^2 - (x_0')^2 \equiv 0 \,(\mathrm{mod}\,p)$$

or

$$(x_0 - x_0')(x_0 + x_0') \equiv 0 \,(\mathrm{mod}\,p).$$

Consequently, by Euclid's first theorem, it follows

$$p \mid (x_0 - x_0') \text{ or } p \mid (x_0 + x_0').$$

Thus, equivalently we have

$$x_0 \equiv x_0' \,(\mathrm{mod}\,p) \text{ or } -x_0 \equiv x_0' \,(\mathrm{mod}\,p).$$

Hence, the solution x_0' is not different than x_0 and $-x_0$, which is a contradiction. $\square$

Theorem 5.1.3. *If p is an odd prime, then there exist exactly $(p-1)/2$ quadratic residues and $(p-1)/2$ quadratic nonresidues* $\mathrm{mod}\,p$.

Proof. It is clear that

$$p - 1 \equiv -1 \,(\mathrm{mod}\,p)$$

$$p - 2 \equiv -2 \,(\mathrm{mod}\,p)$$

$$\vdots$$

$$p - \frac{p-1}{2} \equiv -\frac{p-1}{2} \,(\mathrm{mod}\,p).$$

Therefore, it is evident that

$$(p-1)^2 \equiv 1^2 \,(\mathrm{mod}\,p)$$

$$(p-2)^2 \equiv 2^2 \,(\mathrm{mod}\,p)$$

$$\vdots$$

$$\left(p - \frac{p-1}{2}\right)^2 \equiv \left(\frac{p-1}{2}\right)^2 \,(\mathrm{mod}\,p).$$

So, each of the integers $1^2, 2^2, \ldots, \left(\frac{p-1}{2}\right)^2$ is a quadratic residue mod p. We shall now prove that these are also pairwise not congruent mod p.

Let

$$x_1, x_2 \in \left\{1, 2, \ldots, \frac{p-1}{2}\right\}.$$

Then

$$1 < x_1 + x_2 < p. \tag{1}$$

Therefore, if $x_1^2 \equiv x_2^2 \,(\mathrm{mod}\,p)$, where $x_1 \neq x_2$, it yields $p \mid (x_1 - x_2)(x_1 + x_2)$ and thus $p \mid (x_1 - x_2)$ or $p \mid (x_1 + x_2)$. However, by (1) it follows that

$$p \mid (x_1 - x_2).$$

Since

$$|x_1 - x_2| < p,$$

we get that $x_1 = x_2$, which is a contradiction. Hence, according to the above arguments, it follows that there exist exactly $(p-1)/2$ quadratic residues and $(p-1)/2$ quadratic nonresidues mod p. More specifically, the integers

$$1^2, 2^2, \ldots, \left(\frac{p-1}{2}\right)^2$$

are the quadratic residues. □

Theorem 5.1.4 (Dirichlet's Theorem). *Let p be a prime number and a be an integer such that $1 \leq a \leq p-1$. If the congruence $x^2 \equiv a\,(\mathrm{mod}\,p)$ does not have any solutions, then*

$$p \mid (p-1)! - a^{(p-1)/2}.$$

Else, if $x^2 \equiv a\,(\mathrm{mod}\,p)$ has solutions, then

$$p \mid (p-1)! + a^{(p-1)/2}.$$

Proof. If the prime number p is even, then the validity of the theorem is obvious. Thus, we will examine the case when p is an odd prime number. Let us consider the linear congruence

$$a_1 x \equiv a \,(\mathrm{mod}\, p), \tag{1}$$

where $1 \leq a_1 \leq p - 1$.

By Remark 4.1.6 we know that (1) has a unique solution and by the proof of Theorem 4.1.5 it follows that this solution belongs to the set of integers

$$\{0, 1, \ldots, p - 1\}$$

or the set

$$\left\{ -\frac{p-1}{2}, \ldots, -2, -1, 0, 1, 2, \ldots, \frac{p-1}{2} \right\}.$$

Without loss of generality, we assume that $x \in \{1, 2, \ldots, p - 1\}$. The element 0 is excluded since it must hold $\gcd(a, p) = 1$.

If b is a solution of (1), then

$$a_1 b \equiv a \,(\mathrm{mod}\, p). \tag{2}$$

Therefore, if the congruence

$$x^2 \equiv a \,(\mathrm{mod}\, p) \tag{3}$$

has no solutions, it yields that $a_1 \neq b$. But, since $a_1, b \in \{1, 2, \ldots, p-1\}$, we can partition this set in $(p - 1)/2$ distinct pairs (a_1, b), with $a_1 \neq b$, for which (2) holds true.

Moreover, if we multiply by parts the linear congruences which are derived by those pairs, we obtain

$$(p - 1)! \equiv a^{(p-1)/2} \,(\mathrm{mod}\, p).$$

Thus,

$$p \mid (p - 1)! - a^{(p-1)/2}.$$

If the congruence $x^2 \equiv a \,(\mathrm{mod}\, p)$ has solutions, then by Theorem 5.1.2 it follows that it must have exactly two solutions. Without loss of generality, we assume again that these solutions belong to the set $\{1, 2, \ldots, p - 1\}$.

Let k be one of these two solutions of (3). Then, it is clear that $p - k$ is also a solution and since (3) can only have two solutions, it is obvious that k and $p - k$ are the only ones.

Let us exclude k and $p - k$ from the set $\{1, 2, \ldots, p - 1\}$ and partition the $p - 3$ remaining integers in $(p - 3)/2$ distinct pairs (a_1, b), with $a_1 \neq b$, for which (2) holds true. If we multiply by parts the linear congruences which are derived by those pairs, we obtain

$$\frac{(p - 1)!}{k \cdot (p - k)} \equiv a^{(p-3)/2} \,(\mathrm{mod}\, p),$$

where clearly $N = (p - 1)!/(k \cdot (p - k))$ is a positive integer.

However,

$$k \cdot (p - k) = kp - k^2 \equiv -a \, (\mathrm{mod} \, p).$$

Therefore,

$$N \cdot k \cdot (p - k) \equiv a^{(p-3)/2}(-a) \, (\mathrm{mod} \, p),$$

and hence

$$(p - 1)! \equiv (-a)^{(p-1)/2} \, (\mathrm{mod} \, p).$$

Thus,

$$p \mid (p - 1)! + a^{(p-1)/2}. \qquad \qquad \square$$

Historical Remark. The above theorem was proved by P. Dirichlet in 1828. We shall use Dirichlet's result in order to prove another important theorem, Wilson's theorem.

Wilson's theorem was initially introduced as *Wilson's conjecture* and was announced by his professor Ed. Waring, in 1770. The theorem was proved for the first time by J.L. Lagrange, in 1771. Two years later, in 1773, L. Euler presented a different proof. A third proof is presented in Gauss's book entitled *Disquisitiones Arithmeticae*.

Finally, the theorem was named after Wilson, notwithstanding the fact that G. Leibniz had, almost one hundred years earlier, discovered an equivalent theorem.

Theorem 5.1.5 (Wilson's Theorem). *If p is a prime number, then*

$$p \mid (p - 1)! + 1$$

and conversely if

$$p \mid (p - 1)! + 1,$$

then p is a prime number.

First Proof. We shall first prove that if p is a prime number, then

$$p \mid (p - 1)! + 1.$$

In order to do so, we will use Dirichlet's Theorem.

Consider the congruence

$$x^2 \equiv 1 \, (\mathrm{mod} \, p),$$

which obviously has a solution (for example, $x = p - 1$).

If we apply Dirichlet's Theorem for $a = 1$, we obtain

$$p \mid (p - 1)! + 1^{(p-1)/2}$$

or equivalently

$$p \mid (p - 1)! + 1.$$

In order to prove the converse, we assume that

$$p \mid (p-1)! + 1,$$

and we shall prove that p must be a prime number.

It is clear that none of the integers $2, 3, \ldots, p-1$ divides $(p-1)! + 1$. Thus, the least positive integer which divides $(p-1)! + 1$ is p. However, in Lemma 1.1.4 we proved that the least nontrivial divisor of every positive integer greater than 1 is a prime number. Hence, it is evident that p is a prime number.

Second Proof. (Lagrange). Consider the polynomial

$$f(x) = (x-1)(x-2) \cdots (x-(p-1)) - (x^{p-1} - 1),$$

where $x = 1, 2, \ldots, p-1$ and p is a prime number.

It is evident that $\gcd(x, p) = 1$ and therefore, by Fermat's Little Theorem we have

$$x^{p-1} \equiv 1 \, (\mathrm{mod} \, p).$$

In addition, it is clear that one of the integers $x - 1, x - 2, \ldots, x - (p-1)$ must be equal to zero. Thus,

$$p \mid (x-1)(x-2) \cdots (x-(p-1)).$$

Hence, by the above two relations, we obtain that the polynomial congruence

$$f(x) \equiv 0 \, (\mathrm{mod} \, p)$$

has $p - 1$ solutions. However, since the polynomial $f(x)$ is of degree $p - 2$, by Theorem 4.1.8 it yields that if

$$f(x) = a_{p-2} x^{p-2} + \cdots + a_1 x + a_0,$$

then

$$p \mid a_0, p \mid a_1, \ldots, p \mid a_{p-2}.$$

But $a_0 = (p-1)! + 1$, thus

$$p \mid (p-1)! + 1.$$

In order to prove the converse, we follow the same method as in Proof 1. □

5.2 Legendre's symbol

Definition 5.2.1. *Let p be an odd prime number and a be an integer such that $\gcd(a, p) = 1$. We define **Legendre's symbol** $\left(\frac{a}{p}\right)$ by*

$$\left(\frac{a}{p}\right) = \begin{cases} 1, & \text{if } a \text{ is a quadratic residue } \bmod p \\ -1, & \text{if } a \text{ is a quadratic nonresidue } \bmod p. \end{cases}$$

In case $p \mid a$, Legendre's symbol is defined to be equal to zero.

$$\left(\frac{a}{p}\right) = 0, \text{ if } p \mid a.$$

For example,

$$\left(\frac{11}{7}\right) = 1, \left(\frac{6}{13}\right) = -1, \left(\frac{15}{5}\right) = 0.$$

We shall now prove some basic theorems and properties related to Legendre's symbol.

Theorem 5.2.2 (Euler's Criterion). *Let p be an odd prime number and a an integer such that $\gcd(a, p) = 1$. Then, it holds*

$$\left(\frac{a}{p}\right) \equiv a^{(p-1)/2} \pmod{p}.$$

Proof. Since the hypothesis ensures that $p \nmid a$, by the definition of Legendre's symbol, we obtain that

$$\left(\frac{a}{p}\right) = \pm 1.$$

- If $\left(\frac{a}{p}\right) = 1$, then the integer a is a quadratic residue mod p and thus, there exists an integer x_0, such that

$$x_0^2 \equiv a \pmod{p}.$$

Therefore,

$$(x_0^2)^{(p-1)/2} \equiv a^{(p-1)/2} \pmod{p}$$

or

$$x_0^{p-1} \equiv a^{(p-1)/2} \pmod{p}$$

or

$$a^{(p-1)/2} \equiv x_0^{p-1} \pmod{p}. \tag{1}$$

But, because of the fact that $p \mid (x_0^2 - a)$ and $\gcd(a, p) = 1$, it yields $\gcd(x_0, p) = 1$. Hence, by Fermat's Little Theorem, we get

$$x_0^{p-1} \equiv 1 \pmod{p}. \tag{2}$$

Thus, by (1), (2) it follows

$$1 \equiv a^{(p-1)/2} \pmod{p}$$

or

$$\left(\frac{a}{p}\right) \equiv a^{(p-1)/2} \pmod{p}.$$

- If $\left(\frac{a}{p}\right) = -1$, then the integer a is a quadratic nonresidue $\bmod p$ and thus the congruence $x^2 \equiv a \,(\mathrm{mod}\ p)$ does not have any solutions. However, in this case, Dirichlet's theorem ensures that

$$p \mid (p-1)! - a^{(p-1)/2}$$

and therefore

$$a^{(p-1)/2} \equiv (p-1)! \,(\mathrm{mod}\ p). \qquad (3)$$

But, by Wilson's Theorem, (3) takes the form

$$a^{(p-1)/2} \equiv -1 \,(\mathrm{mod}\ p)$$

or

$$-1 \equiv a^{(p-1)/2} \,(\mathrm{mod}\ p).$$

Hence, we have

$$\left(\frac{a}{p}\right) \equiv a^{(p-1)/2} \,(\mathrm{mod}\ p).$$

This completes the proof of Euler's Criterion.   □

Theorem 5.2.3. *Let p be an odd prime number and a be an integer, such that $\gcd(a,p) = 1$. If $a \equiv b \,(\mathrm{mod}\ p)$, then it holds*

$$\left(\frac{a}{p}\right) = \left(\frac{b}{p}\right).$$

Proof. It is clear that $\gcd(b,p) = 1$, since if $p \mid b$, then we would have $p \mid a$ which contradicts the hypothesis of the theorem. Therefore, because of the fact that $a \equiv b \,(\mathrm{mod}\ p)$, it is evident that a is a quadratic residue (or nonresidue, respectively) $\bmod p$ if and only if b is a quadratic residue (or nonresidue, respectively) $\bmod p$. Hence, by the definition of Legendre's symbol, it follows that

$$\left(\frac{a}{p}\right) = \left(\frac{b}{p}\right).$$ □

Theorem 5.2.4. *Let p be an odd prime number and a, b be integers, such that $\gcd(ab, p) = 1$. Then, it holds*

$$\left(\frac{ab}{p}\right) = \left(\frac{a}{p}\right)\left(\frac{b}{p}\right).$$

Therefore, Legendre's symbol is a completely multiplicative function.

Proof. By Euler's Criterion, we have

$$\left(\frac{ab}{p}\right) \equiv (ab)^{(p-1)/2} \,(\mathrm{mod}\ p).$$

Thus, equivalently we get

$$\left(\frac{ab}{p}\right) \equiv a^{(p-1)/2} b^{(p-1)/2} \pmod{p}$$

$$\equiv \left(\frac{a}{p}\right)\left(\frac{b}{p}\right) \pmod{p}.$$

Thus, equivalently we have

$$p \left| \left(\frac{ab}{p}\right) - \left(\frac{a}{p}\right)\left(\frac{b}{p}\right) \right.. \tag{1}$$

However, the only possible values of $\left(\frac{ab}{p}\right)$, $\left(\frac{a}{p}\right)$ and $\left(\frac{b}{p}\right)$ are $-1, 1$. Hence, the only possible values of the difference

$$D = \left(\frac{ab}{p}\right) - \left(\frac{a}{p}\right)\left(\frac{b}{p}\right)$$

are $0, 2, -2$. But, by (1) and the fact that p is an odd prime number, it yields that $D = 0$. Therefore,

$$\left(\frac{ab}{p}\right) = \left(\frac{a}{p}\right)\left(\frac{b}{p}\right). \qquad \square$$

Lemma 5.2.5. *Let p be an odd prime number. Then, it holds*

$$\left(\frac{-1}{p}\right) = (-1)^{(p-1)/2}.$$

Proof. By Euler's Criterion, we have

$$\left(\frac{-1}{p}\right) \equiv (-1)^{(p-1)/2} \pmod{p}.$$

However, the only possible values of $\left(\frac{-1}{p}\right)$ and $(-1)^{(p-1)/2}$ are $1, -1$. Therefore, since

$$p \left| \left(\frac{-1}{p}\right) - (-1)^{(p-1)/2} \right.$$

and p is an odd prime number, it follows that

$$\left(\frac{-1}{p}\right) = (-1)^{(p-1)/2}. \qquad \square$$

Lemma 5.2.6. *Let p be an odd prime number. Then, it holds*

$$\left(\frac{-1}{p}\right) = \begin{cases} 1, & \text{if } p \equiv 1 \pmod{4} \\ -1, & \text{if } p \equiv 3 \pmod{4}. \end{cases}$$

Proof. The prime number p can either take the form $4n+1$ or $4n+3$, where n is a natural number.

If $p = 4n+1$, then by Lemma 5.2.5, it follows

$$\left(\frac{-1}{p}\right) = (-1)^{(p-1)/2} = (-1)^{2n} = 1.$$

If $p = 4n+3$, then by Lemma 5.2.5, it yields

$$\left(\frac{-1}{p}\right) = (-1)^{(p-1)/2} = (-1)^{2n+1} = -1. \qquad \square$$

Theorem 5.2.7. *Let p be an odd prime number. Then it holds*

$$\left(\frac{2}{p}\right) = (-1)^{(p^2-1)/8} = \begin{cases} 1, & \text{if } p \equiv \pm 1 \,(\text{mod } 8) \\ -1, & \text{if } p \equiv \pm 3 \,(\text{mod } 8). \end{cases}$$

Proof. Consider the following $(p-1)/2$ congruences:

$$p - 1 \equiv 1 \cdot (-1)^1 \,(\text{mod } p)$$

$$2 \equiv 2 \cdot (-1)^2 \,(\text{mod } p)$$

$$p - 3 \equiv 3 \cdot (-1)^3 \,(\text{mod } p)$$

$$4 \equiv 4 \cdot (-1)^4 \,(\text{mod } p)$$

$$\vdots$$

$$k \equiv \frac{p-1}{2} \cdot (-1)^{(p-1)/2} \,(\text{mod } p),$$

where

$$k = \begin{cases} \frac{p-1}{2}, & \text{if the integer } (p-1)/2 \text{ is even} \\ p - \frac{p-1}{2}, & \text{if the integer } (p-1)/2 \text{ is odd.} \end{cases}$$

By multiplying by parts the above congruences, we obtain

$$2 \cdot 4 \cdot 6 \cdots (p-1) \equiv \left(\frac{p-1}{2}\right)!(-1)^{(p^2-1)/8} \,(\text{mod } p). \tag{1}$$

However, it is clear that

$$2 \cdot 4 \cdot 6 \cdots (p-1) \equiv (2 \cdot 1) \cdot (2 \cdot 2) \cdot (2 \cdot 3) \cdots \left(2 \cdot \frac{p-1}{2}\right)$$

$$= 2^{(p-1)/2} \left(\frac{p-1}{2}\right)!.$$

Therefore, (1) takes the form

$$2^{(p-1)/2} \left(\frac{p-1}{2}\right)! \equiv \left(\frac{p-1}{2}\right)!(-1)^{(p^2-1)/2} \pmod{p}. \qquad (2)$$

Moreover, since p does not divide $(\frac{p-1}{2})!$, by (2) we get

$$2^{(p-1)/2} \equiv (-1)^{(p^2-1)/8} \pmod{p}. \qquad (3)$$

Hence, by Euler's Criterion, we have

$$\left(\frac{2}{p}\right) \equiv 2^{(p-1)/2} \pmod{p}. \qquad (4)$$

By relations (3) and (4), we obtain

$$\left(\frac{2}{p}\right) \equiv (-1)^{(p^2-1)/8} \pmod{p}.$$

However, the only possible values of $\left(\frac{2}{p}\right)$ and $(-1)^{(p^2-1)/8}$ are $-1, 1$. Thus, it is evident that the only possible values of the difference

$$D = \left(\frac{2}{p}\right) - (-1)^{(p^2-1)/8}$$

are $0, 2, -2$. But, by the fact that p is an odd prime number, it follows that $D = 0$ and thus

$$\left(\frac{2}{p}\right) = (-1)^{(p^2-1)/8}.$$

The prime number p can take one of the forms

$$8n + 1 \text{ or } 8n + 3 \text{ or } 8n - 3 \text{ or } 8n - 1, \text{ where } n \in \mathbb{N}.$$

In case $p = 8n \pm 1$, it follows that

$$\frac{p^2 - 1}{8} = 8n^2 \pm 2n,$$

which is an even integer.

In case $p = 8n \pm 3$, it follows that

$$\frac{p^2 - 1}{8} = 8n^2 \pm 6n + 1,$$

which is an odd integer. Hence, in conclusion, one has

$$\left(\frac{2}{p}\right) = (-1)^{(p^2-1)/8} = \begin{cases} 1, & \text{if } p \equiv \pm 1 \pmod 8 \\ -1, & \text{if } p \equiv \pm 3 \pmod 8. \end{cases} \qquad \square$$

5.2.1 The law of quadratic reciprocity

Theorem 5.2.8 (Gauss's Lemma). *Let p be an odd prime number and a be an integer, such that $\gcd(a, p) = 1$. Consider the least positive residues* mod p *of the integers*

$$a, 2a, 3a, \ldots, \frac{p-1}{2}a.$$

If s denotes the number of these residues which are greater than $p/2$, it holds

$$\left(\frac{a}{p}\right) = (-1)^s.$$

Proof. It is clear that each of the integers ma, where $m = 1, 2, \ldots,$ $(p-1)/2$, when divided by p leaves a nonzero remainder, since $\gcd(a, p) = 1$ and $\gcd(m, p) = 1$, for every m. Now, we consider a partition of the set of the least positive residues mod p of the integers $a, 2a, 3a, \ldots, ((p-1)/2)a$, in two distinct sets as follows:

$$S_1 = \{r_1, r_2, \ldots, r_\lambda\}, \text{ if } r_i < \frac{p}{2}, \text{ where } i = 1, 2, \ldots, \lambda$$

and

$$S_2 = \{e_1, e_2, \ldots, e_s\}, \text{ if } e_i > \frac{p}{2}, \text{ where } i = 1, 2, \ldots, s.$$

It is evident that $s + \lambda = (p-1)/2$, since $S_1 \cap S_2 = \emptyset$. We shall now try to construct a third set S_3, for which

$$S_1 \cup S_3 = \{1, 2, \ldots, (p-1)/2\}.$$

We can observe that each element $r_i \in S_1$ is different than every w_j with $w_j = p - e_j$, where $e_j \in S_2$. Thus, for every pair (i, j), where $i = 1, 2, \ldots, \lambda$ and $j = 1, 2, \ldots, s$, it holds $r_i \neq w_j$. This is true because if we could determine a pair (i, j), for which $w_j = r_i$, we would have $p = r_i + e_j$. However, by the definition of r_i and e_j, we have

$$ka = k_i p + r_i, \text{ where } 1 \leq k \leq \frac{p-1}{2}, \text{ and } i = 1, 2, \ldots, \lambda, \qquad (1)$$

as well as

$$\nu a = \nu_j p + e_j, \text{ where } 1 \leq \nu \leq \frac{p-1}{2}, \text{ and } j = 1, 2, \ldots, s. \qquad (2)$$

Therefore, we would get

$$(k + \nu)a = (k_i + \nu_j)p + (r_i + e_j)$$

$$= (k_i + \nu_j)p + p$$

and because of the fact that $\gcd(a,p) = 1$, it should hold

$$k + \nu \equiv 0 \,(\mathrm{mod}\,p),$$

which is a contradiction, since

$$2 \le k + \nu \le p - 1.$$

Hence, by the above arguments, it follows that the sets S_1 and $\{w_1, w_2, \ldots, w_s\}$ are mutually disjoint. In addition, it is a fact that $w_j \in \{1, \ldots, \frac{p-1}{2}\}$ for every $j = 1, 2, \ldots, s$, since

$$w_j = p - e_j \quad \text{and} \quad e_j > \frac{p}{2}.$$

Thus, the set S_3 which we were trying to construct, is exactly the set $\{w_1, w_2, \ldots, w_s\}$. Therefore,

$$S_1 \cup S_3 = \{1, \ldots, \frac{p-1}{2}\}.$$

By multiplying the elements of the set $S_1 \cup S_3$, we obtain

$$r_1 r_2 \cdots r_\lambda w_1 w_2 \cdots w_s = 1 \cdot 2 \cdot 3 \cdots \frac{p-1}{2}$$

and equivalently

$$r_1 r_2 \cdots r_\lambda (p - e_1)(p - e_2) \cdots (p - e_s) = \left(\frac{p-1}{2}\right)!.$$

However, there exists an integer c for which

$$r_1 r_2 \cdots r_\lambda (p - e_1)(p - e_2) \cdots (p - e_s) = cp - r_1 r_2 \cdots r_\lambda (-1)^s e_1 e_2 \cdots e_s.$$

But

$$p \left| r_1 r_2 \cdots r_\lambda (p - e_1)(p - e_2) \cdots (p - e_s) - \left(\frac{p-1}{2}\right)! = 0. \right.$$

Thus,

$$p \left| cp - r_1 r_2 \cdots r_\lambda (-1)^s e_1 e_2 \cdots e_s - \left(\frac{p-1}{2}\right)! \right.$$

or

$$p \left| (-1)^s r_1 r_2 \cdots r_\lambda e_1 e_2 \cdots e_s - \left(\frac{p-1}{2}\right)!. \right. \tag{3}$$

By (1), (2) we obtain

$$r_i = ka - k_i p, \text{ where } 1 \le k \le \frac{p-1}{2}, \ i = 1, 2, \ldots, \lambda$$

and

$$e_j = \nu a - \nu_j p, \text{ where } 1 \le \nu \le \frac{p-1}{2}, j = 1, 2, \ldots, s.$$

Hence,

$$r_1 r_2 \cdots r_\lambda e_1 e_2 \cdots e_s \equiv a(2a)(3a) \cdots \left(\frac{p-1}{2}a\right) \pmod{p}. \tag{4}$$

Thus, by (3), (4) we get

$$\left(\frac{p-1}{2}\right)! \equiv (-1)^s a^{(p-1)/2} \left(\frac{p-1}{2}\right)! \pmod{p}.$$

However, by Euler's Criterion we have

$$a^{(p-1)/2} \equiv \left(\frac{a}{p}\right) \pmod{p}.$$

Therefore,

$$\left(\frac{p-1}{2}\right)! \equiv (-1)^s \left(\frac{a}{p}\right) \left(\frac{p-1}{2}\right)! \pmod{p}$$

or

$$1 \equiv (-1)^s \left(\frac{a}{p}\right) \pmod{p}$$

or

$$(-1)^s \equiv \left(\frac{a}{p}\right) \pmod{p},$$

and since the only possible values of $\left(\frac{a}{p}\right) - (-1)^s$ are $0, 2, -2$, it follows that

$$\left(\frac{a}{p}\right) = (-1)^s. \qquad \square$$

Theorem 5.2.9 (The Law of Quadratic Reciprocity). *Let p, q be distinct odd prime numbers. Then it holds*

$$\left(\frac{p}{q}\right)\left(\frac{q}{p}\right) = (-1)^{\frac{(p-1)(q-1)}{4}}.$$

Proof. By Gauss's Lemma, we obtain

$$\left(\frac{p}{q}\right) = (-1)^{s_1} \quad \text{and} \quad \left(\frac{q}{p}\right) = (-1)^{s_2},$$

where s_1 represents the number of positive residues greater than $\frac{q}{2}$, which occur when the integers

$$p, 2p, 3p, \ldots, \frac{q-1}{2}p$$

are divided by q, and s_2 represents the number of positive residues greater than $\frac{p}{2}$, which occur when the integers

$$q, 2q, 3q, \ldots, \frac{p-1}{2}q$$

are divided by p. Therefore,

$$\left(\frac{p}{q}\right)\left(\frac{q}{p}\right) = (-1)^{s_1+s_2}.$$

Step 1. We shall prove that

$$s_1 + s_2 \equiv \sum_{m_1=1}^{(q-1)/2} \left\lfloor \frac{m_1 p}{q} \right\rfloor + \sum_{m_2=1}^{(p-1)/2} \left\lfloor \frac{m_2 q}{p} \right\rfloor \pmod 2.$$

In order to do so, we must first prove that for the number of residues s, which we defined in Gauss's Lemma, it holds

$$s \equiv (a-1)\frac{p^2-1}{8} + \sum_{m=1}^{(p-1)/2} \left\lfloor \frac{ma}{p} \right\rfloor \pmod 2.$$

We have

$$\sum_{m=1}^{(p-1)/2} m = \sum_{i=1}^{\lambda} r_i + \sum_{j=1}^{s} w_j = \sum_{i=1}^{\lambda} r_i + \sum_{j=1}^{s}(p - e_j)$$

$$= \sum_{i=1}^{\lambda} r_i + s \cdot p - \sum_{j=1}^{s} e_j. \tag{1}$$

In addition, it holds

$$\frac{ma}{p} = \left\lfloor \frac{ma}{p} \right\rfloor + \upsilon_m, \text{ where } 0 < \upsilon < 1.$$

Therefore, equivalently, we get

$$ma = \left\lfloor \frac{ma}{p} \right\rfloor p + \upsilon_m p. \tag{2}$$

Set $h_m = \upsilon_m p$. Then, it is clear that $0 < h_m < p$ and that h_m is the least positive residue which occurs when ma is divided by p. Hence,

$$\sum_{i=1}^{\lambda} r_i + \sum_{j=1}^{s} e_j = \sum_{m=1}^{(p-1)/2} h_m$$

and by (2), we obtain

$$a \sum_{m=1}^{(p-1)/2} m - p \sum_{m=1}^{(p-1)/2} \left\lfloor \frac{ma}{p} \right\rfloor = \sum_{i=1}^{\lambda} r_i + \sum_{j=1}^{s} e_j. \tag{3}$$

If we add up the relations (1) and (3), it follows

$$(a+1) \sum_{m=1}^{(p-1)/2} m - p \sum_{m=1}^{(p-1)/2} \left\lfloor \frac{ma}{p} \right\rfloor = 2 \sum_{i=1}^{\lambda} r_i + s \cdot p. \tag{4}$$

However, since $p \equiv 1 \,(\mathrm{mod}\,2)$, it is obvious that

$$sp \equiv s \,(\mathrm{mod}\,2)$$

and

$$p \sum_{m=1}^{(p-1)/2} \left\lfloor \frac{ma}{p} \right\rfloor \equiv \sum_{m=1}^{(p-1)/2} \left\lfloor \frac{ma}{p} \right\rfloor \,(\mathrm{mod}\,2).$$

Furthermore, since $a + 1 \equiv a - 1 \,(\mathrm{mod}\,2)$, we also get

$$(a+1) \sum_{m=1}^{(p-1)/2} m \equiv (a-1) \sum_{m=1}^{(p-1)/2} m \,(\mathrm{mod}\,2).$$

Therefore, by the above relations, we obtain

$$s + p \sum_{m=1}^{(p-1)/2} \left\lfloor \frac{ma}{p} \right\rfloor + (a+1) \sum_{m=1}^{(p-1)/2} m$$

$$\equiv sp + \sum_{m=1}^{(p-1)/2} \left\lfloor \frac{ma}{p} \right\rfloor + (a-1) \sum_{m=1}^{(p-1)/2} m \,(\mathrm{mod}\,2)$$

and thus, by (4) we get

$$s + 2\sum_{i=1}^{\lambda} r_i + sp + 2p \sum_{m=1}^{(p-1)/2} \left\lfloor \frac{ma}{p} \right\rfloor \equiv sp + \sum_{m=1}^{(p-1)/2} \left\lfloor \frac{ma}{p} \right\rfloor + (a-1) \sum_{m=1}^{(p-1)/2} m \,(\mathrm{mod}\,2).$$

Thus,

$$s \equiv \sum_{m=1}^{(p-1)/2} \left\lfloor \frac{ma}{p} \right\rfloor + (a-1) \sum_{m=1}^{(p-1)/2} m \,(\mathrm{mod}\,2)$$

or, equivalently,

$$s \equiv \sum_{m=1}^{(p-1)/2} \left\lfloor \frac{ma}{p} \right\rfloor + (a-1)\frac{p^2-1}{8} \,(\mathrm{mod}\,2).$$

This completes the proof of Step 1. Thus, we have proved that

$$\left(\frac{p}{q}\right)\left(\frac{q}{p}\right) = (-1)^{s_1+s_2},$$

where

$$s_1 + s_2 \equiv \sum_{m_1=1}^{(q-1)/2} \left\lfloor \frac{m_1 p}{q} \right\rfloor + \sum_{m_2=1}^{(p-1)/2} \left\lfloor \frac{m_2 q}{p} \right\rfloor \,(\mathrm{mod}\,2).$$

Step 2. We shall prove that

$$\sum_{m_1=1}^{(q-1)/2} \left\lfloor \frac{m_1 p}{q} \right\rfloor + \sum_{m_2=1}^{(p-1)/2} \left\lfloor \frac{m_2 q}{p} \right\rfloor \equiv \frac{p-1}{2} \cdot \frac{q-1}{2} \pmod{2}.$$

In the Cartesian plane below, let us consider the lattice points (m_2, m_1), where $1 \leq m_1 \leq \frac{q-1}{2}$ and $1 \leq m_2 \leq \frac{p-1}{2}$.

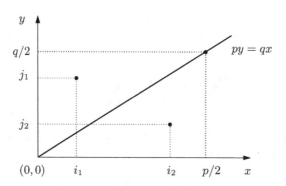

Figure 5.1

None of these lattice points lays upon the line with equation $py = qx$, for if there existed a pair (m_2, m_1) such that $pm_2 = qm_1$, then we would have $qm_1 \equiv 0 \pmod{p}$, which is impossible since $\gcd(q,p) = 1$ and $1 \leq m_1 \leq \frac{q-1}{2}$. Therefore, each lattice point (m_2, m_1) will be either above or below the line with equation $py = qx$. Thus, we shall distinguish two cases.

- *Case 1.* If the lattice point (m_2, m_1) lays above the line $py = qx$, then it is clear that

$$pm_1 > qm_2$$

and thus

$$m_2 < \frac{m_1 p}{q}.$$

Hence, for every fixed value of m_1, there exist $\left\lfloor \frac{m_1 p}{q} \right\rfloor$ lattice points, which lay above the line with equation $py = qx$. Therefore, the total number of lattice points laying above the line is

$$\sum_{m_1=1}^{(q-1)/2} \left\lfloor \frac{m_1 p}{q} \right\rfloor.$$

- *Case 2.* If the lattice point (m_2, m_1) lays below the line with equation $py = qx$, then it is clear that

$$pm_1 < qm_2$$

and thus

$$m_1 < \frac{qm_2}{p}.$$

Thus, similarly to the previous case, it follows that the total number of lattice points laying below the line is

$$\sum_{m_2=1}^{(p-1)/2} \left\lfloor \frac{qm_2}{p} \right\rfloor.$$

Consequently, the total number of lattice points laying above and below the line with equation $py = qx$ is

$$\sum_{m_1=1}^{(q-1)/2} \left\lfloor \frac{m_1 p}{q} \right\rfloor + \sum_{m_2=1}^{(p-1)/2} \left\lfloor \frac{qm_2}{p} \right\rfloor.$$

However, since $1 \le m_1 \le \frac{q-1}{2}$ and $1 \le m_2 \le \frac{p-1}{2}$, it is clear that the total number of lattice points (m_2, m_1) is

$$\frac{p-1}{2} \cdot \frac{q-1}{2}.$$

Therefore, we obtain

$$\sum_{m_1=1}^{(q-1)/2} \left\lfloor \frac{m_1 p}{q} \right\rfloor + \sum_{m_2=1}^{(p-1)/2} \left\lfloor \frac{qm_2}{p} \right\rfloor = \frac{p-1}{2} \cdot \frac{q-1}{2}.$$

This completes the proof of Step 2.

By the results obtained by Steps 1 and 2, we have

$$s_1 + s_2 \equiv \frac{p-1}{2} \cdot \frac{q-1}{2} \pmod 2,$$

or equivalently

$$s_1 + s_2 - \frac{p-1}{2} \cdot \frac{q-1}{2} = 2k, \text{ for some } k \in \mathbb{Z}.$$

Therefore,

$$\left(\frac{p}{q} \right) \left(\frac{q}{p} \right) = (-1)^{s_1+s_2} = (-1)^{\frac{p-1}{2} \cdot \frac{q-1}{2}} (-1)^{2k}$$

$$= (-1)^{\frac{p-1}{2} \cdot \frac{q-1}{2}}.$$

This completes the proof of the law of quadratic reciprocity.   □

Remark 5.2.10. The **law of quadratic reciprocity** relates the solvability of the congruence

$$x^2 \equiv p \,(\mathrm{mod}\, q) \tag{R1}$$

to the solvability of the congruence

$$x^2 \equiv q \,(\mathrm{mod}\, p). \tag{R2}$$

Thus, we have the following two cases:

Case 1. If $p \equiv 1 \,(\mathrm{mod}\, 4)$ or $q \equiv 1 \,(\mathrm{mod}\, 4)$, then we obtain

$$\left(\frac{p}{q}\right)\left(\frac{q}{p}\right) = 1.$$

Hence, the congruence (R1) has a solution if and only if the congruence (R2) has a solution.

Case 2. If $p \equiv 3 \,(\mathrm{mod}\, 4)$ and $q \equiv 3 \,(\mathrm{mod}\, 4)$, then we obtain

$$\left(\frac{p}{q}\right)\left(\frac{q}{p}\right) = -1.$$

Therefore, the congruence (R1) has a solution if and only if the congruence (R2) does not have a solution.

EXAMPLES.

1. It holds $6^2 \equiv 5 \,(\mathrm{mod}\, 31)$, thus we have $\left(\frac{5}{31}\right) = 1$. But, by the law of quadratic reciprocity, we obtain

$$\left(\frac{31}{5}\right)\left(\frac{5}{31}\right) = (-1)^{\frac{31-1}{2} \cdot \frac{5-1}{2}} = (-1)^{15 \cdot 2} = 1.$$

   Therefore,

$$\left(\frac{31}{5}\right) = 1,$$

   which means that 31 is a quadratic residue mod 5.

2. It holds $\left(\frac{29}{17}\right) = -1$. But, by the law of quadratic reciprocity, we get

$$\left(\frac{29}{17}\right)\left(\frac{17}{29}\right) = (-1)^{\frac{29-1}{2} \cdot \frac{17-1}{2}} = (-1)^{14 \cdot 8} = 1.$$

   Therefore,

$$\left(\frac{17}{29}\right) = -1,$$

   which means that 17 is a quadratic nonresidue mod 29.

3. By the law of quadratic reciprocity, we have

$$\left(\frac{7}{29}\right)\left(\frac{29}{7}\right) = (-1)^{\frac{7-1}{2}\cdot\frac{29-1}{2}} = (-1)^{3\cdot 14} = 1.$$

Hence, the congruence $x^2 \equiv 7\,(\mathrm{mod}\,29)$ has a solution if and only if the congruence $x^2 \equiv 29\,(\mathrm{mod}\,7)$ has a solution. However, we have $6^2 \equiv 7\,(\mathrm{mod}\,29)$. Thus, since both congruences have a solution, it yields

$$\left(\frac{7}{29}\right) = \left(\frac{29}{7}\right) = 1.$$

Historical Remark. The law of quadratic reciprocity was discovered for the first time, in a complex form, by L. Euler who published it in his paper entitled "Novae demonstrationes circa divisores numerorum formae $xx + nyy$." On the 20th of November, 1775, he presented his discovery at the Academy of Saint Petersburg, followed by a false proof.

Later, Legendre also discovered independently the law of quadratic reciprocity and in his paper entitled "Recherches d'analyse indéterminée," Hist. Acad. Paris, 1785, p. 465ff, he presented an unsubstantiated proof of the theorem (Legendre proved the law of quadratic reciprocity based on some hypotheses, which he did not prove).

In 1795, Gauss at the age of 18 also discovered this law independently of Euler and Legendre. A year later, he presented the first complete proof. The law of quadratic reciprocity was one of the favorite theorems of Gauss and this is justified by the fact that during his life, he discovered six different proofs of the theorem.

The proof which we presented above is due to a student of Gauss, Gotthold Eisenstein, who presented his proof in his paper "Geometrischer Beweis des Fundamentaltheorems für die quadratischen Reste," J. Reine Angew. Math. 28(1844), 246–248.

5.3 Jacobi's symbol

Jacobi's symbol[2] is a generalization of Legendre's symbol, which we examined in the previous section. Legendre's symbol examines the solvability of the congruence $x^2 \equiv a\,(\mathrm{mod}\,p)$ where p is a prime number and $\gcd(a,p) = 1$. On the other hand, Jacobi's symbol is not strictly referred to a prime number p, but to an arbitrary odd positive integer P and its value does not necessarily provide information related to the solvability of the congruence $x^2 \equiv a\,(\mathrm{mod}\,P)$, where $\gcd(a,P) = 1$. See Remark 5.3.2 for further details. Of course, in case P is a prime number, the Legendre and Jacobi symbols are identical.

---

[2] Jacobi's symbol was named after Carl Gustav Jacobi (1804–1851), who presented it in 1846.

Definition 5.3.1. *Let P be an odd positive integer and a be an integer such that $\gcd(a, P) = 1$. Then, we define **Jacobi's symbol** $\left(\frac{a}{P}\right)$ by*

$$\left(\frac{a}{P}\right) = \begin{cases} 1, & \text{if } P = 1 \\ \left(\frac{a}{p_1}\right)^{m_1} \left(\frac{a}{p_2}\right)^{m_2} \cdots \left(\frac{a}{p_k}\right)^{m_k}, & \text{if } P = p_1^{m_1} p_2^{m_2} \cdots p_k^{m_k}, \end{cases}$$

where $\left(\frac{a}{p_i}\right)$ stands for Legendre's symbol.

Jacobi's symbol is often generalized to obtain a zero value in the case when $\gcd(a, P) > 1$.

EXAMPLES.

1. $\left(\frac{14}{17081}\right) = \left(\frac{14}{19 \cdot 29 \cdot 31}\right) = \left(\frac{14}{19}\right)\left(\frac{14}{29}\right)\left(\frac{14}{31}\right) = (-1) \cdot (-1) \cdot 1 = 1.$

2. $\left(\frac{14}{324539}\right) = \left(\frac{14}{19^2 \cdot 29 \cdot 31}\right) = \left(\frac{14}{19}\right)^2 \left(\frac{14}{29}\right)\left(\frac{14}{31}\right) = (-1)^2 \cdot (-1) \cdot 1 = -1.$

3. $\left(\frac{27}{9}\right) = 0.$

Remark 5.3.2. As mentioned above, Jacobi's symbol does not necessarily provide information whether a is a quadratic residue $\mod P$ or not. This happens because if we assume that

$$P = p_1 p_2 \cdots p_k, \text{ with } k = 2\lambda, \ \lambda \in \mathbb{N}, [3]$$

and $\left(\frac{a}{p_i}\right) = -1$ for every $i = 1, 2, \ldots, k$, then it follows

$$\left(\frac{a}{P}\right) = (-1)^k = 1.$$

However, it is clear that the congruence

$$x^2 \equiv a \, (\mathrm{mod} \, P)$$

does not have solutions because if it did, then each of the congruences

$$x^2 \equiv a \, (\mathrm{mod} \, p_i)$$

would have a solution. In that case, we would have

$$\left(\frac{a}{P}\right) = 1,$$

for every $i = 1, 2, \ldots, k$ which, by the hypothesis, is a contradiction.

But, if a is a quadratic residue $\mod P$, then similarly one has

$$\left(\frac{a}{P}\right) = 1,$$

for every prime divisor p_i of P. Hence, Jacobi's symbol is equal to 1.

---

[3] The prime numbers $p_1, p_2, \ldots, p_k$ are not necessarily distinct.

If $\left(\frac{a}{P}\right) = -1$, then a is not a quadratic residue mod P, because if it was, then every congruence

$$x^2 \equiv a \,(\mathrm{mod}\; p_i)$$

would have a solution and therefore it would follow that $\left(\frac{a}{P}\right) = 1$, which is a contradiction.

Hence, to sum up, we have:

- If $\left(\frac{a}{P}\right) = 1$, then one cannot draw a conclusion on whether a is a quadratic residue mod P.
- If a is a quadratic residue mod P, then it necessarily holds $\left(\frac{a}{P}\right) = 1$.
- If $\left(\frac{a}{P}\right) = -1$, then a is not a quadratic residue mod P.

Theorem 5.3.3. *Let a be an integer, coprime to two odd positive integers P, Q. Then, it holds*

$$\left(\frac{a}{PQ}\right) = \left(\frac{a}{P}\right)\left(\frac{a}{Q}\right).$$

Proof. Let $P = p_1^{m_1} p_2^{m_2} \cdots p_k^{m_k}$ and $Q = q_1^{b_1} q_2^{b_2} \cdots q_\lambda^{b_\lambda}$, where $k, \lambda \in \mathbb{N}$, be the canonical forms of P and Q, respectively. Then, by the definition of Jacobi's symbol, we have

$$\left(\frac{a}{PQ}\right) = \left(\frac{a}{p_1^{m_1} p_2^{m_2} \cdots p_k^{m_k} q_1^{b_1} q_2^{b_2} \cdots q_\lambda^{b_\lambda}}\right)$$

$$= \left(\frac{a}{p_1}\right)^{m_1} \left(\frac{a}{p_2}\right)^{m_2} \cdots \left(\frac{a}{p_k}\right)^{m_k} \left(\frac{a}{q_1}\right)^{b_1} \left(\frac{a}{q_2}\right)^{b_2} \cdots \left(\frac{a}{q_\lambda}\right)^{b_\lambda}$$

$$= \left(\frac{a}{P}\right)\left(\frac{a}{Q}\right). \qquad \square$$

Theorem 5.3.4. *Let a, b be integers, coprime to an odd positive integer P. Then it holds*

$$\left(\frac{a}{P}\right)\left(\frac{b}{P}\right) = \left(\frac{ab}{P}\right).$$

Proof. Let $P = p_1^{m_1} p_2^{m_2} \cdots p_k^{m_k}$, where $k \in \mathbb{N}$, be the canonical form of P. Then, we have

$$\left(\frac{a}{P}\right)\left(\frac{b}{P}\right) = \left(\frac{a}{p_1^{m_1} p_2^{m_2} \cdots p_k^{m_k}}\right)\left(\frac{b}{p_1^{m_1} p_2^{m_2} \cdots p_k^{m_k}}\right)$$

$$= \left(\frac{a}{p_1}\right)^{m_1}\left(\frac{a}{p_2}\right)^{m_2}\cdots\left(\frac{a}{p_k}\right)^{m_k}\left(\frac{b}{p_1}\right)^{m_1}\left(\frac{b}{p_2}\right)^{m_2}\cdots\left(\frac{b}{p_k}\right)^{m_k}$$

$$= \left[\left(\frac{a}{p_1}\right)\left(\frac{b}{p_1}\right)\right]^{m_1} \cdot \left[\left(\frac{a}{p_2}\right)\left(\frac{b}{p_2}\right)\right]^{m_2} \cdots \left[\left(\frac{a}{p_k}\right)\left(\frac{b}{p_k}\right)\right]^{m_k}.$$

$$(1)$$

However, by Theorem 5.2.4 and (1), we obtain

$$\left(\frac{a}{P}\right)\left(\frac{b}{P}\right) = \left(\frac{ab}{p_1}\right)^{m_1}\left(\frac{ab}{p_2}\right)^{m_2}\cdots\left(\frac{ab}{p_k}\right)^{m_k}$$

$$= \left(\frac{ab}{P}\right).$$

□

Corollary 5.3.5. *Let a be an integer, coprime to an odd positive integer P. Then, it holds*

$$\left(\frac{a^2}{P}\right)\left(\frac{a}{P^2}\right) = 1.$$

Proof. By the above two theorems, we have

$$\left(\frac{a}{P^2}\right) = \left(\frac{a}{P}\right)\left(\frac{a}{P}\right)$$

and

$$\left(\frac{a^2}{P}\right) = \left(\frac{a}{P}\right)\left(\frac{a}{P}\right).$$

But, it is clear that

$$\left(\frac{a}{P}\right)\left(\frac{a}{P}\right) = 1$$

and thus

$$\left(\frac{a^2}{P}\right) = \left(\frac{a}{P^2}\right) = 1.$$

□

Theorem 5.3.6. *Let a, b be integers, where a is coprime to an odd positive integer P. If $a \equiv b \,(\mathrm{mod}\, P)$, then it holds*

$$\left(\frac{a}{P}\right) = \left(\frac{b}{P}\right).$$

Proof. Let $P = p_1^{m_1}p_2^{m_2}\cdots p_k^{m_k}$, where $k \in \mathbb{N}$, be the canonical form of P. Then, we have

$$a \equiv b \,(\mathrm{mod}\, p_i), \text{ for every } i = 1, 2, \ldots, k.$$

However, by Theorem 5.2.3, we obtain

$$\left(\frac{a}{p_i}\right) = \left(\frac{b}{p_i}\right), \text{ for every } i = 1, 2, \ldots, k.$$

Hence,

$$\left(\frac{a}{P}\right) = \left(\frac{a}{p_1}\right)^{m_1}\left(\frac{a}{p_2}\right)^{m_2}\cdots\left(\frac{a}{p_k}\right)^{m_k}$$

$$= \left(\frac{b}{p_1}\right)^{m_1}\left(\frac{b}{p_2}\right)^{m_2}\cdots\left(\frac{b}{p_k}\right)^{m_k}$$

$$= \left(\frac{b}{P}\right).$$

□

Theorem 5.3.7. *Let P be an odd positive integer. Then it holds*

$$\left(\frac{-1}{P}\right) = (-1)^{(P-1)/2}.$$

Proof. Let $P = p_1^{m_1} p_2^{m_2} \cdots p_k^{m_k}$, where $k \in \mathbb{N}$, be the canonical form of P. Then, by the definition of Jacobi's symbol, we obtain

$$\left(\frac{-1}{P}\right) = \left(\frac{-1}{p_1}\right)^{m_1} \left(\frac{-1}{p_2}\right)^{m_2} \cdots \left(\frac{-1}{p_k}\right)^{m_k}. \tag{1}$$

But, by Euler's Criterion, we get

$$\left(\frac{-1}{p_i}\right) = (-1)^{(p_i-1)/2} \text{ for every } i = 1, 2, \ldots, k,$$

since each p_i is odd and thus greater than 2. Therefore, by (1) it follows

$$\left(\frac{-1}{P}\right) = (-1)^{\sum_{i=1}^{k}(p_i-1)m_i/2}.$$

Let

$$p_1^{m_1} p_2^{m_2} \cdots p_k^{m_k} = q_1 q_2 \cdots q_\lambda,$$

where $\lambda = m_1 + m_2 + \cdots + m_k$ and $q_1, q_2, \ldots, q_\lambda \in \{p_1, p_2, \ldots, p_k\}$. We obtain

$$\left(\frac{-1}{P}\right) = (-1)^{\sum_{j=1}^{\lambda}(q_j-1)/2}. \tag{2}$$

However, we have

$$P = \prod_{j=1}^{\lambda} q_j = \prod_{j=1}^{\lambda} [1 + (q_j - 1)]$$

$$= 1 + \sum_{j=1}^{\lambda} (q_j - 1) + 4r,$$

for a natural number r, since each term $q_j - 1$ is an even integer. Therefore,

$$\frac{1}{2}(P - 1) = \frac{1}{2} \sum_{j=1}^{\lambda} (q_j - 1) + 2r.$$

Hence, (2) takes the form

$$\left(\frac{-1}{P}\right) = (-1)^{(P-1)/2}(-1)^{-2r} = (-1)^{(P-1)/2}. \qquad \square$$

Theorem 5.3.8. *Let P be an odd positive integer. Then, it holds*

$$\left(\frac{2}{P}\right) = (-1)^{(P^2-1)/8}.$$

Proof. Let us assume that

$$P = p_1^{m_1} p_2^{m_2} \cdots p_k^{m_k} = q_1 q_2 \cdots q_\lambda,$$

where $\lambda = m_1 + m_2 + \cdots + m_k$ and $q_1, q_2, \ldots, q_\lambda \in \{p_1, p_2, \ldots, p_k\}$. Then, we have

$$\left(\frac{2}{P}\right) = \left(\frac{2}{q_1}\right)\left(\frac{2}{q_2}\right)\cdots\left(\frac{2}{q_\lambda}\right). \tag{1}$$

But, by Theorem 5.2.7 we obtain

$$\left(\frac{2}{p}\right) = (-1)^{(p^2-1)/8}.$$

Thus, (1) takes the form

$$\left(\frac{2}{P}\right) = (-1)^{\sum_{i=1}^{\lambda}(q_i^2-1)/8}. \tag{2}$$

Moreover, it holds

$$\frac{P^2-1}{8} = \frac{q_1^2 q_2^2 \cdots q_\lambda^2 - 1}{8}$$

$$= \frac{\left(1+8\cdot\frac{q_1^2-1}{8}\right)\left(1+8\cdot\frac{q_2^2-1}{8}\right)\cdots\left(1+8\cdot\frac{q_\lambda^2-1}{8}\right)-1}{8}. \tag{3}$$

However,

$$\left(1+8\cdot\frac{q_1^2-1}{8}\right)\left(1+8\cdot\frac{q_2^2-1}{8}\right)\cdots\left(1+8\cdot\frac{q_\lambda^2-1}{8}\right)-1$$

$$= 1 + \sum_{i=1}^{\lambda}\frac{8(q_i^2-1)}{8} + \left(\sum_{i\neq j}\frac{8(q_i^2-1)}{8}\cdot\frac{8(q_j^2-1)}{8}\right.$$

$$\left. + \sum_{i\neq j\neq k}\frac{8(q_i^2-1)}{8}\frac{8(q_j^2-1)}{8}\frac{8(q_k^2-1)}{8} + \cdots\right) - 1$$

$$= \sum_{i=1}^{\lambda}(q_i^2-1) + \left(\sum_{i\neq j}8\nu_i\cdot 8\nu_j + \sum_{i\neq j\neq k}8\nu_i\cdot 8\nu_j\cdot 8\nu_k + \cdots\right),$$

since $q_i^2 = 8\nu_i + 1$ for some $\nu_i \in \mathbb{N}$. Therefore, (3) can be written as

$$\frac{P^2 - 1}{8} = \sum_{i=1}^{\lambda} \frac{1}{8}(q_i^2 - 1) + 2r,$$

for some positive integer r. Hence, by (2) we obtain

$$\left(\frac{2}{P}\right) = (-1)^{(P^2-1)/8}(-1)^{-2r} = (-1)^{(P^2-1)/8}. \qquad \square$$

We shall now present an analogue of the law of quadratic reciprocity for Jacobi symbols.

Theorem 5.3.9. *Let P, Q be odd coprime positive integers. Then it holds*

$$\left(\frac{P}{Q}\right)\left(\frac{Q}{P}\right) = (-1)^{\frac{(P-1)(Q-1)}{4}}.$$

Proof. Let $P = p_1 p_2 \cdots p_\lambda$, $\lambda \in \mathbb{N}$, where the prime numbers $p_1, p_2, \ldots, p_\lambda$ are not necessarily pairwise distinct. Similarly, let $Q = q_1 q_2 \cdots q_m$, $m \in \mathbb{N}$, where the prime numbers $q_1, q_2, \ldots, q_m$ are not necessarily pairwise distinct. Then, it follows

$$\left(\frac{P}{Q}\right) = \prod_{i=1}^{m}\left(\frac{P}{q_i}\right) = \prod_{i=1}^{m}\prod_{j=1}^{\lambda}\left(\frac{p_j}{q_i}\right).$$

Similarly, we have

$$\left(\frac{Q}{P}\right) = \prod_{j=1}^{\lambda}\left(\frac{Q}{p_j}\right) = \prod_{j=1}^{\lambda}\prod_{i=1}^{m}\left(\frac{q_i}{p_j}\right).$$

Therefore,

$$\left(\frac{P}{Q}\right)\left(\frac{Q}{P}\right) = \prod_{j=1}^{\lambda}\prod_{i=1}^{m}\left(\frac{p_j}{q_i}\right)\left(\frac{q_i}{p_j}\right). \tag{1}$$

However, by the law of quadratic reciprocity for Legendre symbols, we know that

$$\left(\frac{p_j}{q_i}\right)\left(\frac{q_i}{p_j}\right) = (-1)^{\frac{(p_j-1)(q_i-1)}{4}}.$$

Thus, by (1) we obtain

$$\left(\frac{P}{Q}\right)\left(\frac{Q}{P}\right) = (-1)^{\sum_{j=1}^{\lambda}\sum_{i=1}^{m}\frac{1}{2}(p_j-1)\frac{1}{2}(q_i-1)}$$

$$= (-1)^{\sum_{j=1}^{\lambda}\frac{1}{2}(p_j-1)\sum_{i=1}^{m}\frac{1}{2}(q_i-1)}. \tag{2}'$$

But, in the proof of Theorem 5.3.7, we have shown that

$$\frac{1}{2}(P-1) = \frac{1}{2}\sum_{j=1}^{\lambda}(p_j - 1) + 2r_1,$$

for some positive integer r_1. Thus, similarly, it follows that

$$\frac{1}{2}(Q-1) = \frac{1}{2}\sum_{i=1}^{m}(q_i - 1) + 2r_2,$$

for some positive integer r_2. Hence, by (2) we obtain

$$\left(\frac{P}{Q}\right)\left(\frac{Q}{P}\right) = (-1)^{-2r_1}(-1)^{-2r_2}(-1)^{\frac{1}{2}(P-1)\frac{1}{2}(Q-1)}$$

$$= (-1)^{(P-1)(Q-1)/4}. \qquad \qquad \square$$

5.3.1 An application of the Jacobi symbol to cryptography

The Jacobi symbol was used by Robert M. Solovay and Volker Strassen in their *primality test* algorithm which was introduced in 1982. A primality test algorithm is an algorithm which tests whether a positive integer n is a prime number or not. Before we present the steps of the algorithm, we shall define some basic notions.

Definition 5.3.10. *Let n be an odd composite integer and a be an integer, such that $1 \leq a \leq n - 1$. Then*

- *If*

$$\gcd(a, n) > 1 \ \ or \ \ a^{(n-1)/2} \not\equiv \left(\frac{a}{n}\right) \ (\mathrm{mod}\ n),$$

  *the integer a is said to be an **Euler martyr** for the integer n.*
- *If*

$$\gcd(a, n) = 1 \ \ and \ \ a^{(n-1)/2} \equiv \left(\frac{a}{n}\right) \ (\mathrm{mod}\ n),$$

  *the integer n is said to be an **Euler pseudoprime** to the base a and a is called an **Euler liar**.*

The steps of the algorithm are presented below.

Solovay–Strassen algorithm.

1. Consider an odd positive integer n, with $n \geq 3$, which you want to examine whether it is a prime number.
2. Choose an arbitrary integer from the interval $(1, n - 1)$.
3. Determine the integer x, such that $x \equiv a^{(n-1)/2} \ (\mathrm{mod}\ n)$.
   If $x \neq 1$ and $x \neq n - 1$, then the positive integer n is composite and the algorithm terminates at this step. If that is not the case, then

4. Compute the Jacobi symbol

$$j = \left(\frac{a}{n}\right).$$

If $x \not\equiv j \,(\mathrm{mod}\, n)$, then the positive integer n is composite and the algorithm terminates at this step.

5. If the algorithm has not been terminated in one of the above steps, for several values of a, then the positive integer n is probably a prime number.

According to the Solovay–Strassen algorithm, the only case when we can be led to a false conclusion is when the positive integer n is composite and the outcome of the algorithm is that n is probably a prime. The number of Euler liars is at most

$$\frac{\phi(n)}{2} < \frac{n-1}{2}.$$

Therefore, the probability that an Euler liar occurs is less than $1/2$.

Note. Generally, if the outcome of the algorithm is that n is a composite integer, then this result is undeniably true. In addition, if n is a prime number, then it is certain that the algorithm will verify that fact.

6

The π- and li-functions

Mathematicians have tried in vain to this day to discover some order in the sequence of prime numbers, and we have reason to believe that it is a mystery into which the human mind will never penetrate.

Leonhard Euler (1707–1783)

6.1 Basic notions and historical remarks

Definition 6.1.1. *We define* $\pi(x)$ *to be the number of primes which do not exceed a given real number* x.

If we attempt to find a formula in order to describe $\pi(x)$ we will definitely understand that this is an extremely difficult task. The most important reason why this happens is because we don't know exactly how primes are distributed among the integers. We can understand this by a simple example. We shall prove that the gaps among prime numbers can be arbitrarily large. This is true because we can always find n consecutive composite positive integers, where n is any natural number. For this consider the sequence of consecutive integers

$$(n+1)! + 2, (n+1)! + 3, \ldots, (n+1)! + n, (n+1)! + (n+1).$$

It is easy to see that none of these integers is prime as

$$2|(n+1)! + 2, 3|(n+1)! + 3, \ldots, n|(n+1)! + n, (n+1)|(n+1)! + (n+1).$$

One expects that it might be easier to construct an asymptotic formula for $\pi(x)$. In 1793, Carl Friedrich Gauss (1777–1855), while conducting research in number theory, conjectured that

M.Th. Rassias, *Problem-Solving and Selected Topics in Number Theory: In the Spirit of the Mathematical Olympiads*, DOI 10.1007/978-1-4419-0495-9_6, © Springer Science+Business Media, LLC 2011

$$\pi(x) \sim \frac{x}{\log x}, \text{ [1]}$$

namely, that $\pi(x) \log x/x \to 1$, as $x \to \infty$.

In the same period of time, Adrien-Marie Legendre (1752–1833) formulated an equivalent conjecture. He assumed that there exist constants A, B such that

$$\pi(x) \sim x/(A \log x + B). \text{ [2]}$$

Both Gauss and Legendre tried to prove this conjecture, which is known today as the **Prime Number Theorem**,[3] but none of them succeeded. Some of the most eminent mathematicians of the 19th century failed to give a rigorous proof of the theorem. Among them were Pafnuty Chebyshev (1821–1894) and Georg Bernhard Riemann (1826–1866), who tried to prove it in his very well known paper "Ueber die Anzahl der Primzahlen unter einer gegebenen Grösse" (1859).

We shall outline the basic steps of Riemann's paper in the next chapter.

The first proof of the Prime Number Theorem was given in 1896 by the French mathematician Jacques Hadamard (1865–1963) and the Belgian mathematician Charles-Jean-Gustave-Nicolas de la Vallée-Poussin (1866–1962), who both provided independent proofs. This was the most important single result ever obtained in number theory until that time. An elementary proof of the theorem was given in 1959 by Atle Selberg (1917–2007) and Paul Erdős (1913–1996). In 1980, a simpler proof was given by D. J. Newman in [5]. In his proof, Newman just used basic Complex Analysis. Later on, we are going to analyze Newman's proof step by step.

Apart from the function $x/\log x$, there are other functions which can describe more efficiently the behavior of $\pi(x)$ for large values of x. For example, the function

$$li(x) = \int_2^x \frac{dt}{\log t} \text{ [4]}$$

is a better approximation for $\pi(x)$, since the quotient $\pi(x)/li(x)$ tends to 1 faster than the quotient $\pi(x) \log x/x$, as $x \to +\infty$.

---

[1] Throughout the book we consider log with respect to base e.

[2] In addition, in 1808, Legendre also formulated another conjecture, according to which $\pi(x) \sim x/(\log x - A(x))$, where $\lim_{x \to \infty} A(x) = 1.0836\ldots$. Some progress related to this conjecture has been made by J. B. Rosser and L. Schoenfeld in [49], where they proved that $\lim_{x \to \infty} A(x) = 1$ and for $x < 10^6$ the function $A(x)$ actually obtains values close to $1.0836\ldots$.

[3] An interesting corollary of the Prime Number Theorem is the following: *For any pair of positive real numbers a and b, where $a < b$, there exists a prime number between the real numbers ac and bc, for sufficiently large values of c.*

[4] In the literature, the following notations are also used:

$$Li(x) = \int_2^x \frac{dt}{\log t}, li(x) = \int_0^x \frac{dt}{\log t}.$$

By calculating the values of $\pi(x)$ and $li(x)$ for certain values of x, as is shown in the table below, it seems that $li(x)$ always counts more prime numbers than $\pi(x)$.

x	$\pi(x)$	$li(x) - \pi(x)$
10^8	5761455	753
10^9	50847534	1700
10^{10}	455052511	3103
10^{11}	4118054813	11587
10^{12}	37607912018	38262
10^{13}	346065536839	108970
10^{14}	3204941750802	314889
10^{15}	29844570422669	1052618
10^{16}	279238341033925	3214631
10^{17}	2623557157654233	7956588
10^{18}	24739954287740860	21949554
10^{19}	234057667276344607	99877774
10^{20}	2220819602560918840	222744643

In 1914, J.E. Littlewood (1885–1977) proved that the difference $\pi(x) - li(x)$ changes sign infinitely many times. Some years later, in 1933, Littlewood's student Stanley Skewes [56] proved that the first change of sign of $\pi(x) - li(x)$ should happen for

$$x < 10^{10^{10^{34}}}.$$

However, Skewes in order to prove the above result assumed that the Riemann hypothesis[5] holds true. Twenty-two years later, in 1955, Skewes without being based on Riemann's hypothesis or any other open problem, proved in [57] that the first change of sign of $\pi(x) - li(x)$ should happen for

$$x < 10^{10^{10^{1000}}}.$$

The latest improvement on that bound has been made by C. Bays and R. Hudson, who proved in [10] that the first change of sign should happen for

$$x < 1.3982 \cdot 10^{316}.$$

---

[5] See next chapter for more details on the Riemann hypothesis.

6.2 Open problems concerning prime numbers

1. **Goldbach's conjecture:** Every even integer greater than two can be expressed as the sum of two prime numbers.
   This conjecture appeared for the first time in a letter of Christian Goldbach (1690–1764) to Leonhard Euler (1707–1783), on the 7th of June, 1742.

2. Generally two prime numbers p_1 and p_2 are called **twin prime numbers** if $|p_1 - p_2| = 2$.

   Are there infinitely many twin prime numbers?

3. Consider the sequence $(A_n)_{n \geq 1}$, where

$$A_n = \sqrt{p_{n+1}} - \sqrt{p_n},$$

   and p_n denotes the nth prime number. **Andrica's conjecture** states that the inequality

$$A_n < 1$$

   holds true for every positive integer n. The conjecture has been verified for values of n up to $26 \cdot 10^{10}$.

   Note. From the Prime Number Theorem it follows that there exists an integer $M > 0$, such that Andrica's conjecture is true for all values of n, where $n \geq M$.
   (Dorin Andrica, Problem 34, Newsletter, European Mathematical Society, 67(2008), p. 44)

4. For any prime number p, where $p > 2$, there exist two distinct prime numbers p_1, p_2, with $p_1 < p_2$, such that

$$p = \frac{p_1 + p_2 + 1}{p_1}.$$

   This conjecture can also be stated in the following equivalent form: For any prime number p, where $p > 2$, there exist two distinct prime numbers p_1, p_2, with $p_1 < p_2$, such that the integers $(p - 1)p_1$, p_2 are consecutive.
   (Michael Th. Rassias, Open Problem No. 1825, Octogon Mathematical Magazine, 13(2005), p. 885. See also Problem 25, Newsletter, European Mathematical Society, 65(2007), p. 47)

7

The Riemann zeta function

If I were to awaken after having slept for a thousand years,
my first question would be:
Has the Riemann hypothesis been proven ?
David Hilbert (1862–1943)

7.1 Definition and Riemann's paper

Definition 7.1.1. *The zeta function is defined by*

$$\zeta(s) = \sum_{n=1}^{+\infty} \frac{1}{n^s},$$

for all real values of s with $s > 1$.

This function was defined for the first time in 1737 by Leonhard Euler (1707–1783). More than a century later, in 1859 Riemann rediscovered the zeta function for complex values of s, while he was trying to prove the Prime Number Theorem. In his paper, Riemann formulated six hypotheses. By the use of those hypotheses, he proved the Prime Number Theorem. As of today, only five of those hypotheses have been proved. The sixth hypothesis is the well-known *Riemann hypothesis* and is considered to be one of the most difficult open problems in mathematics.

M.Th. Rassias, *Problem-Solving and Selected Topics in Number Theory: In the Spirit of the Mathematical Olympiads*, DOI 10.1007/978-1-4419-0495-9_7,
© Springer Science+Business Media, LLC 2011

According to the Riemann hypothesis:

the non-trivial zeros[1] of the function $\zeta(s)$, where $s \in \mathbb{C}$, have real part equal to $1/2$.[2]

We mention very briefly the basic steps of Riemann's paper.

- In the beginning, he defined the function

$$H(x) = \sum_{k=1}^{+\infty} \frac{1}{k} \pi(x^{1/k}).$$

- Afterwards, he proved that

$$\pi(x) = \sum_{k=1}^{+\infty} \frac{1}{k} \mu(k) H(x^{1/k}), \tag{a}$$

where $\mu(k)$ is the Möbius function.
However,

$$H(x) = li(x) - \sum_{\rho} li(x^\rho) - \log 2 + \int_{x}^{+\infty} \frac{dt}{t(t^2 - 1) \log t}, \tag{b}$$

where the series $\sum_{\rho} li(x^\rho)$ extends over all nontrivial zeros of $\zeta(s)$. Therefore, from (a) and (b), one can construct a formula for $\pi(x)$ which will be expressed in terms of specific functions and the nontrivial zeros of $\zeta(s)$.

7.2 Some basic properties of the ζ-function

Property 7.2.1 (Euler's Identity).

$$\zeta(s) = \prod_{p} \frac{1}{1 - p^{-s}}, \quad s \in \mathbb{R}, \quad \text{with} \quad s > 1,$$

where the product extends over all prime numbers p.

---

[1] The negative even integers are considered to be the trivial roots of $\zeta(s)$. For further details concerning the trivial roots, see Property 7.2.7.

[2] Riemann in his paper considered the function $\xi(t) = \frac{1}{2}s(s - 1)\pi^{-s/2}\Gamma(\frac{s}{2})\zeta(s)$, where $s = \frac{1}{2} + it$ and $\Gamma(s)$ is the well-known gamma function. He conjectured that all roots of $\xi(t)$ are real. That is the exact statement of the Riemann hypothesis, which is equivalent to the one we have just introduced above.

The basic idea of the proof of the above property is the following:

$$\prod_p \frac{1}{1-p^{-s}} = \frac{1}{1-p_1^{-s}} \cdot \frac{1}{1-p_2^{-s}} \cdots \frac{1}{1-p_k^{-s}}$$

$$= \left(1 + \frac{1}{p_1^s} + \frac{1}{p_1^{2s}} + \cdots\right)\left(1 + \frac{1}{p_2^s} + \frac{1}{p_2^{2s}} + \cdots\right) \cdots$$

$$\times \left(1 + \frac{1}{p_k^s} + \frac{1}{p_k^{2s}} + \cdots\right)\cdots,$$

where $k \in \mathbb{N}$.

Now, if we carry over the calculations in the above product, we will obtain an infinite sum of terms of the form

$$\frac{1}{p_1^{a_1} p_2^{a_2} \cdots p_k^{a_k}},$$

where in the denominator every possible combination of powers of primes will occur. However, by the Fundamental Theorem of Arithmetic, it is known that every positive integer can be represented as the product of prime powers, in a unique way. Hence, it is evident that

$$\prod_p \frac{1}{1-p^{-s}} = \sum_{n \geq 1} \frac{1}{n^s}, \quad s > 1.$$

Therefore,

$$\zeta(s) = \prod_p \frac{1}{1-p^{-s}}, \quad s > 1. \qquad \Box$$

The above identity appeared for the first time in Euler's book entitled *Introductio in Analysin Infinitorum*, which was published in 1748.

Corollary 7.2.2. *By Euler's identity it easily follows that*

$$\frac{\zeta(2s)}{\zeta(s)} = \prod_p \frac{1}{1+p^{-s}}.$$

Property 7.2.3. It holds

$$\frac{1}{s-1} = \int_1^{+\infty} \frac{1}{x^s}dx \leq \zeta(s) \leq 1 + \int_1^{+\infty} \frac{1}{x^s}dx = 1 + \frac{1}{s-1}$$

and therefore the function $\zeta(s)$ takes a real value, for $s > 1$.

Proof. We shall first prove that for any nonnegative, continuous and decreasing function $f(x)$, defined in the interval $[1, +\infty)$, it holds

$$\int_1^{+\infty} f(x) \, dx \le S \le a_1 + \int_1^{+\infty} f(x) \, dx,$$

where $a_n = f(n)$ and

$$S = \lim_{n \to +\infty} S_n = \lim_{n \to +\infty} (f(1) + f(2) + \cdots + f(n)), \quad \text{for } n \in \mathbb{N}.$$

In order to do so, we consider a partition of the interval $[1, n]$, as is shown in the figures below.

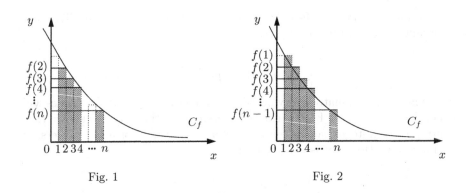

Fig. 1   Fig. 2

In Fig. 1 we can observe that a set of inscribed rectangles is formed. Hence, the area of the shaded region is equal to

$$\sum_{k=2}^{n} f(k).$$

Similarly, in Fig. 2 we can observe that a set of circumscribed rectangles is formed. Hence, the area of the shaded region is equal to

$$\sum_{k=1}^{n-1} f(k).$$

Therefore, by Figs. 1 and 2, it is evident that

$$\sum_{k=2}^{n} f(k) \le \int_1^n f(x) \, dx \le \sum_{k=1}^{n-1} f(k).$$

But

$$S_n = f(1) + f(2) + \cdots + f(n)$$

and thus

$$S_n - f(1) \le \int_1^n f(x) \, dx \le S_{n-1}$$

or

$$\lim_{n\to+\infty} (S_n - f(1)) \leq \lim_{n\to+\infty} \int_1^n f(x)\,dx \leq \lim_{n\to+\infty} S_{n-1}$$

or

$$S - f(1) \leq \int_1^{+\infty} f(x)\,dx \leq S.$$

Therefore,

$$\int_1^{+\infty} f(x)\,dx \leq S \leq f(1) + \int_1^{+\infty} f(x)\,dx$$

or

$$\int_1^{+\infty} f(x)\,dx \leq S \leq a_1 + \int_1^{+\infty} f(x)\,dx.$$

Hence, for the function $\zeta(s)$, we obtain

$$\int_1^{+\infty} \frac{1}{x^s}\,dx \leq \lim_{n\to+\infty} \left(\frac{1}{1^s} + \frac{1}{2^s} + \cdots + \frac{1}{n^s} \right) \leq \frac{1}{1^s} + \int_1^{+\infty} \frac{1}{x^s}\,dx.$$

However,

$$\int_1^{+\infty} \frac{1}{x^s}\,dx = \frac{1}{s-1}$$

and

$$\lim_{n\to+\infty} \left(\frac{1}{1^s} + \frac{1}{2^s} + \cdots + \frac{1}{n^s} \right) = \sum_{n=1}^{+\infty} \frac{1}{n^s}.$$

Thus,

$$\frac{1}{s-1} = \int_1^{+\infty} \frac{1}{x^s}\,dx \leq \zeta(s) \leq 1 + \int_1^{+\infty} \frac{1}{x^s}\,dx = 1 + \frac{1}{s-1}. \qquad \square$$

Property 7.2.4. It holds

$$\zeta(2) = \frac{\pi^2}{6}.$$

(The above property is known as Basel's problem.)

Proof. It is a standard fact in mathematical analysis that the series

$$\sum_{n=1}^{+\infty} \frac{1}{n^2}$$

converges to a real number. However, we shall present a short proof of this fact. It is evident that

$$\sum_{n=1}^{+\infty} \frac{1}{n^2} < 1 + \sum_{n=2}^{+\infty} \frac{1}{(n-1)n} = 1 + \sum_{n=2}^{+\infty} \left(\frac{1}{n-1} - \frac{1}{n} \right) = 2.$$

Therefore,

$$\sum_{n=1}^{+\infty} \frac{1}{n^2} < 2,$$

and thus the series $\sum_{n=1}^{+\infty} 1/n^2$ converges to a real number.

We shall now prove that

$$\sum_{n=1}^{+\infty} \frac{1}{n^2} = \frac{\pi^2}{6}.$$

The strategy which we will follow is to bound the series $\sum_{n=1}^{+\infty} 1/n^2$ from above and below with the same limit. In order to do so, we are going to use the trigonometric inequality

$$\cot^2 x < \frac{1}{x^2} < \csc^2 x, \tag{1}$$

for $0 < x < \frac{\pi}{2}$. However,

$$\frac{\cos(nx) + i\sin(nx)}{(\sin x)^n} = \frac{(\cos x + i\sin x)^n}{(\sin x)^n}$$

and by de Moivre's formula, we obtain

$$\frac{\cos(nx) + i\sin(nx)}{(\sin x)^n}$$

$$= (\cot x + i)^n = \binom{n}{0} \cot^n x + \binom{n}{1} \cot^{n-1} x \cdot i + \binom{n}{2} \cot^{n-2} x \cdot i^2$$

$$+ \binom{n}{3} \cot^{n-3} x \cdot i^3 + \cdots + \binom{n}{n-1} \cot x \cdot i^{n-1} + \binom{n}{n} i^n$$

$$= \left[\binom{n}{0} \cot^n x - \binom{n}{2} \cot^{n-2} x + \cdots \right]$$

$$+ i \cdot \left[\binom{n}{1} \cot^{n-1} x - \binom{n}{3} \cot^{n-3} x + \cdots \right].$$

Therefore,

$$\frac{\sin(nx)}{(\sin x)^n} = \left[\binom{n}{1} \cot^{n-1} x - \binom{n}{3} \cot^{n-3} x + \cdots \right].$$

Set

$$n = 2m+1 \quad \text{and} \quad x = \frac{r\pi}{2m+1}, \quad \text{for } r = 1, 2, \ldots, m. \tag{2}$$

Then, obviously, it holds $nx = r\pi$ and thus

$$\frac{\sin(nx)}{(\sin x)^n} = 0,$$

for all values of x, which verify conditions (2). Hence, we obtain

$$\binom{2m+1}{1}\cot^{2m} x - \binom{2m+1}{3}\cot^{2m-2} x + \cdots = 0.$$

Thus, we have

$$\binom{2m+1}{1}(\cot^2 x)^m - \binom{2m+1}{3}(\cot^2 x)^{m-1} + \cdots = 0,$$

for all values of x, which verify conditions (2). Therefore, the m roots of the polynomial

$$p(t) = \binom{2m+1}{1}t^m - \binom{2m+1}{3}t^{m-1} + \cdots$$

are the values of $\cot^2 x$, for the m different values of x.

Consequently, by Viète's formulae we obtain

$$\cot^2\left(\frac{\pi}{2m+1}\right) + \cot^2\left(\frac{2\pi}{2m+1}\right) + \cdots + \cot^2\left(\frac{m\pi}{2m+1}\right)$$

$$= \frac{\binom{2m+1}{3}}{\binom{2m+1}{1}} = \frac{(2m+1)}{3(2m-2)} \cdot \frac{(2m)}{(2m+1)} = \frac{2m(2m-1)}{6}.$$

Thus,

$$\cot^2\left(\frac{\pi}{2m+1}\right) + \cot^2\left(\frac{2\pi}{2m+1}\right) + \cdots + \cot^2\left(\frac{m\pi}{2m+1}\right) = \frac{2m(2m-1)}{6}. \quad (3)$$

It is a standard fact that $\csc^2 x = \cot^2 x + 1$. Hence, by (3) it follows

$$\csc^2\left(\frac{\pi}{2m+1}\right) + \csc^2\left(\frac{2\pi}{2m+1}\right) + \cdots + \csc^2\left(\frac{m\pi}{2m+1}\right) - m = \frac{2m(2m-1)}{6}$$

or

$$\csc^2\left(\frac{\pi}{2m+1}\right) + \csc^2\left(\frac{2\pi}{2m+1}\right) + \cdots + \csc^2\left(\frac{m\pi}{2m+1}\right)$$

$$= \frac{2m(2m-1)}{6} + \frac{2 \cdot 3m}{6} = \frac{2m(2m+2)}{6}.$$

Thus,

$$\csc^2\left(\frac{\pi}{2m+1}\right) + \csc^2\left(\frac{2\pi}{2m+1}\right) + \cdots + \csc^2\left(\frac{m\pi}{2m+1}\right) = \frac{2m(2m+2)}{6}. \quad (4)$$

By (3), (4) and (1) we obtain

$$\cot^2\left(\frac{\pi}{2m+1}\right) + \cot^2\left(\frac{2\pi}{2m+1}\right) + \cdots + \cot^2\left(\frac{m\pi}{2m+1}\right)$$

$$< \frac{(2m+1)^2}{\pi^2} + \frac{(2m+1)^2}{2^2\pi^2} + \cdots + \frac{(2m+1)^2}{m^2\pi^2}$$

$$< \csc^2\left(\frac{\pi}{2m+1}\right) + \csc^2\left(\frac{2\pi}{2m+1}\right) + \cdots + \csc^2\left(\frac{m\pi}{2m+1}\right).$$

Therefore,

$$\frac{2m(2m-1)}{6} < \frac{(2m+1)^2}{\pi^2}\sum_{n=1}^{m}\frac{1}{n^2} < \frac{2m(2m+2)}{6}$$

and thus

$$\frac{\pi^2}{6}\frac{2m(2m-1)}{(2m+1)^2} < \sum_{n=1}^{m}\frac{1}{n^2} < \frac{\pi^2}{6}\frac{2m(2m+1)}{(2m+1)^2}.$$

Hence,

$$\frac{\pi^2}{6}\lim_{m\to+\infty}\frac{2m(2m-1)}{(2m+1)^2} \leq \sum_{n=1}^{+\infty}\frac{1}{n^2} \leq \frac{\pi^2}{6}\lim_{m\to+\infty}\frac{2m(2m+1)}{(2m+1)^2}$$

or

$$\frac{\pi^2}{6}\lim_{m\to+\infty}\frac{4m^2}{4m^2} \leq \zeta(2) \leq \frac{\pi^2}{6}\lim_{m\to+\infty}\frac{4m^2}{4m^2}$$

or

$$\frac{\pi^2}{6} \leq \zeta(2) \leq \frac{\pi^2}{6}.$$

Therefore, it is evident that

$$\zeta(2) = \frac{\pi^2}{6}. \qquad \square$$

Remark 7.2.5 (L. A. Lyusternik, 1899–1981). By the use of the above result and by Euler's formula Lyusternik presented a new proof of the infinitude of prime numbers. His simple proof is the following:

If the number of primes was finite, then the product

$$\prod_{p}\frac{1}{1-1/p^2}$$

would be a rational number. However,

$$\prod_{p}\frac{1}{1-1/p^2} = \zeta(2) = \frac{\pi^2}{6}.$$

Therefore, $\pi^2/6$ and thus π^2 would be a rational number, which is impossible as we proved in Theorem 1.4.6. $\qquad \square$

Note. The above proof is due to John Papadimitriou and is probably the most elementary proof of this theorem.

The problem of the calculation of the value of $\zeta(2)$ was first posed in 1644 by Pietro Mengoli. The first mathematician to present a solution of the problem was Euler, in 1735. Moreover, Riemann in his paper published in 1859, used some of Euler's ideas concerning generalizations of the problem. The reason why this problem was labeled as Basel's problem is because Basel is the place where Euler was born. We shall now present the basic idea of Euler's proof.

Euler considered the relation

$$\frac{\sin x}{x} = \frac{x - \frac{x^3}{3!} + \frac{x^5}{5!} + \cdots}{x} = 1 - \frac{x^2}{3!} + \frac{x^4}{5!} - \cdots,$$

by which he observed that the coefficient of x^2 is $-1/6$. In addition, the zeros of the function $\sin x/x$ are the real numbers $\pm\pi$, $\pm 2\pi$, $\pm 3\pi$, $\ldots$.

Euler handled the function $\sin x/x$ as a polynomial and therefore, he wrote

$$\frac{\sin x}{x} = \left(1 - \frac{x}{\pi}\right)\left(1 + \frac{x}{\pi}\right)\left(1 - \frac{x}{2\pi}\right)\left(1 + \frac{x}{2\pi}\right)\left(1 - \frac{x}{3\pi}\right)\left(1 + \frac{x}{3\pi}\right)\cdots$$

$$= \left(1 - \frac{x^2}{\pi^2}\right)\left(1 - \frac{x^2}{2^2\pi^2}\right)\left(1 - \frac{x^2}{3^2\pi^2}\right)\cdots\,{}^3 \tag{5}$$

If we carry over the calculations in (5), it follows that the coefficient of x^2 is

$$-\frac{1}{\pi^2}\left(1 + \frac{1}{2^2} + \frac{1}{3^2} + \cdots\right).$$

Hence, Euler claimed that

$$-\frac{1}{\pi^2}\left(1 + \frac{1}{2^2} + \frac{1}{3^2} + \cdots\right) = -\frac{1}{6},$$

by which it is clear that

$$\zeta(2) = \frac{\pi^2}{6}.$$

Property 7.2.6. The Riemann ζ-function $\zeta(s)$ is equal to a rational multiple of π^s, for every even integer s, with $s \geq 2$.

This is true since Euler, in [22], proved that

$$\zeta(2n) = \sum_{k=1}^{+\infty} \frac{1}{k^{2n}} = (-1)^{n-1}\frac{(2\pi)^{2n} \cdot B_{2n}}{2(2n)!}, \quad \text{for } n \in \mathbb{N},$$

---

[3] However, this caused some speculation since some mathematicians claimed that the function $e^x \sin x/x$ has also the same roots, but certainly cannot be expressed in the form (5). This led Euler to justify his assumption, which he did successfully.

where B_n denotes the Bernoulli numbers, which are defined by the following recursive formula:

$$B_0 = 1, \quad B_n = \sum_{s=0}^{n} \binom{n}{s} B_s, \quad \text{for} \ \ n \geq 2.^4$$

Therefore,

$$\zeta(2) = \frac{\pi^2}{6}$$

$$\zeta(4) = \frac{\pi^4}{90}$$

$$\zeta(6) = \frac{\pi^6}{945}$$

$$\zeta(8) = \frac{\pi^8}{9450}$$

$$\zeta(10) = \frac{\pi^{10}}{93555}$$

$$\vdots$$

Property 7.2.7. It holds that

$$\zeta(-n) = (-1)^n \frac{B_{n+1}}{n+1}, \quad \text{for} \ \ n \in \mathbb{N} \cup \{0\}. \tag{6}$$

Previously, we have presented a formula which calculates the value of the zeta function for all positive even integers. However, by formula (6) we can calculate the value of the zeta function for all negative integers and for $s = 0$. Therefore, since $B_{2n+1} = 0$, for $n \in \mathbb{N} \cup \{0\}$, it follows that

$$\zeta(-2k) = 0,$$

for every $k \in \mathbb{N}$.

That is the reason why the negative even integers are considered to be the trivial zeros of $\zeta(s)$.

Property 7.2.8. It holds that $\zeta(3)$ is an irrational number.

In 1979, Roger Apéry proved in [6] that $\zeta(3)$ is an irrational number.

Open Problem. Given an odd integer s, where $s \geq 5$, determine whether $\zeta(s)$ is an irrational number.

---

[4] $B_0 = 1, B_1 = -1/2, B_2 = 1/6, B_4 = -1/30, B_6 = 1/42, B_8 = -1/30, B_{10} = 5/66, \dots$.

We have

$$\zeta(3) = 1,202056903\ldots$$
$$\zeta(5) = 1,036927755\ldots$$
$$\zeta(7) = 1,008349277\ldots$$
$$\zeta(9) = 1,002008392\ldots$$

Property 7.2.9. Let

$$\zeta(s,a) = \sum_{n=0}^{+\infty} \frac{1}{(n+a)^s},\ ^5$$

for real values of s and a, where $s > 1$ and $0 < a \leq 1$. Then

$$\Gamma(s)\zeta(s,a) = \int_0^{+\infty} \frac{x^{s-1}e^{-ax}}{1-e^{-x}}\,dx.$$

The basic idea of the proof of the above property is the following:
By the definition of the Gamma function, we have

$$\Gamma(s) = \int_0^{+\infty} e^{-t}t^{s-1}\,dt.$$

Set $t = (n+a)x$, where $n \in \mathbb{N} \cup \{0\}$. Thus, $dt = (n+a)\,dx$ and therefore

$$\Gamma(s) = (n+a)^{s-1}(n+a)\int_0^{+\infty} e^{-(n+a)x}x^{s-1}\,dx$$

$$= (n+a)^s \int_0^{+\infty} e^{-nx}e^{-ax}x^{s-1}\,dx.$$

Hence,

$$\frac{1}{(n+a)^s}\Gamma(s) = \int_0^{+\infty} e^{-nx}e^{-ax}x^{s-1}\,dx. \tag{1}$$

Therefore, in order to construct the product $\Gamma(s)\zeta(s,a)$, it suffices to sum up all relations of the form (1), for every $n \geq 0$. Thus, by the summation, we obtain

$$\Gamma(s)\sum_{n=0}^{+\infty} \frac{1}{(n+a)^s} = \sum_{n=0}^{+\infty} \int_0^{+\infty} e^{-nx}e^{-ax}x^{s-1}\,dx$$

or

$$\Gamma(s)\zeta(s,a) = \sum_{n=0}^{+\infty} \int_0^{+\infty} e^{-nx}e^{-ax}x^{s-1}\,dx.$$

---

[5] The function $\zeta(s,a)$ is known as the Hurwitz zeta function.

In order to compute the value of the infinite sum

$$\sum_{n=0}^{+\infty} \int_0^{+\infty} e^{-nx} e^{-ax} x^{s-1} \, dx,$$

we shall examine whether the relation

$$\sum_{n=0}^{+\infty} \int_0^{+\infty} e^{-nx} e^{-ax} x^{s-1} \, dx = \int_0^{+\infty} \sum_{n=0}^{+\infty} e^{-nx} e^{-ax} x^{s-1} \, dx \qquad (2)$$

holds true. However, this is an immediate consequence of Tonelli's theorem. Thus, we obtain

$$\Gamma(s)\zeta(s,a) = \sum_{n=0}^{+\infty} \int_0^{+\infty} e^{-nx} e^{-ax} x^{s-1} \, dx = \int_0^{+\infty} \sum_{n=0}^{+\infty} e^{-nx} e^{-ax} x^{s-1} \, dx$$

$$= \int_0^{+\infty} e^{-ax} x^{s-1} \sum_{n=0}^{+\infty} e^{-nx} \, dx$$

$$= \int_0^{+\infty} e^{-ax} x^{s-1} \frac{1}{1-e^{-x}}.$$

Therefore,

$$\Gamma(s)\zeta(s,a) = \int_0^{+\infty} \frac{x^{s-1} e^{-ax}}{1-e^{-x}} \, dx.$$

This completes the proof of the property.   □

Remark 7.2.10. In case $a = 1$ it holds $\zeta(s,1) = \zeta(1)$, since

$$\sum_{n=1}^{+\infty} \frac{1}{n^s} = \sum_{n=0}^{+\infty} \frac{1}{(n+1)^s},$$

and thus

$$\Gamma(s)\zeta(s) = \int_0^{+\infty} \frac{x^{s-1} e^{-x}}{1-e^{-x}} \, dx.$$

The idea of the proof of Property 7.2.9 applies for real values of s. It can also be proved that the same formula holds for $s = a + bi$, for $a > 1$.

Another useful formula which relates the functions $\zeta(s)$ and $\Gamma(s)$ is the following:

$$\pi^{-s/2} \Gamma\left(\frac{s}{2}\right) \zeta(s) = \frac{1}{s(s-1)} + \int_1^{+\infty} W(x)(x^{s/2} + x^{(1-s)/2}) \frac{dx}{x},$$

where

$$W(x) = \frac{w(x) - 1}{2}, \quad w(y) = \theta(iy), \quad \text{with } \theta(z) = \sum_{n \in \mathbb{Z}} e^{\pi i n^2 z}.$$

7.2.1 Applications

Application 7.2.11. Using the function $\zeta(s)$, prove that the number of primes is infinite.

Proof. Let us consider that the number of primes is finite. Then, the product

$$\prod_p \frac{1}{1 - 1/p^s},$$

which extends over all prime numbers, should clearly converge to a real number. Therefore, the same should happen for $s \to 1^+$. But, by Euler's identity we have

$$\lim_{s \to 1+} \prod_p \frac{1}{1 - 1/p^s} = \lim_{s \to 1+} \sum_{n=1}^{+\infty} \frac{1}{n^s} = +\infty,$$

which is a contradiction. $\square$

Application 7.2.12. By the use of the function $\zeta(s)$, prove that the series

$$\sum_p \frac{1}{p},$$

which extends over all prime numbers, diverges (i.e., converges to $+\infty$).

Proof. Because of the fact that $\lim_{s \to 1+} \zeta(s) = +\infty$, it follows that

$$\lim_{s \to 1+} (\log \zeta(s)) = +\infty. \tag{1}$$

However,

$$\log \zeta(s) = \log \left(\sum_{n=1}^{+\infty} \frac{1}{n^s} \right) = \log \left(\prod_p \frac{1}{1 - 1/p^s} \right)$$

$$= \sum_p \log \left(\frac{1}{1 - 1/p^s} \right)$$

$$= \sum_p \sum_{n=1}^{+\infty} \frac{1}{np^{ns}} = \sum_p \left(\frac{1}{p^s} + \sum_{n=2}^{+\infty} \frac{1}{np^{ns}} \right).$$

Hence,

$$\log \zeta(s) = \sum_p \frac{1}{p^s} + \sum_p \sum_{n=2}^{+\infty} \frac{1}{np^{ns}}.$$

Thus,

$$\log \zeta(s) < \sum_p \frac{1}{p^s} + \sum_p \sum_{n=2}^{+\infty} \frac{1}{p^{ns}}.$$

It is true that

$$\sum_{n=2}^{+\infty} \frac{1}{p^{ns}} = \frac{1}{p^{2s} - p^s},$$

since $\sum_{n=2}^{+\infty} 1/p^{ns}$ is the sum of the terms of a decreasing geometric progression. Therefore, for $s > 1$, we obtain that

$$\log \zeta(s) < \sum_p \frac{1}{p^s} + \sum_p \frac{1}{p^{2s} - p^s}$$

$$= \sum_p \frac{1}{p^s} + \sum_p \frac{1}{p^s(p^s - 1)}$$

$$< \sum_p \frac{1}{p^s} + \sum_{n=2}^{+\infty} \frac{1}{n^s(n^s - 1)}$$

$$< \sum_p \frac{1}{p^s} + \sum_{n=2}^{+\infty} \frac{1}{n(n - 1)}$$

$$< \sum_p \frac{1}{p^s} + \sum_{n=2}^{+\infty} \frac{1}{(n - 1)^2}.$$

Thus, it yields

$$\lim_{s \to 1^+} (\log \zeta(s)) \leq \lim_{s \to 1^+} \left(\sum_p \frac{1}{p^s} + \sum_{n=2}^{+\infty} \frac{1}{(n - 1)^2} \right)$$

and thus, by (1), it is evident that

$$\lim_{s \to 1^+} \left(\sum_p \frac{1}{p^s} + \sum_{n=2}^{+\infty} \frac{1}{(n - 1)^2} \right) = +\infty.$$

However, the series

$$\sum_{n=2}^{+\infty} \frac{1}{(n - 1)^2}$$

converges to a real number. Hence,

$$\lim_{s \to 1^+} \sum_p \frac{1}{p^s} = +\infty$$

and therefore

$$\sum_p \frac{1}{p} = +\infty.$$ □

Application 7.2.13. Prove that

$$\frac{1}{\zeta(s)} = \sum_{n=1}^{+\infty} \frac{\mu(n)}{n^s}.$$

The basic idea of the proof of the above property is the following:
    By Euler's identity we have

$$\frac{1}{\zeta(s)} = \frac{1}{\prod_p(1-p^{-s})^{-1}} = \prod_p\left(1 - \frac{1}{p^s}\right)$$

$$= \left(1 - \frac{1}{p_1^s}\right)\left(1 - \frac{1}{p_2^s}\right)\cdots.$$

However, if we perform all possible calculations in the above product, we will obtain an infinite sum of terms of the form

$$\frac{(-1)^k}{(p_1^{a_1} p_2^{a_2} \cdots p_k^{a_k})^s},$$

increased by 1, where $a_i = 0$ or 1, for $i = 1, 2, \ldots, k$ and it is not possible for all the exponents to be zero simultaneously.
    In the denominators of the fractions of the above terms, one encounters all possible combinations of multiples of prime numbers to 0 or 1st powers. Hence, it follows that

$$\frac{1}{\zeta(s)} = 1 + \sum \frac{(-1)^k}{(p_1^{a_1} p_2^{a_2} \cdots p_k^{a_k})^s}$$

$$= \frac{\mu(1)}{1} + \sum \frac{(-1)^k}{(p_1^{a_1} p_2^{a_2} \cdots p_k^{a_k})^s} + 0.$$

But, we can write

$$0 = \sum \frac{\mu(p_1^{q_1} p_2^{q_2} \cdots p_\lambda^{q_\lambda})}{(p_1^{q_1} p_2^{q_2} \cdots p_\lambda^{q_\lambda})^s},$$

where in the denominators of the terms of the above infinite sum, one encounters all possible combinations of multiples of prime numbers with $q_i \geq 2$, $i = 1, 2, \ldots, \lambda$. Therefore, it yields

$$\frac{1}{\zeta(s)} = \frac{\mu(1)}{1} + \sum \frac{(-1)^k}{(p_1^{a_1} p_2^{a_2} \cdots p_k^{a_k})^s} + \sum \frac{\mu(p_1^{q_1} p_2^{q_2} \cdots p_\lambda^{q_\lambda})}{(p_1^{q_1} p_2^{q_2} \cdots p_\lambda^{q_\lambda})^s}.$$

However, each positive integer n with $n > 1$ can be represented either by the term $p_1^{a_1} p_2^{a_2} \cdots p_k^{a_k}$ or by the term $p_1^{q_1} p_2^{q_2} \cdots p_\lambda^{q_\lambda}$. Thus, it follows that

$$\frac{1}{\zeta(s)} = \sum_{n=1}^{+\infty} \frac{\mu(n)}{n^s}. \qquad \qquad \Box$$

8

Dirichlet series

The total number of Dirichlet's publications is not large: jewels are not weighted on a grocery scale.

Carl Friedrich Gauss (1777–1855)

8.1 Basic notions

Definition 8.1.1. *Let f be an arithmetic function. The series*

$$D(f,s) = \sum_{n=1}^{+\infty} \frac{f(n)}{n^s},$$

*where $s \in \mathbb{C}$, is called a **Dirichlet series with coefficients** $f(n)$.*

In this chapter, we will handle Dirichlet series for s being a real number.

Consider now a Dirichlet series, which is absolutely convergent for $s > s_0$.

- If for these values of s it holds

$$\sum_{n=1}^{+\infty} \frac{f(n)}{n^s} = 0,$$

  then $f(n) = 0$, for every integer n with $n \geq 1$.
- If for these values of s it holds

$$D(f,s) = D(g,s),$$

  then by the above argument it holds

$$f(n) = g(n), \text{ for every integer } n \text{ with } n \geq 1.$$

M.Th. Rassias, *Problem-Solving and Selected Topics in Number Theory: In the Spirit of the Mathematical Olympiads*, DOI 10.1007/978-1-4419-0495-9_8,
© Springer Science+Business Media, LLC 2011

Definition 8.1.2. *The **summation** and **multiplication** of Dirichlet series are defined, respectively, as follows:*

$$D(f_1, s) + D(f_2, s) = \sum_{n=1}^{+\infty} \frac{f_1(n) + f_2(n)}{n^s}$$

and

$$D(f_1, s) \cdot D(f_2, s) = \sum_{n=1}^{+\infty} \frac{g(s)}{n^s},$$

where

$$g(n) = \sum_{n_1 n_2 = n} f_1(n_1) f_2(n_2).$$

Theorem 8.1.3. *Let f be a multiplicative function. Then, it holds*

$$D(f, s) = \prod_{p} \left(\sum_{n=0}^{+\infty} \frac{f(p^n)}{p^{ns}} \right),$$

where the product extends over all prime numbers p.

The basic idea of the proof of the theorem is the following:
It is true that

$$\prod_{p} \left(\sum_{n=0}^{+\infty} \frac{f(p^n)}{p^{ns}} \right)$$

$$= \prod_{p} \left(\frac{f(1)}{1} + \frac{f(p)}{p^s} + \frac{f(p^2)}{p^{2s}} + \cdots \right)$$

$$= \left(\frac{f(1)}{1} + \frac{f(p_1)}{p_1^s} + \frac{f(p_1^2)}{p_1^{2s}} + \cdots \right) \left(\frac{f(1)}{1} + \frac{f(p_2)}{p_2^s} + \frac{f(p_2^2)}{p_2^{2s}} + \cdots \right) \cdots \text{[1]}$$

$$= \sum \frac{f(p_1^{a_1}) \cdots f(p_k^{a_k})}{(p_1^{a_1} \cdots p_k^{a_k})^s}, \tag{1}$$

where the sum extends over all possible combinations of multiples of powers of prime numbers. But, since the function $f(n)$ is multiplicative, it is evident that

$$\sum \frac{f(p_1^{a_1}) \cdots f(p_k^{a_k})}{(p_1^{a_1} \cdots p_k^{a_k})^s} = \sum \frac{f(p_1^{a_1} \cdots p_k^{a_k})}{(p_1^{a_1} \cdots p_k^{a_k})^s}$$

$$= \sum_{n=1}^{+\infty} \frac{f(n)}{n^s} = D(f, s). \qquad \square$$

---

[1] Here p_i denotes the ith prime number ($p_1 = 2$, $p_2 = 3, \dots$).

Application 8.1.4. Prove that

$$D(\lambda, s) = \frac{\zeta(2s)}{\zeta(s)},$$

where $\lambda(n)$ stands for the *Liouville function*, which is defined by

$$\lambda(n) = \begin{cases} (-1)^{a_1 + \cdots + a_k}, & \text{for } n = p_1^{a_1} p_2^{a_2} \cdots p_k^{a_k} > 1 \\ 1, & \text{for } n = 1. \end{cases}$$

Proof. By the previous theorem, we obtain

$$D(\lambda, s) = \sum_{n=1}^{+\infty} \frac{(-1)^{a_1 + a_2 + \cdots + a_k}}{n^s} = \prod_p \left(\sum_{n=0}^{+\infty} \frac{(-1)^n}{p^{ns}} \right)$$

$$= \prod_p \left(1 - \frac{1}{p^s} + \frac{1}{p^{2s}} - \frac{1}{p^{3s}} + \cdots \right) = \prod_p \frac{1}{1 - (-1/p^s)}$$

$$= \prod_p \frac{1}{1 + 1/p^s} = \frac{\zeta(2s)}{\zeta(s)}.$$

Therefore,

$$D(\lambda, s) = \frac{\zeta(2s)}{\zeta(s)}. \qquad \square$$

9

Special topics

9.1 The harmonic series of prime numbers

Theorem 9.1.1. *The series*

$$\sum_p \frac{1}{p} = \frac{1}{2} + \frac{1}{3} + \frac{1}{5} + \frac{1}{7} + \cdots$$

diverges (converges to $+\infty$) when p is a prime number.[1]

First Proof. Suppose that the series

$$\sum_p \frac{1}{p}$$

converges to a real number. Then, there exists a positive integer k, such that

$$\frac{1}{p_{k+1}} + \frac{1}{p_{k+2}} + \cdots < \frac{1}{2}.$$

Thus,

$$\frac{x}{p_{k+1}} + \frac{x}{p_{k+2}} + \cdots < \frac{x}{2},$$

where x is a positive integer.

---

[1] We have presented a proof of the fact in the chapter on the function $\zeta(s)$.

M.Th. Rassias, *Problem-Solving and Selected Topics in Number Theory: In the Spirit
of the Mathematical Olympiads*, DOI 10.1007/978-1-4419-0495-9_9,
© Springer Science+Business Media, LLC 2011

Let $N(x, p_k)$ be the number of all positive integers n, where $n \leq x$, which are not divisible by any prime number p with $p > p_k$, where x is a positive integer and p_k is the kth prime number in the sequence of primes ($p_1 = 2$, $p_2 = 3, \ldots$).

We express the arbitrary such positive integer n in the form

$$n = m_1^2 m,$$

where

$$m = 2^{b_1} 3^{b_2} \cdots p_k^{b_k} \quad \text{with} \quad b_i = 0 \quad \text{or} \quad b_i = 1 \quad \text{and} \quad i = 1, 2, \ldots, k.$$

There are 2^k different values of m and

$$\sqrt{n} = m_1 \sqrt{m}, \quad \text{that is,} \quad m_1 \leq \sqrt{n} \leq \sqrt{x}.$$

Therefore, there exist at most $\sqrt{x}$ different values of m_1. This means that there are at most $2^k \sqrt{x}$ different values of n.

Thus,

$$N(x, p_k) \leq 2^k \sqrt{x}.$$

However, $N(x, p_k) > x/2$ because one can choose p_k such that $p_k > x/2$ if x is even or $p_k > (x+1)/2$ if x is odd. Thus, every prime number p with $p > p_k$ does not divide the numbers $1, 2, 3, 4, \ldots, p_k$, which are more than $x/2$ (or $(x+1)/2$ respectively, depending on x).

Therefore, one has

$$\frac{x}{2} < N(x, p_k) \leq 2^k \sqrt{x}$$

and thus

$$x^2 < 4 \cdot 2^{2k} x \Leftrightarrow x < 2^{2k+2},$$

which is not satisfied for $x \geq 2^{2k+2}$.

Hence, the series

$$\sum_p \frac{1}{p}$$

diverges (converges to $+\infty$).   □

Second Proof. Let n be a positive integer with $n > 1$. Then

$$p_1, p_2, \ldots, p_k \leq n \leq p_{k+1}$$

for some positive integer k.

Every positive integer less than or equal to n has prime factors which belong to the set $\{p_1, p_2, \ldots, p_k\}$.

Therefore, every one of these integers can be expressed in the form

$$p_1^{a_1} p_2^{a_2} \cdots p_k^{a_k} \quad \text{with} \quad a_i \geq 0, \quad \text{where} \quad i = 1, 2, \ldots, k.$$

We consider an integer m for which it holds

$$2^m > n.$$

Then it is obvious that $p_i^m > n$ for every prime number p_i, where $i = 1,$ $2, \ldots, k$.

Therefore,

$$p_1^m p_2^m \cdots p_k^m > n \quad \text{or} \quad p_1^{a_1} p_2^{a_2} \cdots p_i^m \cdots p_k^{a_k} > n, \quad \text{where } i = 1, 2, \ldots, k.$$

We know that

$$\frac{1}{1 - \frac{1}{p_i}} = \sum_{j=0}^{+\infty} \frac{1}{p_i^j}.$$

Thus,

$$\frac{1}{1 - \frac{1}{p_i}} > 1 + \frac{1}{p_i} + \frac{1}{p_i^2} + \cdots + \frac{1}{p_i^m}.$$

We obtain

$$\prod_{i=1}^{k} \frac{1}{1 - \frac{1}{p_i}} > \prod_{i=1}^{k} \left(1 + \frac{1}{p_i} + \frac{1}{p_i^2} + \cdots + \frac{1}{p_i^m} \right).$$

If we expand the product,

$$\prod_{i=1}^{k} \left(1 + \frac{1}{p_i} + \frac{1}{p_i^2} + \cdots + \frac{1}{p_i^m} \right),$$

we obtain a summation of terms of the form

$$\frac{1}{p_1^{a_1} p_2^{a_2} \cdots p_k^{a_k}},$$

where $0 \le a_i \le m$, for $i = 1, 2, \ldots, k$. We have shown that

$$p_1^{a_1} p_2^{a_2} \cdots p_i^m \cdots p_k^{a_k} > n, \quad \text{for } i = 1, 2, \ldots, k.$$

Therefore, in the denominators of the terms of the summation, all positive integers $1, 2, \ldots, n$ will appear and in addition, some more positive integers greater than n will appear too.

Therefore,

$$\prod_{i=1}^{k} \left(1 + \frac{1}{p_i} + \frac{1}{p_i^2} + \cdots + \frac{1}{p_i^m} \right) > \sum_{l=1}^{n} \frac{1}{l},$$

that is,

$$\prod_{i=1}^{k} \frac{1}{1 - \frac{1}{p_i}} > \sum_{l=1}^{n} \frac{1}{l}. \tag{1}$$

We know that

$$\ln(1 + x) = \frac{x}{1} - \frac{x^2}{2} + \cdots + (-1)^{n+1} \frac{x^n}{n} + \cdots, \quad \text{if } |x| < 1.$$

Thus,

$$\ln \frac{1}{1 - \frac{1}{p_i}} = -\ln \left(1 - \frac{1}{p_i}\right)$$

$$= -\left(-\frac{1}{p_i} - \frac{1}{2\,p_i^2} - \frac{1}{3\,p_i^3} - \frac{1}{4\,p_i^4} - \cdots\right)$$

$$= \frac{1}{p_i} + \frac{1}{2\,p_i^2} + \frac{1}{3\,p_i^3} + \frac{1}{4\,p_i^4} + \cdots$$

$$< \frac{1}{p_i} + \frac{1}{2\,p_i^2} + \frac{1}{2\,p_i^3} + \frac{1}{2\,p_i^4} + \cdots$$

$$= \frac{1}{p_i} + \frac{1}{2\,p_i^2} \left(1 + \frac{1}{p_i} + \frac{1}{p_i^2} + \cdots\right)$$

$$= \frac{1}{p_i} + \frac{1}{2\,p_i^2} \cdot \frac{1}{1 - \frac{1}{p_i}}.$$

However,

$$p_i \geq 2 \Rightarrow \frac{1}{p_i} \leq \frac{1}{2} \Rightarrow -\frac{1}{p_i} \geq -\frac{1}{2} \Rightarrow \frac{1}{1 - \frac{1}{p_i}} \leq \frac{1}{1 - \frac{1}{2}},$$

thus

$$\ln \frac{1}{1 - \frac{1}{p_i}} < \frac{1}{p_i} + \frac{1}{2\,p_i^2} \cdot \frac{1}{1 - \frac{1}{p_i}} \leq \frac{1}{p_i} + \frac{1}{2\,p_i^2} \cdot \frac{1}{1 - \frac{1}{2}} = \frac{1}{p_i} + \frac{1}{p_i^2}.$$

Therefore,

$$\ln \frac{1}{1 - \frac{1}{p_i}} < \frac{1}{p_i} + \frac{1}{p_i^2}, \quad \text{for } i = 1, 2, \ldots, k. \tag{2}$$

Applying relation (2) we obtain

$$\sum_{i=1}^{k} \ln \frac{1}{1 - \frac{1}{p_i}} < \sum_{i=1}^{k} \frac{1}{p_i} + \sum_{i=1}^{k} \frac{1}{p_i^2} < \sum_{i=1}^{k} \frac{1}{p_i} + \sum_{l=1}^{+\infty} \frac{1}{l^2}.$$

But

$$\sum_{i=1}^{k} \ln \frac{1}{1 - \frac{1}{p_i}} = \ln \prod_{i=1}^{k} \frac{1}{1 - \frac{1}{p_i}}.$$

Thus, because of inequality (1) we get

$$\sum_{i=1}^{k} \frac{1}{p_i} + \sum_{l=1}^{+\infty} \frac{1}{l^2} > \ln \prod_{i=1}^{k} \frac{1}{1 - \frac{1}{p_i}} > \ln \left(\sum_{l=1}^{n} \frac{1}{l}\right).$$

Therefore,

$$\sum_{i=1}^{k} \frac{1}{p_i} > \ln\left(\sum_{l=1}^{n} \frac{1}{l}\right) - \sum_{l=1}^{+\infty} \frac{1}{l^2}.$$

If $n \to +\infty$, then obviously $k \to +\infty$.

However, we know that

$$\sum_{l=1}^{+\infty} \frac{1}{l} = +\infty \quad \text{and} \quad \sum_{l=1}^{+\infty} \frac{1}{l^2} \quad \text{converges to a real number.}$$

Hence, the series

$$\sum_{i=1}^{+\infty} \frac{1}{p_i}$$

diverges (converges to $+\infty$).   □

Third Proof. Suppose that the series

$$\sum_{p} \frac{1}{p}$$

converges to a real number.

Then, there exists a positive integer n such that

$$\frac{1}{p_{n+1}} + \frac{1}{p_{n+2}} + \cdots < \frac{1}{2},$$

which means

$$\sum_{k=n+1}^{+\infty} \frac{1}{p_k} < \frac{1}{2}.$$

Consider the numbers

$$Q_m = 1 + mN, \quad \text{where} \quad m = 1, 2, \ldots \quad \text{and} \quad N = p_1 p_2 \cdots p_n.$$

It is evident that none of the numbers Q_m is divisible by some of the prime numbers $p_1, p_2, \ldots, p_n$. Therefore, the prime factors of the numbers of the form $1 + mN$ belong to the set $\{p_{n+1}, p_{n+2}, \ldots\}$.

In the infinite summation

$$\left(\frac{1}{p_{n+1}} + \frac{1}{p_{n+2}} + \cdots\right) + \left(\frac{1}{p_{n+1}} + \frac{1}{p_{n+2}} + \cdots\right)^2 + \cdots$$

$$+ \left(\frac{1}{p_{n+1}} + \frac{1}{p_{n+2}} + \cdots\right)^t + \cdots$$

there exists every number of the form

$$\frac{1}{p_{n+1}^{m_1} p_{n+2}^{m_2} \cdots p_{n+\lambda}^{m_\lambda}},$$

but there also exist infinitely many other different terms.

Thus, we obtain

$$\frac{1}{Q_1} + \frac{1}{Q_2} + \cdots + \frac{1}{Q_m} \leq \sum_{t=1}^{+\infty} \left(\sum_{k=n+1}^{+\infty} \frac{1}{p_k} \right)^t,$$

that is,

$$\sum_{k=1}^{m} \frac{1}{Q_k} \leq \sum_{t=1}^{+\infty} \left(\sum_{k=n+1}^{+\infty} \frac{1}{p_k} \right)^t.$$

However,

$$\sum_{k=n+1}^{+\infty} \frac{1}{p_k} < \frac{1}{2},$$

thus

$$\sum_{k=1}^{m} \frac{1}{Q_k} < \sum_{t=1}^{+\infty} \left(\frac{1}{2} \right)^t.$$

But, the series

$$\sum_{t=1}^{+\infty} \left(\frac{1}{2} \right)^t$$

converges to a real number, since it is an infinite summation of a decreasing geometric progression. Therefore, the series

$$\sum_{k=1}^{+\infty} \frac{1}{Q_k}$$

is bounded, which is not possible because

$$\frac{1}{Q_m} = \frac{1}{1 + mN} > \frac{1}{N + mN} = \frac{1}{N(m+1)},$$

and the series

$$\sum_{m=1}^{+\infty} \frac{1}{N(m+1)} = \frac{1}{N} \sum_{m=1}^{+\infty} \frac{1}{m+1}$$

diverges (converges to $+\infty$). Therefore, the assumption that the series

$$\sum_{p} \frac{1}{p}$$

converges to a real number has led to a contradiction. Hence, the series

$$\sum_{p} \frac{1}{p}$$

diverges (converges to $+\infty$).   □

Note. The above proof is due to J. A. Clarkson, On the series of prime reciprocals, Proc. Amer. Math. Soc. 17(1966), 541.

Fourth Proof. Consider the summation

$$S_N = \sum_{n=1}^{N} \frac{1}{p_n}, \quad N \in \mathbb{N}.$$

Then, one has

$$e^{S_N} = e^{\frac{1}{p_1} + \frac{1}{p_2} + \cdots + \frac{1}{p_N}} = \prod_{n=1}^{N} e^{\frac{1}{p_n}}. \tag{1}$$

The Maclaurin's expansion of e^x is given by the formula

$$e^x = 1 + \sum_{n=1}^{+\infty} \frac{x^n}{n!}, \quad \text{where } x \in \mathbb{R}.$$

Thus,

$$e^{\frac{1}{p_n}} = 1 + \frac{1}{p_n} + \frac{1}{2! \, p_n^2} + \cdots$$

and therefore

$$e^{\frac{1}{p_n}} > 1 + \frac{1}{p_n}.$$

From (1), we obtain

$$e^{S_N} > \prod_{n=1}^{N} \left(1 + \frac{1}{p_n} \right).$$

After the calculation of the product

$$\prod_{n=1}^{N} \left(1 + \frac{1}{p_n} \right)$$

the result will get the representation

$$1 + \sum \frac{1}{p_{\lambda_1} p_{\lambda_2} \cdots p_{\lambda_\mu}}, \quad \text{with } p_{\lambda_1} \neq p_{\lambda_2} \neq \cdots \neq p_{\lambda_\mu}.$$

The greatest value of the denominator of a term, which appears in the summation

$$\sum \frac{1}{p_{\lambda_1} p_{\lambda_2} \cdots p_{\lambda_\mu}},$$

is the number

$$p_1 p_2 \cdots p_n.$$

Thus, we obtain

$$\prod_{n=1}^{N} \left(1 + \frac{1}{p_n}\right) > \sum_{q \leq N} \frac{1}{q},$$

where q is an arbitrary integer less than or equal to N, which is not divisible by the square of any prime number (i.e., it is a squarefree integer).

Therefore,

$$s^{S_N} > \sum_{q \leq N} \frac{1}{q}.$$

Suppose that

$$\lim_{N \to +\infty} S_N = S,$$

where $S \in \mathbb{R}$. It is evident that

$$S > S_n, \quad \text{for every } n \in \mathbb{N}$$

and thus

$$e^S > \sum_{q \leq N} \frac{1}{q}, \quad \text{for every } N \in \mathbb{N}. \tag{2}$$

However, by Lemma 1.3.4, we know that every positive integer can be represented in the form $a^2 q$, where q is a squarefree integer. Therefore, it holds

$$\left(\sum_{q \leq N} \frac{1}{q}\right) \cdot \left(\sum_{a=1}^{N} \frac{1}{a^2}\right) > \sum_{n=1}^{N} \frac{1}{n}, \quad \text{where } N \neq 1.$$

It is a standard fact from mathematical analysis that

$$\sum_{a=1}^{+\infty} \frac{1}{a^2}$$

converges to a real number.

Let D be that real number. Then

$$D > \sum_{a=1}^{N} \frac{1}{a^2}.$$

Hence,

$$\sum_{q \leq N} \frac{1}{q} > \frac{1}{D} \sum_{n=1}^{N} \frac{1}{n}.$$

In addition, from mathematical analysis we also know that

$$\sum_{n=1}^{+\infty} \frac{1}{n}$$

diverges (converges to $+\infty$).

Thus, we can choose N to be sufficiently large, so that

$$\sum_{n=1}^{N} \frac{1}{n} > De^S.$$

But then,

$$\sum_{q \le N} \frac{1}{q} > e^S,$$

which is impossible because of (2).

Therefore, the hypothesis that the series

$$\sum_{n=1}^{+\infty} \frac{1}{p_n}$$

converges to a real number leads to a contradiction. Consequently, the series diverges. □

Historical Remark. The question whether the series

$$\sum_{k=1}^{+\infty} \frac{1}{p_k}$$

diverges was answered for the first time in 1737, by L. Euler.

9.2 Lagrange's four-square theorem

> *The highest form of pure thought is in mathematics.*
> Plato (428 BC–348 BC)

Leonhard Euler (1707–1783) proved that a positive integer can be expressed as the sum of two squares of integers, only when certain conditions are being satisfied. In addition, it has also been proved that it is not possible to express every positive integer as the sum of three squares of integers.

Joseph-Louis Lagrange (1736–1813) was the first mathematician to prove that every positive integer can be expressed as the sum of four squares of integers.

At the beginning of the 20th century, David Hilbert (1862–1943) proved that

> for every positive integer n there exists a number K_n, such that each positive integer m can be expressed as the sum of no more than K_n positive nth powers.

Here we shall present the proof of Lagrange's theorem.

Theorem 9.2.1 (Lagrange's Theorem). *Every positive integer can be expressed as the sum of four squares of integers.*

Proof. By Lagrange's identity, we have

$$(x_1^2 + x_2^2 + x_3^2 + x_4^2)(y_1^2 + y_2^2 + y_3^2 + y_4^2) - (x_1y_1 + x_2y_2 + x_3y_3 + x_4y_4)^2$$

$$= \begin{vmatrix} x_1 & x_2 \\ y_1 & y_2 \end{vmatrix}^2 + \begin{vmatrix} x_1 & x_3 \\ y_1 & y_3 \end{vmatrix}^2 + \begin{vmatrix} x_1 & x_4 \\ y_1 & y_4 \end{vmatrix}^2 + \begin{vmatrix} x_2 & x_3 \\ y_2 & y_3 \end{vmatrix}^2 + \begin{vmatrix} x_2 & x_4 \\ y_2 & y_4 \end{vmatrix}^2 + \begin{vmatrix} x_3 & x_4 \\ y_3 & y_4 \end{vmatrix}^2$$

$$= (x_1y_2 - x_2y_1)^2 + (x_1y_3 - x_3y_1)^2 + (x_1y_4 - x_4y_1)^2 + (x_2y_3 - x_3y_2)^2$$
$$+ (x_2y_4 - x_4y_2)^2 + (x_3y_4 - x_4y_3)^2$$

$$= (x_1y_2 - x_2y_1 + x_3y_4 - x_4y_3)^2 - 2(x_1y_2 - x_2y_1)(x_3y_4 - x_4y_3)$$
$$+ (x_1y_3 - x_3y_1 + x_4y_2 - x_2y_4)^2 + 2(x_1y_3 - x_3y_1)(x_2y_4 - x_4y_2)$$
$$+ (x_1y_4 - x_4y_1 + x_2y_3 - x_3y_2)^2 - 2(x_1y_4 - x_4y_1)(x_2y_3 - x_3y_2)$$

$$= (x_1y_2 - x_2y_1 + x_3y_4 - x_4y_3)^2 + (x_1y_3 - x_3y_1 + x_4y_2 - x_2y_4)^2$$
$$+ (x_1y_4 - x_4y_1 + x_2y_3 - x_3y_2)^2 - 2[(x_1y_2 - x_2y_1)(x_3y_4 - x_4y_3)$$
$$- (x_1y_3 - x_3y_1)(x_2y_4 - x_4y_2) + (x_1y_4 - x_4y_1)(x_2y_3 - x_3y_2)].$$

Thus, we obtain

$$(x_1^2 + x_2^2 + x_3^2 + x_4^2)(y_1^2 + y_2^2 + y_3^2 + y_4^2)$$
$$= (x_1y_1 + x_2y_2 + x_3y_3 + x_4y_4)^2 + (x_1y_2 - x_2y_1 + x_3y_4 - x_4y_3)^2$$
$$+ (x_1y_3 - x_3y_1 + x_4y_2 - x_2y_4)^2 + (x_1y_4 - x_4y_1 + x_2y_3 - x_3y_2)^2.$$

Therefore, by the above relation, it follows that the product of two positive integers, each of which can be expressed as the sum of four squares of integers, can always be expressed as the sum of four squares of integers.

However, by the Fundamental Theorem of Arithmetic, we know that every integer can be either a prime number or a product of powers of prime numbers. Hence, it is evident that it suffices to prove that every prime number p can be expressed as the sum of four squares of integers.

In the special case when $p = 2$, it holds

$$2 = 1^2 + 1^2 + 0^2 + 0^2.$$

We shall now prove that for every prime number p, with $p > 2$, there exist integers x, y, such that
$$x^2 + y^2 + 1 = mp,$$
where $0 < m < p$.

Consider the sets

$$A = \left\{ x^2 : x = 0, 1, \ldots, \frac{p-1}{2} \right\}$$

and

$$B = \left\{ -1 - y^2 : y = 0, 1, \ldots, \frac{p-1}{2} \right\}.$$

There does not exist a pair (x_1^2, x_2^2) of elements of A, for which

$$x_1^2 \equiv x_2^2 \, (\text{mod } p).$$

This happens because if such a pair existed, then we would have

$$x_1^2 - x_2^2 = kp$$

or

$$(x_1 - x_2)(x_1 + x_2) = kp, \quad \text{for some } k \in \mathbb{Z}$$

and thus

$$p \mid (x_1 - x_2) \quad \text{or} \quad p \mid (x_1 + x_2).$$

But that is a contradiction, since if $p \mid (x_1 - x_2)$, then

$$|x_1 - x_2| \geq p.$$

However,

$$\left.\begin{array}{l} x_1 < \dfrac{p}{2} \\[2mm] x_2 < \dfrac{p}{2} \end{array}\right\} \Rightarrow x_1 + x_2 < p \Rightarrow |x_1 - x_2| < p.$$

Moreover, if $p \mid (x_1 + x_2)$, then

$$x_1 + x_2 \geq p.$$

But

$$\left.\begin{array}{l} x_1 < \dfrac{p}{2} \\[2mm] x_2 < \dfrac{p}{2} \end{array}\right\} \Rightarrow x_1 + x_2 < p.$$

Similarly, there does not exist a pair of congruent elements $(\bmod\, p)$ of B.

The cardinality of the set $A \cup B$ is $p + 1$, since each of the sets A, B has cardinality $(p+1)/2$. These $p+1$ elements are clearly pairwise distinct. Thus, if we divide each of the elements of $A \cup B$ by p, we obtain $p + 1$ pairwise distinct residues.[2]

However, if we divide any positive integer by p, then the only possible residues are the integers $0, 1, 2, \ldots, p - 1$. Thus, we have at most p pairwise distinct residues. Hence, by the *pigeonhole principle*,[3] it follows that there exists at least one pair of congruent $(\bmod\, p)$ elements in $A \cup B$. Let (u, v) be such a pair. Then, according to the above arguments it yields $u \in A$ and $v \in B$, since it is not possible for both u and v to belong in the same set. Hence, we obtain that

$$u \equiv v \,(\bmod\, p)$$

or

$$x^2 \equiv (-1 - y^2)\,(\bmod\, p)$$

or

$$x^2 + y^2 + 1 = mp, \quad \text{for some } m \in \mathbb{Z}.$$

But

$$x^2 \leq \left(\frac{p-1}{2}\right)^2 \quad \text{and} \quad y^2 \leq \left(\frac{p-1}{2}\right)^2.$$

Therefore,

$$x^2 + y^2 + 1 \leq 2\left(\frac{p-1}{2}\right)^2 + 1 = \frac{(p-1)^2}{2} + 1 < p^2.$$

---

[2] If the element is an integer less than p, the residue is the integer itself.

[3] The *pigeonhole principle* (also known as *Dirichlet's box principle*) is just the obvious remark, that states if we place m pigeons in n pigeonholes, where $m > n$, then there must be at least one pigeonhole with more than one pigeon. The pigeonhole principle was first introduced by Dirichlet, in 1834.

Therefore, it holds $0 < m < p$. Thus, there exist integers x, y, where

$$0 \leq x < \frac{p}{2} \quad \text{and} \quad 0 \leq y < \frac{p}{2},$$

such that

$$x^2 + y^2 + 1 \equiv 0 \,(\text{mod}\, p).$$

Hence, it follows that there exist integers

$$m, x_1, x_2, x_3, x_4, {}^4$$

where $0 < m < p$, such that

$$mp = x_1^2 + x_2^2 + x_3^2 + x_4^2.$$

The integers x_1^2, x_2^2, x_3^2, x_4^2 are not all divisible by p, since if that was the case, then we would have

$$mp = k_1^2 p^2 + k_2^2 p^2 + k_3^2 p^2 + k_4^2 p^2$$
$$= (k_1^2 + k_2^2 + k_3^2 + k_4^2)p^2,$$

and thus

$$m = (k_1^2 + k_2^2 + k_3^2 + k_4^2)p > p,$$

which is a contradiction.

It suffices to prove that the least possible value of m is 1. Thus, let $m_0 p$ be the least multiple of p, which satisfies the property

$$m_0 p = x_1^2 + x_2^2 + x_3^2 + x_4^2,$$

where $m_0 > 1$ and $0 < m_0 < p$.

- If m_0 is an even integer, then the sum

$$x_1 + x_2 + x_3 + x_4$$

is also an even integer, for if it was an odd integer we would have

$$(x_1 + x_2 + x_3 + x_4)^2$$
$$= x_1^2 + x_2^2 + x_3^2 + x_4^2 + 2(x_1 x_2 + x_1 x_3 + x_1 x_4 + x_2 x_3 + x_2 x_4 + x_3 x_4)$$
$$= m_0 p + 2(x_1 x_2 + x_1 x_3 + x_1 x_4 + x_2 x_3 + x_2 x_4 + x_3 x_4),$$

which is a contradiction, since the result is an even integer.

Therefore, concerning the integers x_1, x_2, x_3 and x_4, one of the following cases must hold:

---

[4] $x_1 = x$, $x_2 = y$, $x_3 = 1$, $x_4 = 0$.

(i) All four integers are even.

(ii) All four integers are odd.

(iii) Only two of the integers are even and the other two are odd.

It can be verified that cases (i) and (ii) lead to a contradiction. Hence, let us suppose that (iii) holds true and without loss of generality that x_1, x_2 are even and x_3, x_4 are odd integers. Then, clearly the integers

$$x_1 + x_2, x_1 - x_2, x_3 + x_4, x_3 - x_4$$

are even numbers.

Furthermore, by the identity

$$\frac{a^2 + b^2}{2} = \left(\frac{a+b}{2}\right)^2 + \left(\frac{a-b}{2}\right)^2,$$

it yields

$$\frac{1}{2}m_0 p = \left(\frac{x_1 + x_2}{2}\right)^2 + \left(\frac{x_1 - x_2}{2}\right)^2 + \left(\frac{x_3 + x_4}{2}\right)^2 + \left(\frac{x_3 - x_4}{2}\right)^2.$$

The squares

$$\left(\frac{x_1 + x_2}{2}\right)^2, \left(\frac{x_1 - x_2}{2}\right)^2, \left(\frac{x_3 + x_4}{2}\right)^2, \left(\frac{x_3 - x_4}{2}\right)^2$$

are not all divisible by p, since if that was the case, we would have

$$p \left| \left(\frac{x_1 + x_2}{2}\right) + \left(\frac{x_1 - x_2}{2}\right) \right.$$

or equivalently

$$p \mid x_1.$$

Similarly, we would have $p \mid x_2$, $p \mid x_3$ and $p \mid x_4$ which, as we have shown previously, leads to a contradiction.

Since we have assumed that m_0 is an even integer, it follows that there exists an integer ν, such that

$$\frac{1}{2} 2\nu p = \left(\frac{x_1 + x_2}{2}\right)^2 + \left(\frac{x_1 - x_2}{2}\right)^2 + \left(\frac{x_3 + x_4}{2}\right)^2 + \left(\frac{x_3 - x_4}{2}\right)^2$$

or

$$\nu p = \left(\frac{x_1 + x_2}{2}\right)^2 + \left(\frac{x_1 - x_2}{2}\right)^2 + \left(\frac{x_3 + x_4}{2}\right)^2 + \left(\frac{x_3 - x_4}{2}\right)^2,$$

where $\nu < m_0$, which is obviously a contradiction, since we have assumed that the integer $m_0 p$ is the least multiple of p which can be expressed as the sum of four squares of integers.

- If m_0 is an odd integer, then it is not possible for all four integers x_1, x_2, x_3 and x_4 to be divisible by m_0, because in that case we would have

$$m_0 p = \lambda_1^2 m_0^2 + \lambda_2^2 m_0^2 + \lambda_3^2 m_0^2 + \lambda_4^2 m_0^2$$
$$= (\lambda_1^2 + \lambda_2^2 + \lambda_3^2 + \lambda_4^2) m_0^2$$

and thus $m_0 \mid p$ which is impossible, since p is a prime number.

Let us assume that $m_0 \geq 3$. We can choose integers b_1, b_2, b_3, b_4 such that

$$x_i = b_i m_0 + y_i,$$

for $i = 1, 2, 3, 4$ where

$$|y_i| < \frac{1}{2} m_0$$

and

$$y_1^2 + y_2^2 + y_3^2 + y_4^2 > 0.[5]$$

Hence,

$$y_1^2 + y_2^2 + y_3^2 + y_4^2 < 4 \left(\frac{1}{4} m_0^2 \right) = m_0^2. \qquad (1)$$

However,

$$y_1^2 + y_2^2 + y_3^2 + y_4^2 = (x_1 - b_1 m_0)^2 + (x_2 - b_2 m_0)^2$$
$$+ (x_3 - b_3 m_0)^2 + (x_4 - b_4 m_0)^2$$
$$= (x_1^2 + x_2^2 + x_3^2 + x_4^2) - 2 x_1 b_1 m_0$$
$$- 2 x_2 b_2 m_0 - 2 x_3 b_3 m_0 - 2 x_4 b_4 m_0$$
$$+ (b_1^2 + b_2^2 + b_3^2 + b_4^2) m_0^2$$
$$= m_1 m_0,$$

for some integer m_1.

---

[5] Let $[a]$ be the set of all integers, which are congruent to $a \,(\bmod \, m_0)$. It is evident that each of the integers x_1, x_2, x_3, x_4 belongs to a set $[a_i]$, where $i = 1, 2, 3, 4$ are not necessarily the same. Hence, if $x_i \in [a_i]$, then there exists $y_i \in [a_i]$. Thus, it holds $x_i \equiv y_i \,(\bmod \, m_0)$. Therefore, we can choose integers b_i, such that $x_i = b_i m_0 + y_i$, where

$$|y_i| < \frac{1}{2} m_0 \quad \text{and} \quad y_1^2 + y_2^2 + y_3^2 + y_4^2 > 0.$$

Here, we make use of the fact that m_0 is an odd integer, because if it was an even integer then $m_0 \geq 2$ and thus for $m_0 = 2$ we would have $|y_i| < 1$ or equivalently $|y_i| = 0$. But, this is a contradiction, since in that case we would have $x_i = b_i m_0$ or equivalently $m_0 \mid x_i$, which is impossible.

Since m_0 is an odd integer, its least value is 3. Therefore, for $m_0 = 3$, it follows that $|y_i| < 3/2$. Hence, it is possible for $|y_i|$ to be equal to 1, which is an acceptable value since $0 < 1 < 3/2$.

It holds $0 < m_1 < m_0$, since if $m_1 \geq m_0$, then $m_1 m_0 \geq m_0^2$ which, due to (1), is impossible. Therefore, we have

$$x_1^2 + x_2^2 + x_3^2 + x_4^2 = m_0 p, \tag{2}$$

with $m_0 < p$ and

$$y_1^2 + y_2^2 + y_3^2 + y_4^2 = m_1 m_0. \tag{3}$$

Hence, by (2) and (3), it yields

$$m_0^2 m_1 p = (x_1^2 + x_2^2 + x_3^2 + x_4^2)(y_1^2 + y_2^2 + y_3^2 + y_4^2) = z_1^2 + z_2^2 + z_3^2 + z_4^2.$$

But

$$z_1 = x_1 y_1 + x_2 y_2 + x_3 y_3 + x_4 y_4$$
$$= x_1(x_1 - b_1 m_0) + x_2(x_2 - b_2 m_0) + x_3(x_3 - b_3 m_0) + x_4(x_4 - b_4 m_0)$$
$$= (x_1^2 + x_2^2 + x_3^2 + x_4^2) - (x_1 b_1 + x_2 b_2 + x_3 b_3 + x_4 b_4) m_0$$
$$= m_0 p - (x_1 b_1 + x_2 b_2 + x_3 b_3 + x_4 b_4) m_0$$
$$\equiv 0 \, (\mathrm{mod}\, m_0).$$

Similarly, we get

$$z_2, z_3, z_4 \equiv 0 \, (\mathrm{mod}\, m_0).$$

Therefore, there exist integers t_1, t_2, t_3, t_4, such that

$$m_0^2 m_1 p = t_1^2 m_0^2 + t_2^2 m_0^2 + t_3^2 m_0^2 + t_4^2 m_0^2$$

or

$$m_1 p = t_1^2 + t_2^2 + t_3^2 + t_4^2,$$

where p does not divide all four integers t_1, t_2, t_3, t_4, since if it did we would have

$$m_1 p = (\xi_1^2 + \xi_2^2 + \xi_3^2 + \xi_4^2) p^2$$

or

$$m_1 = (\xi_1^2 + \xi_2^2 + \xi_3^2 + \xi_4^2) p > p,$$

which is a contradiction, since $0 < m_1 < m_0 < p$.

Hence, $m_1 p$ can be expressed as the sum of four squares of integers and $m_1 < m_0$, which contradicts the property of $m_0 p$, being the least multiple of p which can be expressed as the sum of four squares of integers. Thus, the assumption $m_0 > 1$ leads in every case to a contradiction.

Consequently, $m = 1$ is the least integer for which mp has the property

$$mp = x_1^2 + x_2^2 + x_3^2 + x_4^2,$$

for every prime number p. Thus,

$$p = x_1^2 + x_2^2 + x_3^2 + x_4^2.$$

Therefore, every positive integer can be expressed as the sum of four squares of integers. $\qquad\square$

Definition 9.2.2. *Let $r_k(n)$ denote the number of ways a positive integer n can be expressed as the sum of k squares of integers. We must note that permutations and sign changes, count as different ways of representation.*

We shall now present two very interesting results concerning the sum of four squares.

Theorem 9.2.3. *It holds*

$$r_4(2n) = \begin{cases} 3r_4(n), & \text{if } n \text{ is an odd integer} \\ r_4(n), & \text{if } n \text{ is an even integer.} \end{cases}$$

Theorem 9.2.4 (Jacobi). *It holds*

$$r_4(n) = \begin{cases} 8\sigma(n), & \text{if } n \text{ is an odd integer} \\ 24\sigma(d), & \text{if } n \text{ is an even integer and } d \text{ is its largest divisor.} \end{cases}$$

9.3 Bertrand's postulate

I'll tell you once and I'll tell you again.
There's always a prime between n and 2n.
Paul Erdős (1913–1996)

The theorem whose proof is presented below is known as *Bertrand's Postulate*.

In 1845, the French mathematician Joseph Louis François Bertrand (1822–1900) formulated the conjecture that there is always a prime number p, such that $n < p \leq 2n$, for every positive integer n, with $n \geq 1$. Bertrand verified his conjecture for every positive integer n up to 3,000,000. A number of mathematicians tried to prove Bertrand's Postulate, but the Russian mathematician Pafnuty Lvovich Chebyshev (1821–1894) was the first to give a proof, in 1852. However, Landau in his book entitled *Handbuch der Lehre von der Verteinlung der Primzahlen*, (1909), pp. 89–92, presented a proof of the Postulate, which is very similar to that of Chebyshev's. Later, the Indian mathematician Srinivasa Ramanujan (1887–1920) presented a simpler proof using properties of the Gamma function $\Gamma(s)$ and in 1919, he even proved a more general form of the theorem. The proof presented below is due to the Hungarian mathematician Paul Erdős (1913–1996), who published it in 1932 (Acta Litt. Ac. Sci. (Szeged), 5(1932), 194–198), at the age of 19. It is worth mentioning that in 1892, Bertrand's Postulate was generalized by James Joseph Sylvester (1814–1897), who proved that

> If m, n are positive integers, such that $m > n$, then at least one of the positive integers m, $m + 1$, $m + 2, \ldots, m + n - 1$ has a prime divisor greater than n.[6]

Theorem 9.3.1 (Bertrand's Postulate). *For every positive integer n, with $n \geq 1$, there exists a prime number p, such that $n < p \leq 2n$.*

Proof. We shall first prove the theorem for every positive integer n, with $n \geq 4000$. For positive integers up to 4000 the proof is elementary and we shall present it toward the end.

Thus, let us assume that for some positive integer $n \geq 4000$ there does not exist a prime number p, such that $n < p \leq 2n$.

By Legendre's theorem we know that the highest power of p which divides the integer $n!$ is[7]

$$\sum_{k=1}^{+\infty} \left\lfloor \frac{n}{p^k} \right\rfloor.$$

---

[6] Bertrand's Postulate follows if we set $m = n + 1$.

[7] With $\lfloor x \rfloor$ we denote the integer part of x and with $\lceil x \rceil$ the least integer, greater than or equal to x.

This is true, since the number of terms of $n!$ which are divisible by p is $\lfloor n/p \rfloor$. More specifically, these terms are the integers

$$1 \cdot p, 2 \cdot p, \ldots, \left\lfloor \frac{n}{p} \right\rfloor \cdot p.$$

However, some terms of $n!$ are divisible at least by the second power of p, namely, contain p^2 at least one time. These terms are the integers

$$1 \cdot p^2, 2 \cdot p^2, \ldots, \left\lfloor \frac{n}{p^2} \right\rfloor \cdot p^2,$$

which are exactly

$$\left\lfloor \frac{n}{p^2} \right\rfloor$$

in number.

If we continue similarly for higher powers of p, it follows that the integer $n!$ contains the prime number p exactly

$$\left\lfloor \frac{n}{p} \right\rfloor + \left\lfloor \frac{n}{p^2} \right\rfloor + \cdots + \left\lfloor \frac{n}{p^k} \right\rfloor + \cdots$$

times and therefore that is exactly the highest power of p which divides $n!$.

The above sum is finite since for $k > r$, where $p^r \geq n$, it holds

$$\left\lfloor \frac{n}{p^k} \right\rfloor = 0.$$

Hence, according to Legendre's theorem, it yields that the integer

$$\binom{2n}{n} = \frac{(2n)!}{n!n!}$$

contains[8] the prime number p, as many times as p is contained in $(2n)!$ minus the number of times which it is contained in $n!n!$. This happens because

$$\binom{2n}{n} = \frac{(2n)!}{n!n!} = \frac{p_1^{a_1} p_2^{a_2} \cdots p^a \cdots p_\lambda^{a_\lambda}}{(p_1^{q_1} p_2^{q_2} \cdots p^q \cdots p_\lambda^{q_\lambda}) \cdot (p_1^{q_1} p_2^{q_2} \cdots p^q \cdots p_\lambda^{q_\lambda})},$$

where

$$a_1 \geq 2q_1, a_2 \geq 2q_2, \ldots, a \geq 2q, \ldots, a_\lambda \geq 2q_\lambda.$$

Therefore,

$$\binom{2n}{n} = \frac{(2n)!}{n!n!} = p_1^{a_1 - 2q_1} p_2^{a_2 - 2q_2} \cdots p^{a - 2q} \cdots p_\lambda^{a_\lambda - 2q_\lambda},$$

---

[8] We must clarify that "p is contained in m exactly k times" means that the highest power of p which divides m is k.

where a is the number of times that p is contained in the numerator and $2q$ is the number of times that p is contained in the denumerator. Thus, it is evident that the prime number p is contained exactly

$$S = \sum_{k=1}^{+\infty} \left(\left\lfloor \frac{2n}{p^k} \right\rfloor - 2 \left\lfloor \frac{n}{p^k} \right\rfloor \right)$$

times in

$$\binom{2n}{n}.$$

However, it holds

$$\left\lfloor \frac{2n}{p^k} \right\rfloor - 2 \left\lfloor \frac{n}{p^k} \right\rfloor < \frac{2n}{p^k} - 2 \left(\frac{n}{p^k} - 1 \right) = 2.$$

Thus, clearly

$$\left\lfloor \frac{2n}{p^k} \right\rfloor - 2 \left\lfloor \frac{n}{p^k} \right\rfloor = 0 \text{ or } 1.$$

Hence, according to the above arguments, the prime numbers p with $p > \sqrt{2n}$ are contained in $\binom{2n}{n}$ at most once. This is true, since

- If $k = 1$, then $S = 0$ or 1.
- If $k \geq 2$, then $S = 0$.

In addition, the prime numbers p for which

$$\frac{2}{3} n < p \leq n, \quad \text{where } n \geq 3,$$

do not divide the integer

$$\binom{2n}{n},$$

since $3p > 2n$ and thus only p, $2p$ could possibly divide $2n$. However, both p and $2p$ are factors of $n!$ and therefore are eliminated by the numerator of the fraction $(2n)!/(n!n!)$. The prime number p is a factor of $n!$ since $p \leq n$ and $2p$ is a factor of $n!$ because of the fact that

$$2p \leq 2n \leq n!, \quad \text{for } n \geq 3.$$

We shall now prove that

$$\frac{4^n}{2n} \leq \binom{2n}{n}. \tag{1}$$

By the binomial coefficients

$$\binom{n}{0}, \binom{n}{1}, \ldots, \binom{n}{\lfloor n/2 \rfloor}, \binom{n}{\lceil n/2 \rceil}, \ldots, \binom{n}{n},$$

the integers

$$\binom{n}{\lfloor n/2 \rfloor}, \binom{n}{\lceil n/2 \rceil}$$

obtain the greatest value and it holds

$$\binom{n}{\lfloor n/2 \rfloor} = \binom{n}{\lceil n/2 \rceil}.$$

This happens because

$$\binom{n}{k} = \frac{n!}{k!(n-k)!} = \frac{n!}{(k-1)!(n-k)!k}$$

$$= \frac{n-k+1}{k} \frac{n!}{(k-1)!(n-k)!(n-k+1)}$$

$$= \frac{n-k+1}{k} \frac{n!}{(k-1)!(n-(k-1))!} = \frac{n-k+1}{k} \binom{n}{k-1}.$$

Hence,

$$\binom{n}{k} = \frac{n-k+1}{k} \binom{n}{k-1}.$$

Therefore, if

$$\frac{n-k+1}{k} < 1,$$

then

$$\binom{n}{k} < \binom{n}{k-1}.$$

But, by

$$\frac{n-k+1}{k} < 1,$$

we obtain equivalently

$$k \geq \frac{n+2}{2} = \frac{n}{2} + 1 > \left\lfloor \frac{n}{2} \right\rfloor.$$

Hence,

- If $k > \left\lfloor \frac{n}{2} \right\rfloor$, then

$$\binom{n}{k} < \binom{n}{k-1}.$$

Similarly

- If $\frac{n-k+1}{k} > 1$, then

$$\binom{n}{k-1} < \binom{n}{k}.$$

However, the inequality $(n - k + 1)/k > 1$ is equivalent to

$$k \leq \frac{n}{2} \leq \left\lceil \frac{n}{2} \right\rceil.$$

Thus, evidently

- If $k \leq \left\lceil \frac{n}{2} \right\rceil$, then

$$\binom{n}{k-1} < \binom{n}{k}.$$

Therefore, by the above arguments it yields

$$\binom{n}{0} < \binom{n}{1} < \cdots < \binom{n}{\lfloor n/2 \rfloor} = \binom{n}{\lceil n/2 \rceil} > \cdots > \binom{n}{n}.$$

If we consider the sum

$$A = \binom{n}{0} + \binom{n}{1} + \cdots + \binom{n}{n},$$

then

$$A = 2^n$$

and thus

$$\frac{A}{n} = \frac{2^n}{n}$$

or

$$\bar{A} = \frac{2^n}{n},$$

where $\bar{A}$ stands for the mean value of the integers

$$\binom{n}{0} + \binom{n}{n}, \binom{n}{1}, \binom{n}{2}, \ldots, \binom{n}{n-1}.$$

Thus, it is obvious that

$$\binom{n}{\lfloor n/2 \rfloor} \geq \bar{A} \quad \text{or} \quad \binom{n}{\lfloor n/2 \rfloor} \geq \frac{2^n}{n}, \tag{2}$$

since the integer $\binom{n}{\lfloor n/2 \rfloor}$ is greater than or equal to each of the integers

$$\binom{n}{0} + \binom{n}{n}, \binom{n}{1}, \binom{n}{2}, \ldots, \binom{n}{n-1}.$$

In (2), if we substitute n by $2n$ we obtain

$$\binom{2n}{n} \geq \frac{4^n}{2n}.$$

We shall now prove that

$$\prod_{p \leq x} p \leq 4^{x-1}, \quad \text{for every } x \geq 2, \tag{3}$$

where the product extends over all prime numbers p, such that $p \leq x$. $\square$

Proof.

- For $x = 2$, the inequality (3) holds true. Let us suppose that (3) holds true for every x, such that $x < n$ and $n \geq 3$.
- For $n = 2r$, we obtain

$$\prod_{p \leq n} p = \prod_{p \leq 2r} p = \prod_{p \leq 2r-1} p \leq 4^{2r-2} < 4^{2r-1}.$$

- For $n = 2r + 1$, we have

$$\prod_{p \leq n} p = \prod_{p \leq 2r+1} p = \prod_{p \leq r+1} p \cdot \prod_{r+1 < p \leq 2r+1} p \leq 4^r \prod_{r+1 < p \leq 2r+1} p.$$

However,

$$\prod_{r+1 < p \leq 2r+1} p \leq \binom{2r+1}{r},$$

since the prime numbers p, where $p \in (r+1, 2r+1]$, appear in the numerator of the fraction

$$\binom{2r+1}{r} = \frac{(2r+1)!}{r!(r+1)!},$$

but not in the denominator. But

$$\binom{2r+1}{r} \leq 2^{2r},$$

since

$$2^{2r+1} = (1+1)^{2r+1} = 1 + \binom{2r+1}{1} + \cdots + \binom{2r+1}{r} + \binom{2r+1}{r+1} + \cdots + 1.$$

We have

$$\binom{2r+1}{r} = \binom{2r+1}{r+1},$$

and thus

$$2\binom{2r+1}{r} < 2^{2r+1}.$$

This completes the proof of inequality (3). Therefore, from the above, it follows that

$$\frac{4^n}{2n} \leq \binom{2n}{n} \leq \prod_{p \leq \sqrt{2n}} 2n \cdot \prod_{\sqrt{2n} < p \leq \frac{2}{3}n} p \cdot \prod_{\frac{2}{3}n < p \leq 2n} p, \quad \text{where } n \geq 3, \quad (4)$$

since the prime numbers p such that $p > \sqrt{2n}$ divide $\binom{2n}{n}$ at most once and the prime numbers p such that $2n/3 < p \leq n$ where $n \geq 3$ do not divide $\binom{2n}{n}$.

However, we have assumed that there does not exist a prime number in the interval $[n + 1, 2n]$. Hence,

$$\prod_{\frac{2}{3}n < p \leq 2n} p = 1.$$

In addition, it holds

$$\prod_{p \leq \sqrt{2n}} 2n \leq (2n)^{\sqrt{2n}}$$

since the number of primes p for which $p \leq \sqrt{2n}$ does not exceed $\sqrt{2n}$. Therefore, by (4) we obtain that

$$\frac{4^n}{2n} \leq (2n)^{\sqrt{2n}} \cdot \prod_{\sqrt{2n} < p \leq \frac{2}{3}n} p.$$

Thus,

$$4^n \leq (2n)^{1+\sqrt{2n}} \cdot \prod_{\sqrt{2n} < p \leq \frac{2}{3}n} p. \tag{5}$$

By (3), relation (5) takes the form

$$4^n \leq (2n)^{1+\sqrt{2n}} \cdot 4^{2n/3}$$

or

$$4^{n/3} \leq (2n)^{1+\sqrt{2n}}. \tag{6}$$

However, the above inequality does not hold for appropriate value of n. This happens because by (6) we have

$$4^n \leq (2n)^{3(1+\sqrt{2n})}$$

and therefore

$$2^{2n} \leq (2n)^{3(1+\sqrt{2n})}. \tag{7}$$

But, by applying mathematical induction we can easily prove that

$$k + 1 < 2^k, \quad \text{for} \ \ k \in \mathbb{N} \ \ \text{with} \ \ k \geq 2.$$

Thus,

$$2n = ((2n)^{1/6})^6 < (\lfloor (2n)^{1/6} \rfloor + 1)^6 < (2^{\lfloor (2n)^{1/6} \rfloor})^6$$
$$\leq (2^{(2n)^{1/6}})^6.$$

Hence,

$$2n < 2^{6(2n)^{1/6}}.$$

Therefore, (7) takes the form

$$2^{2n} \leq 2^{(6(2n)^{1/6})3(1+\sqrt{2n})} = 2^{(2n)^{1/6}(18+18\sqrt{2n})}.$$

But, for $n \geq 50$ it holds

$$18 < 2\sqrt{2n},$$

thus

$$2^{2n} < 2^{(2n)^{1/6}(2\sqrt{2n}+18\sqrt{2n})} = 2^{20(2n)^{1/6}\sqrt{2n}} = 2^{20(2n)^{2/3}}.$$

One has

$$2n < 20(2n)^{2/3}$$

or

$$(2n)^{1/3} < 20$$

or

$$n < 4000,$$

which is a contradiction. Hence, the initial hypothesis that there exists a positive integer n, with $n \geq 4000$, for which there is no prime number p in the interval $(n, 2n]$, leads to a contradiction. Thus, Bertrand's Postulate holds true for all positive integers $n \geq 4000$.

Therefore, it suffices to prove the theorem for all positive integers $n < 4000$. In order to do so, we are going to use a method which is due to Landau. The integers

$$2, 3, 5, 7, 13, 23, 43, 83, 163, 317, 631, 1259, 2503, 4001$$

form a sequence of prime numbers, each of which is less than the double of the previous prime in the sequence. Therefore, it is clear that for every positive integer $n < 4000$, the interval between n and $2n$ contains at least one prime number from the above sequence.

This completes the proof of the theorem.   □

Note. The proof of argument (3) is due to Professor N. G. Tzanakis.

Remarks 9.3.2. We must mention that there have been determined much smaller intervals than $(n, 2n]$, in which at least one prime number exists.

• J. Nagura in his paper: *On the interval containing at least one prime number*, Proc. Japan. Acad., 28(1952), 177–181, proved that for any real value of x, where $x \geq 25$, the interval $[x, 6x/5]$ contains at least one prime number.
• H. Rohrbach and J. Weis in their paper: *Zum finiten Fall des Bertrandschen Postulats*, J. Reine Angew. Math., 214/215(1964), 432–440, proved that for every positive integer n with $n \geq 118$, the interval $(n, 14n/13]$ contains at least one prime number.
• N. Costa Pereira in his paper: Elementary estimate for the Chebyshev function $\psi(x)$ and the Möbius function $\mu(x)$, Acta Arith. 52(1989), 307–337, proved that for every prime number x with $x > 485492$, the interval $[x, 258x/257)$ contains at least one prime number.

Historical Remark. Pafnuty Lvovich Chebyshev was born on the 16th of May, 1821, at Okatovo of the former Russian empire and died on the 8th of December, 1894, in Saint Petersburg.

In 1837, Chebyshev begun his studies at the University of Moscow, from which he graduated in 1841. Six years later he became a lecturer at the University of Saint Petersburg. In 1853, he was elected a member of the Academy of Sciences of Saint Petersburg, where he was bestowed the chair of Applied Mathematics. The chairs for Pure Mathematics at the Academy were held at that period by the mathematicians P. H. Fuss (1798–1855),[9] M. V. Ostrogradsky (1801–1862) and V. Ya. Bunyakovsky (1804–1889).

Chebyshev conducted research in several areas of mathematics, such as number theory, probability theory, approximation theory, numerical analysis, real analysis, differential geometry and kinematics. Due to his vast and profound contribution in pure and applied mathematics, he was elected member of the Academies of Sciences of several countries, including the Paris Academy of Sciences (1860), the Berlin Academy of Sciences (1871), the Academy of Bolonia (1873), London's Royal Academy (1877), the Italian Royal Academy (1880) and the Swedish Academy of Sciences (1893).

---

[9] Fuss was a great-grandson of L. Euler.

9.4 An inequality for the π-function

There are three reasons for the study of inequalities:
practical, theoretical and aesthetic.
On the aesthetic aspects, as has been pointed out,
beauty is in the eyes of the beholder.
However, it is generally agreed that
certain pieces of music, art, or mathematics are beautiful.
There is an elegance to inequalities that makes them very attractive.
Richard E. Bellman (1920–1984)

Theorem 9.4.1. *For every positive integer n, where $n \geq 2$, the following inequality holds:*

$$\frac{1}{6} \cdot \frac{n}{\log n} < \pi(n) < 6 \cdot \frac{n}{\log n}.^{10}$$

Proof.

- We claim that

$$2^n \leq \binom{2n}{n} < 4^n. \tag{1}$$

The inequality

$$2^n \leq \binom{2n}{n}$$

follows by mathematical induction.

For $n = 2$ one has

$$4 \leq \binom{4}{2} = 6,$$

which holds. Suppose that (1) is valid for n, i.e.,

$$2^n \leq \binom{2n}{n}.$$

It suffices to prove (1) for $n + 1$.

---

[10] This inequality is known as Chebyshev's inequality for the function $\pi(n)$. For further relative results, the reader is referred to the book by T. Apostol [7].

It is clear that

$$\binom{2n+2}{n+1} = \frac{(2n+2)!}{(n+1)!(n+1)!} = \frac{(2n)!}{n!\,n!} \frac{(2n+1)(2n+2)}{(n+1)^2}$$

$$\geq 2^n \frac{(2n+1)(2n+2)}{(n+1)^2}.$$

It is enough to prove that

$$\frac{(2n+1)(2n+2)}{(n+1)^2} \geq 2, \text{ for } n \geq 2.$$

However,

$$\frac{(2n+1)(2n+2)}{(n+1)^2} \geq 2 \Leftrightarrow 2n \geq 0,$$

which obviously holds.
    Thus,

$$\binom{2n+2}{n+1} \geq 2^{n+1}$$

and therefore we have proved that

$$2^n \leq \binom{2n}{n}$$

for every positive integer n, where $n \geq 2$.
    The proof of the right-hand side of inequality (1) follows from the fact that

$$\binom{2n}{n} < \binom{2n}{0} + \binom{2n}{1} + \cdots + \binom{2n}{2n} = 2^{2n} = 4^n.$$

From (1) we get that

$$\log 2^n \leq \log \frac{(2n)!}{n!n!} < \log 4^n$$

and therefore

$$n \log 2 \leq \log(2n)! - 2 \log n! < n \log 4. \qquad (2)$$

However, from Legendre's theorem,[11] it easily follows that

$$n! = \prod_{p \leq n} p^{j(n,p)}, \qquad (3)$$

---

[11] The proof of Legendre's theorem appears in the proof of Bertrand's postulate.

where

$$j(n,p) = \sum_{k=1}^{+\infty} \left\lfloor \frac{n}{p^k} \right\rfloor.$$

From (3), it follows that

$$\log n! = \log \prod_{p \le n} p^{j(n,p)}$$

$$= \sum_{p \le n} \log p^{j(n,p)}$$

$$= \sum_{p \le n} j(n,p) \log p.$$

By applying this result we get

$$\log(2n)! - 2\log n! = \sum_{p \le 2n} j(n,p) \log p - 2 \sum_{p \le n} j(n,p) \log p$$

$$= \sum_{p \le 2n} \left(\sum_{k=1}^{+\infty} \left\lfloor \frac{2n}{p^k} \right\rfloor \right) \log p - 2 \sum_{p \le n} \left(\sum_{k=1}^{+\infty} \left\lfloor \frac{n}{p^k} \right\rfloor \right) \log p.$$

However,

$$\sum_{p \le n} \left(\sum_{k=1}^{+\infty} \left\lfloor \frac{n}{p^k} \right\rfloor \right) \log p = \sum_{p \le 2n} \left(\sum_{k=1}^{+\infty} \left\lfloor \frac{n}{p^k} \right\rfloor \right) \log p$$

since for $p > n$ it is true that

$$\left\lfloor \frac{n}{p^k} \right\rfloor = 0.$$

Therefore,

$$\log(2n)! - 2\log n! = \sum_{p \le 2n} \left(\sum_{k=1}^{+\infty} \left\lfloor \frac{2n}{p^k} \right\rfloor - 2 \sum_{k=1}^{+\infty} \left\lfloor \frac{n}{p^k} \right\rfloor \right) \log p$$

$$= \sum_{p \le 2n} \left[\sum_{k=1}^{+\infty} \left(\left\lfloor \frac{2n}{p^k} \right\rfloor - 2 \left\lfloor \frac{n}{p^k} \right\rfloor \right) \right] \log p.$$

We have proved (see Bertrand's postulate) that

$$\left\lfloor \frac{2n}{p^k} \right\rfloor - 2 \left\lfloor \frac{n}{p^k} \right\rfloor = 0 \text{ or } 1.$$

The terms of the infinite summation

$$\sum_{k=1}^{+\infty}\left(\left\lfloor\frac{2n}{p^k}\right\rfloor - 2\left\lfloor\frac{n}{p^k}\right\rfloor\right)$$

obtain the value zero for k such that $p^k > 2n$, that means for

$$k > \frac{\log 2n}{\log p}.$$

Thus,

$$\sum_{k=1}^{+\infty}\left(\left\lfloor\frac{2n}{p^k}\right\rfloor - 2\left\lfloor\frac{n}{p^k}\right\rfloor\right) = \sum_{k=1}^{\left\lfloor\frac{\log 2n}{\log p}\right\rfloor}\left(\left\lfloor\frac{2n}{p^k}\right\rfloor - 2\left\lfloor\frac{n}{p^k}\right\rfloor\right)$$

$$\leq \sum_{k=1}^{\left\lfloor\frac{\log 2n}{\log p}\right\rfloor} 1.$$

Hence,

$$\log(2n)! - 2\log n! \leq \sum_{p\leq 2n}\left(\sum_{k=1}^{\left\lfloor\frac{\log 2n}{\log p}\right\rfloor} 1\right)\log p$$

$$\leq \sum_{p\leq 2n}\frac{\log 2n}{\log p}\log p$$

$$= \sum_{p\leq 2n}\log 2n$$

$$= \pi(2n)\log 2n.$$

From this relation and inequality (2), it follows that

$$n\log 2 \leq \pi(2n)\log 2n$$

$$\Leftrightarrow \pi(2n) \geq \frac{n\log 2}{\log 2n} > \frac{n/2}{\log 2n} = \frac{2n}{4\log 2n}$$

$$\Leftrightarrow \pi(2n) > \frac{1}{4}\cdot\frac{2n}{\log 2n} > \frac{1}{6}\cdot\frac{2n}{\log 2n}. \tag{4}$$

Therefore, the inequality

$$\frac{1}{6}\cdot\frac{n}{\log n} < \pi(n)$$

is satisfied if n is an even number. It remains to examine the case when n is an odd number.

It is true that

$$\pi(2n+1) \geq \pi(2n)\frac{1}{4} \cdot \frac{2n}{\log 2n} = \frac{1}{4} \cdot \frac{2n}{2n+1} \cdot \frac{2n+1}{\log 2n}$$

$$\frac{1}{4} \cdot \frac{2n}{2n+1} \cdot \frac{2n+1}{\log(2n+1)}.$$

It is evident that

$$\frac{2n}{2n+1} \geq \frac{2}{3}$$

for every positive integer n.

Therefore,

$$\pi(2n+1)\frac{1}{4} \cdot \frac{2}{3} \cdot \frac{2n+1}{\log(2n+1)}$$

$$= \frac{1}{6} \cdot \frac{2n+1}{\log(2n+1)}.$$

Hence, the inequality

$$\frac{1}{6} \cdot \frac{n}{\log n} < \pi(n)$$

is also satisfied in the case where n is an odd number.

Thus,

$$\frac{1}{6} \cdot \frac{n}{\log n} < \pi(n),$$

for every positive integer n, with $n \geq 2$.

- We will prove the inequality

$$\pi(n) < 6 \cdot \frac{n}{\log n}$$

for every positive integer n with $n \geq 2$.

We have already proved that

$$\log(2n)! - 2\log n! = \sum_{p \leq 2n} \left[\sum_{k=1}^{+\infty} \left(\left\lfloor \frac{2n}{p^k} \right\rfloor - 2 \left\lfloor \frac{n}{p^k} \right\rfloor \right) \right] \log p,$$

where none of the terms

$$\left\lfloor \frac{2n}{p^k} \right\rfloor - 2 \left\lfloor \frac{n}{p^k} \right\rfloor$$

is negative.

Therefore, it is clear that

$$\left\lfloor \frac{2n}{p} \right\rfloor - 2 \left\lfloor \frac{n}{p} \right\rfloor \leq \sum_{k=1}^{+\infty} \left(\left\lfloor \frac{2n}{p^k} \right\rfloor - 2 \left\lfloor \frac{n}{p^k} \right\rfloor \right).$$

Thus,

$$\log(2n)! - 2\log n! \geq \sum_{p \leq 2n} \left(\left\lfloor \frac{2n}{p} \right\rfloor - 2 \left\lfloor \frac{n}{p} \right\rfloor \right) \log p$$

$$\geq \sum_{n < p \leq 2n} \left(\left\lfloor \frac{2n}{p} \right\rfloor - 2 \left\lfloor \frac{n}{p} \right\rfloor \right) \log p.$$

However, for the prime numbers p, such that $n < p \leq 2n$ one has

$$\left\lfloor \frac{2n}{p} \right\rfloor - 2 \left\lfloor \frac{n}{p} \right\rfloor = 1,$$

since

$$\left\lfloor \frac{2n}{p} \right\rfloor = 1 \text{ and } \left\lfloor \frac{n}{p} \right\rfloor = 0.$$

Hence,

$$\log(2n)! - 2\log n! \geq \sum_{n < p \leq 2n} \log p. \tag{5}$$

By the definition of Chebyshev's function $\vartheta(x)$, one has

$$\vartheta(x) = \sum_{p \leq x} \log p.$$

Therefore, (5) can be written as follows:

$$\log(2n)! - 2\log n! \geq \vartheta(2n) - \vartheta(n).$$

Thus, by means of (2), we obtain

$$\vartheta(2n) - \vartheta(n) < n \log 4. \tag{6}$$

Suppose that the positive integer n can be expressed as an exact power of 2. Then from (6) it follows

$$\vartheta(2 \cdot 2^m) - \vartheta(2^m) < 2^m \log 2^2$$

and therefore

$$\vartheta(2^{m+1}) - \vartheta(2^m) < 2^{m+1} \log 2.$$

For $m = 1, 2, \ldots, \lambda - 1, \lambda$ the above inequality, respectively, yields

$$\left. \begin{array}{l} \vartheta(2^2) - \vartheta(2) < 2^2 \log 2 \\ \vartheta(2^3) - \vartheta(2^2) < 2^3 \log 2 \\ \qquad \vdots \\ \vartheta(2^\lambda) - \vartheta(2^{\lambda-1}) < 2^\lambda \log 2 \\ \vartheta(2^{\lambda+1}) - \vartheta(2^\lambda) < 2^{\lambda+1} \log 2 \end{array} \right\}.$$

Adding up the above inequalities we get

$$\vartheta(2^{\lambda+1}) - \vartheta(2) < (2^2 + 2^3 + \cdots + 2^\lambda + 2^{\lambda+1}) \log 2.$$

But $\vartheta(2) = \log 2$, therefore

$$\vartheta(2^{\lambda+1}) < (1 + 2^2 + 2^3 + \cdots + 2^\lambda + 2^{\lambda+1}) \log 2$$
$$= (2^{\lambda+1} - 1) \log 2.$$

Hence,

$$\vartheta(2^{\lambda+1}) < 2^{\lambda+2} \log 2. \tag{7}$$

For every positive integer n we can choose a suitable integer m such that

$$2^m \le n \le 2^{m+1}.$$

Then

$$\vartheta(n) = \sum_{p \le n} \log p \le \sum_{p \le 2^{m+1}} \log p = \vartheta(2^{m+1})$$

and by means of (7) it follows that

$$\vartheta(n) < 2^{m+2} \log 2 = 2^2 \cdot 2^m \log 2 \le 4n \log 2. \tag{8}$$

Let N be the number of primes p_i, such that

$$n^r < p_i \le n,$$

where $0 < r < 1$, for $i = 1, 2, \ldots, N$. Then

$$\left.\begin{array}{l} \log n^r < \log p_1 \\ \log n^r < \log p_2 \\ \vdots \\ \log n^r < \log p_N \end{array}\right\} \Rightarrow N \log n^r < \sum_{n^r < p \le n} \log p,$$

and therefore

$$[\pi(n) - \pi(n^r)] \log n^r < \sum_{n^r < p \le n} \log p. \tag{9}$$

It is obvious that

$$\vartheta(n) \ge \sum_{n^r < p \le n} \log p. \tag{10}$$

Therefore, by means of (8), (9) and (10) one has

$$[\pi(n) - \pi(n^r)] \log n^r < 4n \log 2$$

$$\Leftrightarrow \pi(n) \log n^r < 4n \log 2 + \pi(n^r) \log n^r$$

$$\Leftrightarrow \pi(n) < \frac{4n \log 2}{\log n^r} + \pi(n^r)$$

$$< \frac{4n \log 2}{r \log n} + n^r.$$

Thus, equivalently we obtain

$$\pi(n) < \frac{n}{\log n}\left(\frac{4\log 2}{r} + n^{r-1}\log n\right). \tag{11}$$

□

Consider the function defined by the formula

$$f(x) = \frac{\log x}{x^{1-r}}, x \in \mathbb{R}^+.$$

Then

$$f'(x) = \frac{\frac{1}{x}x^{1-r} - (1-r)x^{-r}\log x}{(x^{1-r})^2}.$$

It is clear that

$$f'(x) = 0$$

if

$$x^{-r} = (1-r)x^{-r}\log x \iff \log x = \frac{1}{1-r},$$

which means

$$x = e^{1/(1-r)}.$$

For $x = e^{1/(1-r)}$ the function $f(x)$ attains its maximal value.
    Thus,

$$f(x) \le \frac{1}{e(1-r)} \Rightarrow f(n) \le \frac{1}{e(1-r)},$$

and therefore

$$n^{r-1}\log n \le \frac{1}{e(1-r)}. \tag{12}$$

From (11) and (12) it follows

$$\pi(n) < \frac{n}{\log n}\left(\frac{4\log 2}{r} + \frac{1}{e(1-r)}\right).$$

Set $r = \frac{2}{3}$. Then

$$\pi(n) < \frac{n}{\log n}\left(6\log 2 + \frac{3}{e}\right).$$

However, it holds

$$6\log 2 + \frac{3}{e} < 6 \text{ and thus } \pi(n) < 6 \cdot \frac{n}{\log n}.$$

Hence, for every positive integer n, where $n \ge 2$, the following inequality holds:

$$\frac{1}{6} \cdot \frac{n}{\log n} < \pi(n) < 6 \cdot \frac{n}{\log n}.$$

□

9.5 Some diophantine equations

A) *Determine the solution of the diophantine equation*

$$x^2 + y^2 = z^2,$$

where $x, y, z \in \mathbb{Z}^+$ *and* $\gcd(x, y, z) = 1$.

Solution. The reason why we consider the diophantine equation

$$x^2 + y^2 = z^2, \quad \text{with} \ \gcd(x, y, z) = 1$$

is because if we considered the equation

$$(x')^2 + (y')^2 = (z')^2,$$

with the positive integers x', y', z' not being pairwise relatively prime, then we would obtain the general solution of the diophantine equation

$$(x')^2 + (y')^2 = (z')^2$$

by multiplying the general solution of the equation

$$x^2 + y^2 = z^2$$

by the $\gcd(x', y', z')$.

It is impossible for both integers x, y to be odd. This happens because if

$$x = 2k + 1, y = 2\lambda + 1, \quad \text{for some} \ k, \lambda \in \mathbb{Z}^+,$$

then we would obtain

$$(2k + 1)^2 + (2\lambda + 1)^2 = 4(k^2 + k + \lambda^2 + \lambda) + 2 = z^2.$$

Consequently the integer $4(k^2 + k + \lambda^2 + \lambda) + 2$ should be a square of an integer.

However, for every square of an integer, such as z^2, the congruence

$$z^2 \equiv 0 \ \text{or} \ 1 \,(\text{mod}\,4)$$

always holds true, but the congruence

$$z^2 \equiv 2 \,(\text{mod}\,4)$$

never holds true.

Hence, the positive integers x, y will either be both even or one will be even and the other will be odd. However, the first case is impossible, since if both integers x and y were even, then we would have $\gcd(x, y, z) \neq 1$, which contradicts the hypothesis.

Without loss of generality we consider x to be even and y to be odd. Thus, by the fact that $x^2 + y^2 = z^2$, it follows that z is an odd integer. We have that

$$x^2 + y^2 = z^2 \Leftrightarrow x^2 = (z - y)(z + y). \tag{1}$$

By the fact that z, y are odd integers, it follows that $z - y, z + y$ must be even integers. Thus,

$$(1) \Leftrightarrow \frac{x^2}{4} = \frac{z - y}{2} \cdot \frac{z + y}{2}. \tag{2}$$

The integers z, y are relatively prime, since if there existed a prime number p such that $p \mid z$ and $p \mid y$, then we would have

$$p \mid (z^2 - y^2)$$

and hence

$$p \mid x.$$

In other words, p would divide x, y and z at the same time, which contradicts the hypothesis.

Therefore, since z, y are relatively prime, the integers

$$\frac{z - y}{2}, \frac{z + y}{2}$$

must also be relatively prime. This happens because if there existed a prime number which divided both

$$\frac{z - y}{2}, \frac{z + y}{2},$$

then the same prime number would divide their sum and their difference and thus would divide z and y, which is impossible.

Since $x^2/4$ is a square of an integer, by a well-known theorem (which we will prove at the end of the solution) it follows that each of the integers

$$\frac{z - y}{2}, \frac{z + y}{2}$$

is also a square of an integer. Therefore, we obtain that

$$\frac{z - y}{2} = \alpha^2, \frac{z + y}{2} = \beta^2, \text{ where } \alpha, \beta \in \mathbb{Z}^+.$$

Hence,

$$z = \alpha^2 + \beta^2, \ y = \beta^2 - \alpha^2$$

and

$$\frac{x^2}{4} = \alpha^2 \cdot \beta^2,$$

due to (2). That is, $x = 2\alpha\beta$. Hence, a general representation of the solution of the diophantine equation $x^2 + y^2 = z^2$ is

$$\begin{cases} x = 2\alpha\beta \\ y = \beta^2 - \alpha^2 \\ z = \beta^2 + \alpha^2, \end{cases}$$

where $\beta > \alpha$ and $\gcd(\alpha, \beta) = 1$.[12]   □

We shall now present the statement and a proof of the theorem which was mentioned in the above solution.

Theorem 9.5.1. *If $q_1 q_2 \cdots q_\mu = A^n$, where $q_1, q_2, \ldots, q_\mu$ are pairwise relatively prime integers, then each of the integers $q_1, q_2, \ldots, q_\mu$ can be represented as the nth power of an integer. That is, $q_i = a^n$, for every $i = 1, 2, \ldots, \mu$, where $a \in \mathbb{Z}$.*

Proof. By the canonical form of A, we obtain

$$A = p_1^{a_1} p_2^{a_2} \cdots p_k^{a_k}$$

and thus

$$A^n = p_1^{na_1} p_2^{na_2} \cdots p_k^{na_k}.$$

Therefore,

$$q_1 q_2 \cdots q_\mu = p_1^{na_1} p_2^{na_2} \cdots p_k^{na_k}, \text{ where } \mu \leq k.$$

Because of the fact that the integers $q_1, q_2, \ldots, q_\mu$ are relatively prime, it is clear that they do not have any common prime divisors. In addition, all the prime divisors of $q_1, q_2, \ldots, q_\mu$ belong in the set $\{p_1, p_2, \ldots, p_k\}$. Hence, it follows that we can express each q_i in the form

$$q_i = p_{k_1}^{na_{k_1}} p_{k_2}^{na_{k_2}} \cdots p_{k_p}^{na_{k_p}}, \text{ where } p \leq k.$$

Thus,

$$q_i = a^n$$

for every $i = 1, 2, \ldots, \mu$, where $a \in \mathbb{Z}$.   □

B) *Prove that the diophantine equation*

$$x^4 + y^4 = z^4$$

does not accept any nonzero integer solutions.

---

[12] We must have $\gcd(\alpha, \beta) = 1$, since $\gcd(\alpha, \beta) \leq \gcd(\alpha^2, \beta^2) = 1$.

Proof. It suffices to prove that the equation $x^4 + y^4 = z^2$ does not accept any nonzero integer solutions. This happens because if the equation $x^4 + y^4 = z^4$ accepted even one nonzero integer solution (x_k, y_k, z_k), then the equation $x^4 + y^4 = z^2$ would also accept the nonzero integer solution (x_k, y_k, z_k^2).

Let us suppose that the diophantine equation $x^4 + y^4 = z^2$ has at least one nonzero solution (x, y, z). Consider the integer S, such that

$$S = xyz.$$

Let S_0 be the least of the values that S obtains for the different integer solutions (x, y, z). Thus, S_0 can be expressed in the form

$$S_0 = x_0 y_0 z_0.$$

We shall show that $\gcd(x_0, y_0, z_0) = 1$. In order to do so, it suffices to show that $\gcd(x_0, y_0) = 1$, because if there existed a prime number p, such that $p \mid x_0, p \mid y_0$ and $p \mid z_0$, then clearly it would hold $\gcd(x_0, y_0) \neq 1$.

Hence, we assume that there exists a prime number p, such that $p \mid x_0$ and $p \mid y_0$. However, we have assumed that the equation $x^4 + y^4 = z^2$ has at least one nonzero integer solution (x, y, z). Therefore,

$$\left(\frac{x_0}{p}\right)^4 + \left(\frac{y_0}{p}\right)^4 = \left(\frac{z_0}{p^2}\right)^2 \in \mathbb{Z}.$$

Consequently, $p^2 \mid z_0$ and thus

$$x_1 = \frac{x_0}{p}, y_1 = \frac{y_0}{p}, z_1 = \frac{z_0}{p^2}.$$

Hence, we obtain

$$S_1 = x_1 y_1 z_1 = \frac{x_0}{p} \frac{y_0}{p} \frac{z_0}{p^2} < x_0 y_0 z_0 = S_0.$$

Thus,

$$S_1 < S_0,$$

which is impossible, due to the definition of S_0.

According to the above arguments, it follows that for the greatest common divisor of the integers $x_0, y_0\ z_0$, it holds

$$\gcd(x_0, y_0, z_0) = 1$$

and thus

$$\gcd(x_0^2, y_0^2, z_0) = 1.$$

Furthermore, the integers x_0^2, y_0^2 and z_0 form a Pythagorean triple, since

$$(x_0^2)^2 + (y_0^2)^2 = z_0^2.$$

However, we have already shown that a general representation of the solution of the diophantine equation $x^2 + y^2 = z^2$ is the following:

$$x = 2\alpha\beta, y = \beta^2 - \alpha^2, z = \beta^2 + \alpha^2,$$

where $\beta > \alpha$ and $\gcd(\alpha, \beta) = 1$.

Since one of the numbers x_0^2, y_0^2 must be an even integer, without loss of generality we can assume that x_0^2 is even. In that case, it yields

$$x_0^2 = 2\gamma\delta, \, y_0^2 = \delta^2 - \gamma^2, \, z_0 = \delta^2 + \gamma^2,$$

where $\delta > \gamma$ and $\gcd(\gamma, \delta) = 1$. Therefore,

$$\delta^2 = y_0^2 + \gamma^2,$$

where $\gcd(\delta, y_0, \gamma) = 1$ since $\gcd(\gamma, \delta) = 1$. Thus, the triple (δ, y_0, γ) is a Pythagorean triple.

Because of the fact that we have set x_0^2 to be even, it is clear that y_0^2 and thus y_0 must be an odd integer. Hence, γ is an even integer. Consequently,

$$\gamma = 2cd, \, y_0 = d^2 - c^2, \, \delta = d^2 + c^2,$$

where $d > c$ and $\gcd(c, d) = 1$. However, we have shown that

$$x_0^2 = 2\gamma\delta.$$

Therefore,

$$x_0^2 = 2(2cd)(d^2 + c^2).$$

Hence,

$$\left(\frac{x_0}{2}\right)^2 = cd(d^2 + c^2).$$

Clearly $x_0/2 \in \mathbb{Z}$, since x_0 is an even integer. Thus,

$$\left(\frac{x_0}{2}\right)^2$$

is a square of an integer.

We can easily prove that the integers $c, d, d^2 + c^2$ are pairwise relatively prime. This happens because:
- If there existed a prime number p_1, such that $p_1 \mid c$ and $p_1 \mid (d^2 + c^2)$, then we would clearly have $p_1 \mid d^2$ and thus $p_1 \mid d$, which is impossible since $\gcd(c, d) = 1$.
- If there existed a prime number p_2, such that $p_2 \mid d$ and $p_2 \mid (d^2 + c^2)$, then similarly we would have $p_2 \mid c^2$ and thus $p_2 \mid c$, which is impossible.

Therefore, according to Theorem 9.5.1, it follows that

$$c = x_m^2, d = y_m^2, d^2 + c^2 = z_m^2,$$

or

$$x_m^4 + y_m^4 = z_m^2.$$

Hence, the triple (x_m, y_m, z_m) is an integer solution of the equation

$$x^4 + y^4 = z^2.$$

Thus, we obtain

$$S_m = x_m y_m z_m = \sqrt{x_m^2 y_m^2 z_m^2} = \sqrt{cd(d^2 + c^2)}$$

$$= \frac{x_0}{2} < x_0 y_0 z_0 = S_0$$

or equivalently

$$S_m < S_0,$$

which is impossible.

Therefore, the assumption that the equation $x^4 + y^4 = z^2$ has at least one nonzero integer solution leads to a contradiction. Hence, the equation $x^4 + y^4 = z^4$ does not have any nonzero integer solutions.   □

Note. The above proof is due to the German mathematician Ernst Eduard Kummer (1810–1893).

Historical Remark. Ernst Eduard Kummer was born in Germany on the 29th of January, 1810. In 1828, he entered the University of Halle in order to study theology. However, during his studies he was deeply influenced by the mathematician Heinrich Ferdinand Scherk (1798–1885), who introduced him to the areas of algebra and number theory. Thus, Kummer started studying mathematics and on the 10th of September, 1831 he obtained his Ph.D. In 1842 he was elected professor of mathematics at the University of Breslau, where he taught for 13 years. In 1855, he was appointed a professor at the War School of Berlin.

Kummer worked in several areas of mathematics, such as number theory, algebra, analysis, geometry and applied physics. One of his most significant contributions in number theory was the proof of the fact that the diophantine equation

$$x^p + y^p = z^p$$

does not have any positive integer solutions for all prime exponents p, with $p \leq 100$. In other words, he proved Fermat's Last Theorem for exponents less than or equal to 100.

Ernst Kummer died on the 14th of May, 1893, in Berlin. He is considered to be the father of algebraic number theory.

9.6 Fermat's two-square theorem

Theorem 9.6.1. *Every prime number p of the form $4n + 1, n \in \mathbb{N}$, can be represented as the sum of two squares of integers.*

Proof. Firstly, we shall prove that there exists a positive integer k such that the integer kp can be represented as the sum of two squares of integers. In other words, we shall prove that there exists a positive integer k, such that

$$kp = a^2 + b^2,$$

where $a, b \in \mathbb{Z}$.

By Lemma 5.2.6, we know that

$$\left(\frac{-1}{p}\right) = \begin{cases} 1, & \text{if } p \equiv 1 \,(\text{mod } 4) \\ -1, & \text{if } p \equiv 3 \,(\text{mod } 4). \end{cases}$$

Therefore, in the case when $p = 4n + 1$ it follows that the integer -1 is a quadratic residue mod p. Hence, there exists an integer a such that

$$a^2 \equiv -1 \,(\text{mod } p),$$

and thus there exist integers a, k, such that

$$kp = a^2 + 1^2.$$

Thus, we have shown that there exists $k \in \mathbb{N}$ such that

$$kp = a^2 + b^2.$$

It suffices now to prove that the least possible value of the positive integer k is 1.

Let k_0 be the least possible value of k. If we assume that $k_0 > 1$, then

$$k_0 p = a_0^2 + b_0^2. \tag{1}$$

If we divide the integers a_0 and b_0 by k_0, then the remainders r_1 and r_2, which will occur respectively, will belong to the interval

$$\left(-\frac{k_0}{2}, \frac{k_0}{2}\right].$$

Therefore, we can consider integers r_1 and r_2 to have the property

$$r_1 \equiv a_0 \,(\text{mod } k_0) \tag{2a}$$

$$r_2 \equiv b_0 \,(\text{mod } k_0), \tag{2b}$$

where

$$r_1, r_2 \in \left(-\frac{k_0}{2}, \frac{k_0}{2} \right].$$

Thus,

$$r_1^2 \equiv a_0^2 \,(\mathrm{mod}\; k_0)$$

$$r_2^2 \equiv b_0^2 \,(\mathrm{mod}\; k_0),$$

from which it follows that

$$r_1^2 + r_2^2 \equiv (a_0^2 + b_0^2) \,(\mathrm{mod}\; k_0).$$

Hence,

$$k_0 \,|\, [r_1^2 + r_2^2 - (a_0^2 + b_0^2)]$$

and since

$$k_0 \,|\, (a_0^2 + b_0^2),$$

we get

$$k_0 \,|\, (r_1^2 + r_2^2).$$

Therefore, there exists $\lambda \in \mathbb{N} \cup \{0\}$, such that

$$r_1^2 + r_2^2 = \lambda k_0. \tag{3}$$

By (1) and (3), we obtain that

$$(a_0^2 + b_0^2)(r_1^2 + r_2^2) = k_0^2 \lambda p.$$

However, it holds that

$$(a_0 r_1 + b_0 r_2)^2 + (a_0 r_2 - b_0 r_1)^2 = (a_0^2 + b_0^2)(r_1^2 + r_2^2) = k_0^2 \lambda p. \tag{4}$$

By the above identity it follows that the product of two integers which can be represented as the sum of two squares of integers can also be represented as the sum of two squares of integers.

By (2a) and (2b), we obtain

$$r_1^2 \equiv a_0 r_1 \,(\mathrm{mod}\; k_0) \tag{5a}$$

$$r_2^2 \equiv b_0 r_2 \,(\mathrm{mod}\; k_0) \tag{5b}$$

and

$$r_1 r_2 \equiv a_0 r_2 \,(\mathrm{mod}\; k_0) \tag{6a}$$

$$r_1 r_2 \equiv b_0 r_1 \,(\mathrm{mod}\; k_0). \tag{6b}$$

By (5a), (5b), (6a) and (6b), we get

$$a_0 r_1 + b_0 r_2 \equiv (r_1^2 + r_2^2) \,(\mathrm{mod}\, k_0)$$

and

$$a_0 r_2 - b_0 r_1 \equiv (r_1 r_2 - r_1 r_2) \,(\mathrm{mod}\, k_0).$$

Thus, since $r_1^2 + r_2^2 = \lambda k_0$ and $r_1 r_2 - r_1 r_2 = 0$, we obtain

$$k_0 \mid (a_0 r_1 + b_0 r_2)$$

and

$$k_0 \mid (a_0 r_2 - b_0 r_1).$$

Hence, by the identity (4), it holds

$$\left(\frac{a_0 r_1 + b_0 r_2}{k_0} \right)^2 + \left(\frac{a_0 r_2 - b_0 r_1}{k_0} \right)^2 = \lambda p.$$

In other words, the integer λp can be represented as the sum of two squares of integers. However, we have previously assumed that

$$-\frac{k_0}{2} < r_1, r_2 \leq \frac{k_0}{2}.$$

Therefore,

$$r_1^2 + r_2^2 \leq 2 \cdot \frac{k_0^2}{4} = \frac{k_0^2}{2}.$$

By (3) and the above inequality, we get

$$\lambda k_0 \leq \frac{k_0^2}{2} \Leftrightarrow \lambda \leq \frac{k_0}{2} < k_0. \tag{7}$$

• If $\lambda = 0$, then

$$r_1^2 + r_2^2 = 0 \Leftrightarrow r_1 = r_2 = 0.$$

Hence, by (2a) and (2b) we obtain

$$k_0^2 \mid a_0^2$$

and

$$k_0^2 \mid b_0^2.$$

By (1) and the above results, it follows that

$$k_0 p = m k_0^2,$$

for some positive integer m. Thus, $p = m k_0$, which is impossible since p is a prime number.

It follows that (7) contradicts the property of k_0 being the least positive integer, such that $k_0 p$ can be represented as the sum of two squares of positive integers. Thus, the assumption that $k_0 > 1$ leads to a contradiction, thus $k_0 = 1$.

Therefore, every prime number p of the form $4n + 1, n \in \mathbb{N}$, can be represented as the sum of two squares of integers. $\qquad \square$

10

Problems

Problems worthy of attack prove their worth by fighting back.
Paul Erdős (1913–1996)

1) If a positive integer n is perfect, prove that

$$\sum_{d|n} \frac{1}{d} = 2.$$

2) Let $a \equiv b \pmod{k}$ and d be an arbitrary common divisor of a and b. Suppose $\gcd(k, d) = g$, where k is a positive integer. Prove that

$$\frac{a}{d} \equiv \frac{b}{d} \left(\mod \frac{k}{g}\right).$$

Applying the above property prove that if

$$185\, c \equiv 1295 \pmod{259},$$

then $c \equiv 0 \pmod 7$.

3) Prove that there are no integers a, b, c, d such that

$$abcd - a = 111\ldots1 \text{ (The digit 1 appears } k \text{ times)} \qquad \text{(a)}$$
$$abcd - b = 111\ldots1 \text{ (The digit 1 appears } k \text{ times)} \qquad \text{(b)}$$
$$abcd - c = 111\ldots1 \text{ (The digit 1 appears } k \text{ times)} \qquad \text{(c)}$$
$$abcd - d = 111\ldots1 \text{ (The digit 1 appears } k \text{ times)} \qquad \text{(d)}$$

where $k \in \mathbb{N} - \{1\}$.

M.Th. Rassias, *Problem-Solving and Selected Topics in Number Theory: In the Spirit of the Mathematical Olympiads*, DOI 10.1007/978-1-4419-0495-9_10, © Springer Science+Business Media, LLC 2011

4) Prove that there does not exist any prime number in the sequence of integers

$$10001, \quad 100010001, \quad 1000100010001, \ldots.$$

5) Find the last three digits of the integer 7^{9999}.

(N.M.M., 1937–38, p. 415, Problem 216. Proposed by Victor Thebault, Le Mans, France)

6) Determine the last three digits of the integer

$$2003^{2002^{2001}}.$$

(Canada, 2003)

7) Prove that if the integers $a_1, a_2, \ldots, a_9$ are not divisible by 3, then

$$a_1^2 + a_2^2 + \cdots + a_9^2 \equiv 0 \,(\mathrm{mod}\ 3).$$

8) Let a be an integer. Prove that there are no integers b, c with $c > 1$, such that

$$(a+1)^2 + (a+2)^2 + \cdots + (a+99)^2 = b^c. \tag{a}$$

(1998 Hungarian Mathematical Olympiad)

9) If m, n, m_1, n_1 are positive integers such that

$$(m+n)(m+n-1) + 2m = (m_1+n_1)(m_1+n_1-1) + 2m_1, \tag{a}$$

prove that $m = m_1$ and $n = n_1$.

10) Prove that if m, n are integers, then the expression

$$E = m^5 + 3m^4n - 5m^3n^2 - 15m^2n^3 + 4mn^4 + 12n^5$$

cannot take the value 33.

11) Prove that

$$(2m+1)^{2^n} = 2^{n+2}\lambda_n + 1,$$

for every positive integer n, with $\lambda_n \in \mathbb{Z}$.

(Elias Karakitsos, Sparta, Greece)

12) Consider $m, n \in \mathbb{N}$, such that $m+n$ is odd. Prove that there is no $A \subseteq \mathbb{N}$ such that for all $x, y \in \mathbb{N}$, if $|x-y| = m$ then $x \in A$ or $y \in A$, and if $|x-y| = n$ then $x \notin A$ or $y \notin A$.

(Dimiter Skordev, Problem No. 11074, Amer. Math. Monthly, 2004)

13) Let p be an odd prime number. If r_j is the remainder when the integer $j^{p-1} - 1/p$ is divided by p, where $j = 1, 2, \ldots, p - 1$, prove that

$$r_1 + 2r_2 + \cdots + (p-1)r_{p-1} \equiv \frac{p+1}{2} \pmod{p}.$$

(Dorin Andrica, "Babes-Bolyai" University, Cluj-Napoca, Romania)

14) Find all possible decimal digits a such that, for a given n, the decimal expansions of 2^n and 5^n both begin by a, and give a necessary and sufficient condition to determine all such integers n.

(Konstantinos Drakakis, University College Dublin, Ireland; Newsletter of the European Mathematical Society, Issue 73, 2009, Problem 48, p. 54)

15) Let a, n be positive integers such that a^n is a perfect number. Prove that

$$a^{n/\mu} > \frac{\mu}{2},$$

where μ denotes the number of distinct prime divisors of a^n.

(M. Th. Rassias, Proposed problem W. 27, Octogon Mathematical Magazine, 17(1)(2009), p. 311)

16) Prove that the sum

$$S = \frac{1}{2} + \frac{1}{3} + \cdots + \frac{1}{n},$$

where $n > 1$ cannot be an integer.

17) Let $n \geq 3$ be an odd positive integer. Prove that the set

$$A = \left\{ \binom{n}{1}, \binom{n}{2}, \ldots, \binom{n}{\frac{n-1}{2}} \right\}$$

contains an odd number of odd integers.

(Revista Matematică Timisoara, No.2 (1984), Problem 5346)

18) Prove that every positive rational number can be expressed in the form

$$\frac{a^3 + b^3}{c^3 + d^3},$$

where a, b, c, d are positive integers.

19) Prove that every composite positive integer can be represented in the form

$$xy + xz + yz + 1,$$

where x, y, z are positive integers.

(Problem 1, The Forty-Ninth William Lowell Putnam Mathematical Competition, 1988)

20) If for the rational number x the value of the expression

$$2x^4 + 3x + 1$$

is an integer, prove that x is also an integer.

(School of Aviation Engineers of Greece, Entrance Examinations, 1968)

21) Consider the sequence (x_n) of real numbers, which is defined by the recursive formula

$$x_1 = 0, \; x_{n+1} = 5x_n + \sqrt{24x_n^2 + 1},$$

where $n = 1, 2, 3, \ldots$.
Prove that all the terms of the sequence are integers.

22) Let $a \in \mathbb{Z}$, $n, k \in \mathbb{N}$ with $k \equiv -a^2 \pmod{2n}$. Prove that

$$\sqrt{2}\sqrt{k + a^2} \le \frac{k + a^2}{2n} + n \le \frac{1}{2}(k + a^2) + 1.$$

23) Let $n_1 = \overline{abcabc}$ and $n_2 = \overline{d00d}$ be positive integers represented in the decimal system, where $a, b, c, d \in \{0, 1, 2, \ldots, 9\}$ with $a \ne 0$ and $d \ne 0$.

i) Prove that $\sqrt{n_1}$ cannot be an integer.
ii) Find all positive integers n_1 and n_2 such that $\sqrt{n_1 + n_2}$ is an integer.
iii) From all the pairs (n_1, n_2) such that $\sqrt{n_1 n_2}$ is an integer find those for which $\sqrt{n_1 n_2}$ has the greatest possible value.

(48th National Mathematical Olympiad, Suceava, 1997)

24) Determine the number of real solutions a of the equation

$$\left\lfloor \frac{a}{2} \right\rfloor + \left\lfloor \frac{a}{3} \right\rfloor + \left\lfloor \frac{a}{5} \right\rfloor = a. \tag{a}$$

(Canadian Mathematical Olympiad, 1998)

25) Let $a, b \in \mathbb{N}$. Prove that

$$\left\lfloor \frac{2a}{b} \right\rfloor - 2\left\lfloor \frac{a}{b} \right\rfloor = \begin{cases} 0, & \text{if } \left\lfloor \frac{2a}{b} \right\rfloor \text{ is an even integer} \\ 1, & \text{if } \left\lfloor \frac{2a}{b} \right\rfloor \text{ is an odd integer.} \end{cases}$$

26) Prove that

$$\lfloor \sqrt{n} + \sqrt{n+1} + \sqrt{n+2} \rfloor = \lfloor \sqrt{9n+8} \rfloor \tag{1}$$

for $n = 0, 1, 2, \ldots$.

(Crux Mathematicorum 28(1)(2002). See also Amer. Math. Monthly, 1988, pp. 133–134)

27) Solve the equation

$$\lfloor 3x - 2 \rfloor - \lfloor 2x - 1 \rfloor = 2x - 6, \ x \in \mathbb{R}.$$

(Elias Karakitsos, Sparta, Greece)

28) Solve the equation

$$\lfloor x \rfloor^2 = \lfloor 3x - 2 \rfloor, \quad x \in \mathbb{R}. \tag{1}$$

(Elias Karakitsos, Sparta, Greece)

29) Solve the equation

$$\lfloor x^2 - 3x + 2 \rfloor = 3x - 7, \quad x \in \mathbb{R}.$$

(Elias Karakitsos, Sparta, Greece)

30) Prove that for every real number x and a given positive integer n it holds

$$\lfloor x \rfloor + \left\lfloor x + \frac{1}{n} \right\rfloor + \left\lfloor x + \frac{2}{n} \right\rfloor + \cdots + \left\lfloor x + \frac{n-1}{n} \right\rfloor = \lfloor nx \rfloor.$$

(Charles Hermite, 1822–1901)

31) Let k be a positive integer. Prove that there exist polynomials $P_0(n)$, $P_1(n), \ldots, P_{k-1}(n)$ (which may depend on k) such that for any integer n,

$$\left\lfloor \frac{n}{k} \right\rfloor^k = P_0(n) + P_1(n) \left\lfloor \frac{n}{k} \right\rfloor + \cdots + P_{k-1}(n) \left\lfloor \frac{n}{k} \right\rfloor^{k-1}.$$

(Problem B5, The Sixty-Eighth William Lowell Putnam Mathematical Competition, 2007. Amer. Math. Monthly, 115(2008), pp. 732, 737)

32) A rational number $r = a/b$, where a, b are coprime positive integers, is called *good* if and only if $r > 1$ and there exist integers N, c, such that for every positive integer $n \geq N$, it holds

$$|\{r^n\} - c| \leq \frac{1}{2(a+b)},$$

where $\{r\} = r - \lfloor r \rfloor$.
Prove that every *good* rational number is an integer.

(Chinese National Team Selection Contest, 2007)

33) Determine the integer part of

$$\sum_{n=1}^{10^9} \frac{1}{\sqrt[3]{n^2}},$$

where $n \in \mathbb{N}$.

34) Calculate the integer part of

$$\sum_{n=1}^{+\infty} \sum_{m=1}^{+\infty} \frac{1}{m^2n + mn^2 + 2mn}.$$

35) Find all positive integers a, b such that

$$a^4 + 4b^4$$

is a prime number.

36) Let n and $8n^2 + 1$ be two prime numbers. Prove that the number $8n^2 - 1$ is also a prime number.

37) Prove that there does not exist a nonconstant polynomial $p(n)$ with integer coefficients, such that for every natural number n, the number $p(n)$ is prime.

38) Let n be an odd integer greater than or equal to 5. Prove that

$$\binom{n}{1} - 5\binom{n}{2} + 5^2\binom{n}{3} - \cdots + 5^{n-1}\binom{n}{n}$$

is not a prime number.

(Titu Andreescu, Korean Mathematical Competition, 2001)

39) Prove that there are infinitely many prime numbers of the form $4n + 3$, where $n \in \mathbb{N}$.

40) Let $y \in \mathbb{Z}^* = \mathbb{Z} - \{0\}$. If $x_1, x_2, \ldots, x_n \in \mathbb{Z}^* - \{1\}$ with $n \in \mathbb{N}$ and

$$(x_1x_2 \cdots x_n)^2 y \le 2^{2(n+1)},$$

as well as

$$x_1x_2 \cdots x_n y = z + 1 \quad z \in \mathbb{N},$$

prove that at least one of the integers $x_1, x_2, \ldots, x_n, z$ is a prime number.

(M. Th. Rassias, Proposed problem W.3, Octogon Mathematical Magazine
15 (1) (2007), p. 291. See also Proposed problem No. 109, Euclid
Mathematical Magazine B', Greek Math. Soc. 66 (2007), p. 71)

41) Let p be a prime number. Let $h(x)$ be a polynomial with integer coefficients such that $h(0)$, $h(1)$,..., $h(p^2 - 1)$ are distinct modulo p^2. Prove that $h(0)$, $h(1)$,..., $h(p^3 - 1)$ are distinct modulo p^3.

(Problem B4, The Sixty-Ninth William Lowell Putnam Mathematical Competition, 2008, Amer. Math. Monthly 116(2009), pp. 722, 725)

42) Prove that every odd perfect number has at least three distinct prime factors.

43) Let (a_n) be a sequence of positive integers, such that $(a_i, a_j) = 1$ for every $i \neq j$. If

$$\sum_{n=0}^{+\infty} \frac{1}{a_n} = +\infty,$$

prove that the sequence (a_n) contains infinitely many prime numbers.

(K. Gaitanas, student of the School of Applied Mathematics and Physical Sciences, NTUA, Greece, 2005)

44) Let p_i denote the ith prime number. Prove that

$$p_1^k + p_2^k + \cdots + p_n^k > n^{k+1},$$

for every pair of positive integers n, k.

(Dorin Andrica, Revista Matematică Timisoara, No. 2(1978), p. 45, Problem 3483)

45) Prove that 7 divides the number

$$1^{47} + 2^{47} + 3^{47} + 4^{47} + 5^{47} + 6^{47}.$$

46) Prove that if $3 \nmid n$, then

$$13 \mid 3^{2n} + 3^n + 1,$$

where $n \in \mathbb{N}$.

47) Prove that for every positive integer n the value of the expression

$$2^{4n+1} - 2^{2n} - 1$$

is divisible by 9.

48) Prove that 7 divides the number

$$2222^{5555} + 5555^{2222}.$$

49) If p is a prime number and a, λ are two positive integers such that $p^\lambda \mid (a-1)$, prove that

$$p^{n+\lambda} \mid (a^{p^n} - 1)$$

for every $n \in \mathbb{N} \cup \{0\}$.

(Crux Mathematicorum, 1992, p. 84, Problem 1617. Proposed by Stanley Rabinowitz, Westford, Massachusetts)

50) Prove that for any prime number p greater than 3, the number

$$\frac{2^p + 1}{3}$$

is not divisible by 3.

51) Determine all positive integers n for which the number $n^8 - n^2$ is not divisible by 72.

(38th National Mathematical Olympiad, Slovenia, 1997)

52) Prove that for every positive integer n the number $3^n + n^3$ is a multiple of 7 if and only if the number $3^n \cdot n^3 + 1$ is a multiple of 7.

(Bulgarian Mathematical Competition, 1995)

53) Find the sum of all positive integers that are less than 10,000 whose squares divided by 17 leave remainder 9.

54) What is the largest positive integer m with the property that, for any positive integer n, m divides $n^{241} - n$? What is the new value of m if n is restricted to be odd?

(Konstantinos Drakakis, University College Dublin, Ireland; Newsletter of the European Mathematical Society, Issue 77, 2010, Problem 69)

55) Let f be a nonconstant polynomial with positive integer coefficients. Prove that if n is a positive integer, then $f(n)$ divides $f(f(n)+1)$ if and only if $n = 1$.

(Problem B1, The Sixty-Eighth William Lowell Putnam Mathematical Competition, 2007. Amer. Math. Monthly, 115(2008), pp. 731, 735)

56) Let N_n and D_n be two relatively prime positive integers. If

$$1 + \frac{1}{2} + \frac{1}{3} + \cdots + \frac{1}{n} = \frac{N_n}{D_n},$$

find all prime numbers p with $p \geq 5$, such that

$$p \mid N_{p-4}.$$

(Crux Mathematicorum, 1989, p. 62, Problem 1310. Proposed by Robert E. Shafer, Berkeley, California)

57) Given the positive integer n and the prime number p such that $p^p \mid n!$, prove that

$$p^{p+1} \mid n!.$$

(Proposed by D. Beckwith, Amer. Math. Monthly, Problem No. 11158, 2005)

58) Prove that there are no integer values of x, y, z, where x is of the form $4k + 3 \in \mathbb{Z}$, such that

$$x^n = y^n + z^n,$$

where $n \in \mathbb{N} - \{1\}$.

59) Let P_n denote the product of all distinct prime numbers $p_1, p_2, \ldots, p_k$, which are less than or equal to n (where $k < n$). Prove that P_n divides the integer

$$n^k \sum_{\lambda=0}^{p_1-1} (-1)^\lambda \binom{n}{\lambda} \cdot \sum_{\lambda=0}^{p_2-1} (-1)^\lambda \binom{n}{\lambda} \cdots \sum_{\lambda=0}^{p_k-1} (-1)^\lambda \binom{n}{\lambda}.$$

60) Determine all pairs of positive integers (a, b), such that the number

$$\frac{a^2}{2ab^2 - b^3 + 1}$$

is a positive integer.

(44th IMO, Tokyo, Japan)

61) Prove that for every integer $m \geq 2$ we have

$$F_m^{(F_{m+1}-1)} \equiv 1 \pmod{F_{m+1}},$$

where F_m denotes the mth Fermat number.

62) Prove that

$$\phi(n) \geq \frac{\sqrt{n}}{2}$$

for every positive integer n.

63) Let n be a perfect even number. Prove that the integer

$$n - \phi(n)$$

is a square of an integer and determine an infinity of integer values of k, such that the integer

$$k - \phi(k)$$

is a square of an integer.

(Crux Mathematicorum, 1988, p. 93, Problem 1204. Proposed by Thomas E. Moore, Bridgewater State College, Bridgewater, Massachusetts)

64) Let n be an integer greater than one. If $n = p_1^{k_1} p_2^{k_2} \cdots p_r^{k_r}$ is the canonical form of n, then prove that

$$\sum_{d|n} d\phi(d) = \frac{p_1^{2k_1+1} + 1}{p_1 + 1} \cdot \frac{p_2^{2k_2+1} + 1}{p_2 + 1} \cdots \frac{p_r^{2k_r+1} + 1}{p_r + 1}.$$

65) Let n be an integer greater than one. If $n = p_1^{k_1} p_2^{k_2} \cdots p_r^{k_r}$ is the canonical form of n, then prove that

$$\sum_{d|n} \mu(d)\phi(d) = (2 - p_1)(2 - p_2) \cdots (2 - p_r).$$

66) Let $n, \lambda \in \mathbb{N}$ with $\lambda > 1$ and $4 \mid n$. Solve the diophantine equation

$$\Phi(n)x + \phi(n)y = \phi(n)^\lambda, \tag{a}$$

where

$$\Phi(n) = \sum_{\substack{1 \le q < n \\ \gcd(n,q)=1}} q$$

and $\phi(n)$ is the Euler function.

67) Prove that

$$\sigma_1(n!) < \frac{(n+1)!}{2}$$

for all positive integers n, where $n \ge 8$.

(Crux Mathematicorum, 1990, Problem 1399, p. 58. Proposed by Sydney Bulman-Fleming and Edward T.H. Wang, Wilfried Laurier, University of Waterloo, Ontario)

68) Prove that

$$\sum_{n=1}^{+\infty} \frac{\tau(n)}{2^n} = \sum_{n=1}^{+\infty} \frac{1}{\phi(2^{n+1}) - 1}.$$

69) If f is a multiplicative arithmetic function, then

(α) Prove that

$$\sum_{d|n} f(d) = \prod_{p^a \| n} (1 + f(p) + f(p^2) + \cdots + f(p^a)),$$

where $p^a \| n$ denotes the greatest power of the prime number p which divides n.

(β) Prove that the function

$$g(n) = \sum_{d|n} f(d)$$

is multiplicative.

(γ) Prove that

$$\left(\sum_{d|n} \tau(d) \right)^2 = \sum_{d|n} \tau^3(d).$$

70) Consider two arithmetic functions f, g, such that

$$A(n) = \sum_{d|n} f(d) g \left(\frac{n}{d} \right)$$

and g are multiplicative.

Prove that f must also be multiplicative.

71) Prove that

$$\sum_{n=2}^{+\infty} f(\zeta(n)) = 1,$$

where $f(x) = x - \lfloor x \rfloor$ denotes the fractional part of $x \in \mathbb{R}$ and $\zeta(s)$ is the Riemann zeta function.

(H. M. Srivastava, University of Victoria, Canada)

72) Prove that

$$\pi(x) \geq \log \log x,$$

where $x \geq 2$.

(Hint: Prove first the inequality

$$p_n < 2^{2^n},$$

where p_n denotes the nth prime number.)

73) Prove that any integer can be expressed as the sum of the cubes of five integers not necessarily distinct.

(T. Andreescu, D. Andrica and Z. Feng)

74) Let n be an integer. An integer A is formed by $2n$ digits each of which is 4; however, another integer B is formed by n digits each of which is 8. Prove that the integer

$$A + 2B + 4$$

is a perfect square of an integer.

(7th Balcan Mathematical Olympiad, Kusadasi, Turkey)

75) Find the integer values of x for which the expression $x^2 + 6x$ is a square of an integer.

76) Express the integer 459 as the sum of four squares of integers.

77) Find the three smallest positive consecutive natural numbers, whose sum is a perfect square and a perfect cube of a natural number.

(M. Th. Rassias, Proposed problem No. 94, Euclid Mathematical Magazine
B', Greek Math. Soc., 62(2006), p. 80)

78) Find all prime numbers p such that the number

$$\frac{2^{p-1} - 1}{p}$$

is a square of an integer.

(S. E. Louridas, Athens, Greece)

79) Let n be a positive integer, such that the $\gcd(n, 6) = 1$. Prove that the sum of n squares of consecutive integers is a multiple of n.

80) Prove that for every $m \in \mathbb{N} - \{1, 2\}$, such that the integer $7 \cdot 4^m$ can be expressed as a sum of four squares of nonnegative integers a, b, c, d, each of the numbers a, b, c, d is at least equal to 2^{m-1}.

(W. Sierpiński, 250 Problèmes de Théorie Élémentaire des Nombres, P.W.,
Warsaw, 1970)

81) Let n be a positive integer and d_1, d_2, d_3, d_4 the smallest positive integer divisors of n with $d_1 < d_2 < d_3 < d_4$. Find all integer values of n, such that

$$n = d_1^2 + d_2^2 + d_3^2 + d_4^2.$$

82) Let a, b be two positive integers, such that

$$ab + 1 \mid a^2 + b^2.$$

Prove that the integer

$$\frac{a^2 + b^2}{ab + 1}$$

is a perfect square of a positive integer.

(Shortlist, 29th International Mathematical Olympiad, 1988)

83) Let k be an integer, which can be expressed as a sum of two squares of integers, that is,

$$k = a^2 + b^2 \quad \text{with} \quad a, b \in \mathbb{Z}.$$

If p is a prime number greater than 2, which can be expressed as a sum of two squares of integers c, d for which it holds

$$(c^2 + d^2) \mid (a^2 + b^2) \quad \text{and} \quad (c^2 + d^2) \nmid (a + b),$$

prove that the integer

$$\frac{a^2 + b^2}{c^2 + d^2} = \frac{k}{p}$$

can be expressed as a sum of two squares of integers.

84) Prove that the integer

$$p_1 p_2 \cdots p_n - 1, \text{ where } n \in \mathbb{N} \text{ with } n > 1,$$

cannot be represented as a perfect power of an integer.
(By $p_1, p_2, \ldots, p_n$ we denote, respectively, the 1st, 2nd, 3rd, ..., nth prime number.)

(M. Le, The perfect powers in $\{p_1 p_2 \cdots p_n\}_{n=1}^{+\infty}$, Octogon Mathematical Magazine 13(2)(2005), pp. 1101–1102)

85) Let $f(x) = ax^2 + bx + c$ be a quadratic polynomial with integer coefficients such that $f(0)$ and $f(1)$ are odd integers. Prove that the equation $f(x) = 0$ does not accept an integer solution.

86) A function $f : \mathbb{N} \to \mathbb{N}$ is defined as follows: writing a number $x \in \mathbb{N}$ in its decimal expansion and replacing each digit by its square we obtain the decimal expansion of the number $f(x)$. For example, $f(2) = 4$, $f(35) = 925$, $f(708) = 49064$. Solve the equation $f(x) = 29x$.

(Vladimir Protasov, Moscow State University; Newsletter of the European Mathematical Society, Issue 77, 2010, Problem 67)

87) Prove that the only integer solution of the equation

$$y^2 = x^3 + x$$

is $x = 0$, $y = 0$.

88) Prove that the equation $7x^3 - 13y = 5$ does not have any integer solutions.

(S. E. Louridas, Athens, Greece)

89) Show that for any $n \in \mathbb{N}$, the equation $q = 2p^{2n} + 1$, where p and q are prime numbers, has at most one solution.

(Konstantinos Drakakis, University College Dublin, Ireland; Newsletter of the European Mathematical Society, Issue 67, 2008, Problem 23, p. 46)

90) Find all positive integers x, y, z such that

$$x^3 + y^3 + z^3 - 3xyz = p,$$

where p is a prime number with $p > 3$.

(Titu Andreescu and Dorin Andrica, Problem 27, Newsletter of the European Mathematical Society, 69(2008), p. 24)

91) Prove that there exists an integer n such that

$$p^{(p+3)/2} \left| \left[(p-1)^{p-1} - p - 1 + \sum_{i=0}^{2a} (p^2 - a + i)^m \right] \right.$$

$$\times [(2 \cdot 4 \cdots (p-1))^{p-1} - n(-1)^{(p+1)/2}],$$

where m is an odd positive integer, $a \in \mathbb{N}$ and p is an odd prime number.

92) Find the minimum value of the product xyz over all triples of positive integers x, y, z for which 2010 divides $x^2 + y^2 + z^2 - xy - yz - zx$.

(Titu Andreescu, The University of Texas at Dallas, USA; Newsletter of the European Mathematical Society, Issue 77, 2010, Problem 70)

93) Find all pairs (x, y) of positive integers x, y for which it holds

$$\frac{1}{x} + \frac{1}{y} = \frac{1}{pq},$$

where p, q are prime numbers.

94) Let n be a positive integer. Prove that the equation

$$x + y + \frac{1}{x} + \frac{1}{y} = 3n$$

does not accept solutions in the set of positive rational numbers.

(66th Panhellenic Mathematical Competition, "ARCHIMEDES")

95) Find all integers n, $n \geq 2$, for which it holds

$$1^n + 2^n + \cdots + (n-1)^n \equiv 0 \pmod{n}.$$

96) Prove that for every positive integer k, the equation

$$x_1^3 + x_2^3 + \cdots + x_k^3 + x_{k+1}^2 = x_{k+2}^4$$

has infinitely many solutions in positive integers, such that $x_1 < x_2 < \cdots < x_{k+1}$.

(Dorin Andrica, "Babes-Bolyai" University, Cluj-Napoca, Romania; Newsletter of the European Mathematical Society, Issue 77, 2010, Problem 71)

97) Prove that for every prime number p, the equation

$$2^p + 3^p = a^n$$

does not have integer solutions for all a, n with $a, n \in \mathbb{N} - \{1\}$.

98) Let p_1, p_2 be two odd prime numbers and a, n integers such that $a > 1$ and $n > 1$. Prove that the equation

$$\left(\frac{p_2 - 1}{2}\right)^{p_1} + \left(\frac{p_2 + 1}{2}\right)^{p_1} = a^n$$

accepts integer solutions for a, n only in the case $p_1 = p_2$.

(M. Th. Rassias, Proposed problem W. 5, Octogon Mathematical Magazine, 17(1)(2009), p. 307)

99) Find all integer solutions of the equation

$$\frac{a^7 - 1}{a - 1} = b^5 - 1.$$

(Shortlisted, 47th IMO, Slovenia, 2006)

100) Find all integer solutions of the system

$$x + 4y + 24z + 120w = 782 \tag{a}$$
$$0 \le x \le 4 \tag{b}$$
$$0 \le y \le 6 \tag{c}$$
$$0 \le z \le 5. \tag{d}$$

101) Find all integer solutions of the system

$$35x + 63y + 45z = 1 \tag{a}$$
$$|x| < 9 \tag{b}$$
$$|y| < 5c \tag{c}$$
$$|z| < 7. \tag{d}$$

102) Find the integer solutions of the system

$$x^2 + 2yz < 36 \tag{a}$$

$$y^2 + 2zx = -16 \tag{b}$$

$$z^2 + 2xy = -16. \tag{c}$$

(National Technical University of Athens, Entrance Examinations, 1946)

103) Let a, b, c be real numbers which are not all equal. Prove that positive integer solutions of the system

$$(b - a)x - (c - b)z = 3b \tag{a}$$

$$(c - b)y - (a - c)x = 3c \tag{b}$$

$$(a - c)z - (b - a)y = 3a \tag{c}$$

do not exist, except the trivial solution

$$(x, y, z) = (1, 1, 1),$$

which occurs only when $a + b + c = 0$.

104) Show that, for any $n \in \mathbb{N}$, any $k \in \mathbb{N}$ which is not equal to a power of 10, and any sequence of (decimal) digits $x_0, x_1, \ldots, x_{n-1}$ in $\{0, 1, \ldots, 9\}$, there exists an $m \in \mathbb{N} \cup \{0\}$ such that the first n decimal digits of the power k^m are, from left to right, $x_{n-1}x_{n-2} \cdots x_1 x_0$. As an example, a power of 2 beginning with the digits 409 is $2^{12} = 4096$.

(Konstantinos Drakakis, University College Dublin, Ireland; Newsletter of the European Mathematical Society, Issue 69, 2008, Problem 37, p. 23)

105) In order to file a collection of n books, each book needs a number label from 1 to n. To form this number, digit stickers are used: for example, the number 123 will be formed by the three stickers 1, 2, and 3 side by side (unnecessary zeros in the beginning, such as 00123, are not added, as this would be a terrible waste).

These stickers are sold in sets of 10, and each decimal digit $\{0, 1, 2, \ldots, 9\}$ appears exactly once in the set. How many sets of stickers are needed? As an example, for $n = 21$ books, digit 1 appears 13 times (in numbers 1, 10–19, and 21—note that it appears twice in 11!), 2 appears 4 times (2, 12, 20, and 21), and every other digit from 3 to 9 appears exactly twice, so overall 13 sets are needed.

(Konstantinos Drakakis, University College Dublin, Ireland; Newsletter of the European Mathematical Society, Issue 73, 2009, Problem 45, p. 52)

11

Solutions

1) If a positive integer n is perfect, prove that

$$\sum_{d|n} \frac{1}{d} = 2.$$

Proof. We know that the positive integer n is perfect if and only if

$$\sigma_1(n) = 2n.$$

Therefore,

$$\sum_{d|n} \frac{1}{d} = \sum_{d|n} \frac{1}{n/d} = \frac{1}{n} \sum_{d|n} d = \frac{1}{n} \sigma_1(n)$$

$$= \frac{1}{n} \cdot 2n = 2. \qquad \square$$

2) Let $a \equiv b \pmod{k}$ and d be an arbitrary common divisor of a and b. Suppose $\gcd(k,d) = g$, where k is a positive integer. Prove that

$$\frac{a}{d} \equiv \frac{b}{d} \left(\bmod \, \frac{k}{g} \right).$$

Applying the above property prove that if

$$185\,c \equiv 1295 \pmod{259},$$

then $c \equiv 0 \pmod 7$.

M.Th. Rassias, *Problem-Solving and Selected Topics in Number Theory: In the Spirit of the Mathematical Olympiads*, DOI 10.1007/978-1-4419-0495-9_11,
© Springer Science+Business Media, LLC 2011

Proof.

• Since $a \equiv b \,(\text{mod}, k)$, there exists $\lambda \in \mathbb{Z}$ such that

$$a - b = \lambda k \Leftrightarrow \frac{a-b}{d}d = \lambda k \Leftrightarrow \left(\frac{a}{d} \cdot \frac{b}{d}\right) d = \lambda k$$

$$\Leftrightarrow \left(\frac{a}{d} - \frac{b}{d}\right)\frac{d}{g} = \lambda\frac{k}{g}. \tag{1}$$

However, we know that

$$\gcd(k, d) = g.$$

Thus,

$$\left(\frac{k}{g}, \frac{d}{g}\right) = 1. \tag{2}$$

From (1), (2) we get

$$\frac{k}{g}\left|\left(\frac{a}{d} - \frac{b}{d}\right)\right..$$

That means

$$\frac{a}{d} \equiv \frac{b}{d}\left(\text{mod}, \frac{k}{g}\right).$$

• For the case when $a = 185\,c$, $b = 1295$, $k = 259$ it holds

$$185 = 5 \cdot 37, \quad 1295 = 5 \cdot 7 \cdot 37 \quad \text{and} \quad \gcd(37, 259) = 37.$$

Therefore,

$$\frac{185}{37}c \equiv \frac{1295}{37}\left(\text{mod}, \frac{259}{37}\right)$$

or

$$5c = 35\,(\text{mod}, 7),$$

which means

$$5c - 35 = \text{mult. } 7 \Rightarrow 5(c - 7) = \text{mult. } 7.$$

However, $7 \nmid 5$, thus $7 | (c - 7)$ and therefore

$$c - 7 = \text{mult. } 7,$$

that is,

$$c \equiv 7\,(\text{mod } 7).$$

Hence,

$$c \equiv 0\,(\text{mod } 7). \qquad \square$$

3) Prove that there are no integers a, b, c, d such that

$$abcd - a = 111\ldots1 \text{ (The digit 1 appears } k \text{ times)} \qquad (a)$$

$$abcd - b = 111\ldots1 \text{ (The digit 1 appears } k \text{ times)} \qquad (b)$$

$$abcd - c = 111\ldots1 \text{ (The digit 1 appears } k \text{ times)} \qquad (c)$$

$$abcd - d = 111\ldots1 \text{ (The digit 1 appears } k \text{ times)} \qquad (d)$$

where $k \in \mathbb{N} - \{1\}$.

Proof. One has

$$(a) \Leftrightarrow a(bcd - 1) = 111\cdots1 \text{ (The digit 1 appears } k \text{ times)}.$$

The integer $111\cdots1$ (the digit 1 appears k times) is odd. Therefore, the integer a must necessarily be odd. Similarly, b, c, d must also be odd integers. However, in this case the integers

$$abcd - a, \ abcd - b, \ abcd - c, \ abcd - d$$

are even, which is not possible, since the integer

$$111\cdots1 \text{ (the digit 1 appears } k \text{ times)}$$

is odd.   □

4) Prove that there does not exist any prime number in the sequence of integers
$$10001, \quad 100010001, \quad 1000100010001, \ldots.$$

Proof. The above sequence of integers can be expressed by the following representation:

$$1 + 10^4, 1 + 10^4 + 10^8, \ldots, 1 + 10^4 + 10^8 + \cdots + 10^{4n}, \ldots.$$

This sequence is a special case of the sequence

$$1 + x^4, 1 + x^4 + x^8, \ldots, 1 + x^4 + x^8 + \cdots + x^{4n}, \ldots,$$

where x is an integer with $x > 1$, $n \in \mathbb{N}$. We consider two cases:

Case 1. If $n = 2k$ and $k \in \mathbb{N}$, then

$$1 + x^4 + x^8 + \cdots + x^{4n} = 1 + x^4 + x^8 + \cdots + x^{4(2k)}$$

$$= 1 + x^4 + (x^4)^2 + \cdots + (x^4)^{2k}$$

$$= \frac{1 - (x^4)^{2k+1}}{1 - x^4}$$

$$= \frac{1 - (x^{2k+1})^4}{1 - x^4}$$

$$= \frac{1 - (x^{2k+1})^2}{1 - x^2} \cdot \frac{1 + (x^{2k+1})^2}{1 + x^2}$$

$$= \frac{1 - (x^2)^{2k+1}}{1 - x^2} \cdot \frac{1 + (x^2)^{2k+1}}{1 + x^2}.$$

Thus,

$$1 + x^4 + x^8 + \cdots + x^{4n} = \left(1 + x^2 + \cdots + (x^2)^{2k}\right)\left(1 - x^2 + \cdots + (x^2)^{2k}\right).$$

Therefore, for every $n \in \mathbb{N}$ with $n = 2k$, $k \in \mathbb{N}$, the number

$$1 + x^4 + x^8 + \cdots + x^{4n}$$

is composite.

Case 2. If $n = 2k + 1$ and $k \in \mathbb{N}$, then

$$1 + x^4 + x^8 + \cdots + x^{4n} = 1 + x^4 + x^8 + \cdots + x^{4(2k+1)}$$

$$= (1 + x^4) + (x^8 + x^{12}) + \cdots + (x^{8k} + x^{8k+4})$$

$$= (1 + x^4) + x^8(1 + x^4) + \cdots + x^{8k}(1 + x^4)$$

$$= (1 + x^4)(1 + x^8 + x^{12} + \cdots + x^{8k}).$$

Therefore, for every $n \in \mathbb{N}$ with $n = 2k + 1$, $k \in \mathbb{N}$, the number

$$1 + x^4 + x^8 + \cdots + x^{4n}$$

is composite.

In the special case when $x = 10$ and $n = 1$ we obtain

$$1 + 10^4 = 10001 = 73 \times 137.$$

Therefore, in every case, the sequence of integers

$$1 + x^4, \ 1 + x^4 + x^8, \ldots, \ 1 + x^4 + x^8 + \cdots + x^{4n}, \ldots,$$

where x is an integer with $x > 1$ and $n \in \mathbb{N}$, does not contain prime numbers. □

5) Find the last three digits of the integer 7^{9999}.

(N.M.M., 1937–38, p. 415, Problem 216. Proposed by Victor Thebault, Le Mans, France. Solved by D.P. Richardson, University of Arkansas)

Proof. We note that $7^4 = 2401$. Therefore, we obtain

$$7^{4n} = (2401)^n = (1 + 2400)^n = 1 + n \cdot 2400 + \binom{n}{2} \cdot 2400^2 + \cdots .$$

In the above expression from the third term onwards, all terms end with at least four zero digits and therefore do not influence the three final digits of the number 7^{4n}, where $n \in \mathbb{N}$.

In order to determine the last three digits of the integer 7^{4n}, it is enough to determine the last three digits of the integer $1 + n \cdot 2400$.

However,

$$1 + n \cdot 2400 = 24n \cdot 100 + 1.$$

Consider the integer m to be the last digit of $24n$.

Then

$$24n \cdot 100 + 1 = (\cdots m)100 + 1 = \cdots m01,$$

which means that the integers $m, 0, 1$ are the last three digits of the integer

$$24n \cdot 100 + 1.$$

For $n = 2499$ one has $24n = 59976$ which ends up with 6. Thus, the number

$$7^{4n} = 7^{9996}$$

ends up with 601.

However, $7^3 = 343$ and therefore

$$7^{9999} = 7^{9996} \cdot 7^3 = (\cdots 601)(343)$$

$$= \cdots 143,$$

where $(\cdots 143)$ is easily derived if one multiplies the numbers $(\cdots 601)$ and $(\cdots 343)$.

Therefore, the last three digits of the integer 7^{9999} are the numbers 1, 4, 3.   □

6) Determine the last three digits of the integer

$$2003^{2002^{2001}} .$$

(Canada, 2003)

Solution. It is evident that

$$2003^{2002^{2001}} \equiv 3^{2002^{2001}} \pmod{1000}. \tag{1}$$

However, we observe that $\phi(1000) = 400$. Therefore, we shall compute

$$2002^{2001} \pmod{400}.$$

But, $2000 \equiv 0 \pmod{400}$. Thus,

$$\begin{aligned}
2002^{2001} &\equiv 2^{2001} \pmod{400} \\
&\equiv 2^{1997} \cdot 16 \pmod{400} \\
&\equiv 16\,m \pmod{400},
\end{aligned} \tag{2}$$

for some integer m.

Of course, generally we know that if

$$a \equiv b \pmod{k}$$

and d is any common divisor of a and b, with $g = \gcd(k, d)$, then

$$\frac{a}{d} \equiv \frac{b}{d} \left(\bmod \frac{k}{g} \right).$$

Hence, in our case, since $400 = 16 \cdot 25$, we get

$$2^{1997} \equiv m \pmod{25}. \tag{3}$$

However, we also observe that $\phi(25) = 20$. Hence, we have

$$2^{20} \equiv 1 \pmod{25}.$$

Thus,

$$2^{2000} \equiv 1 \pmod{25}.$$

Therefore, by (3) we get

$$\frac{2^{2000}}{2^3} \equiv m \pmod{25}$$

or

$$\frac{1}{2^3} \equiv m \pmod{25}$$

or

$$22 \equiv m \pmod{25}.$$

By the above congruence and relation (2), we obtain

$$2002^{2001} \equiv 16 \cdot 22 \pmod{400}.$$

So, by (1) we get

$$2003^{2002^{2001}} \equiv 3^{2002^{2001}} \equiv 3^{16 \cdot 22} \pmod{1000}$$

$$\equiv 9^{176} \equiv (10-1)^{176} \pmod{1000}.$$

But

$$(10-1)^{176} = 10^{176} - \binom{176}{1}10^{175} + \cdots + \binom{176}{2}10^2 - \binom{176}{1}10 + 1$$

$$= 1000r + \frac{175 \cdot 176}{2}10^2 - 176 \cdot 10 + 1,$$

for some integer r_1.

Hence,

$$(10-1)^{176} = 15,400 \cdot 10^2 - 1,760 + 1$$

and then

$$2003^{2002^{2001}} \equiv 1,540 \cdot 10^3 - 1,760 + 1 \equiv 241 \pmod{1000}.$$

Therefore, the last three digits of the integer $2003^{2002^{2001}}$ are 2, 4, 1.   □

7) Prove that if the integers $a_1, a_2, \ldots, a_9$ are not divisible by 3, then

$$a_1^2 + a_2^2 + \cdots + a_9^2 \equiv 0 \pmod{3}.$$

Proof. Since the integers $a_1, a_2, \ldots, a_9$ are not divisible by 3, it follows that

$$a_i \equiv 1 \pmod{3} \quad \text{or} \quad a_i \equiv 2 \pmod{3}, \quad \text{where} \quad i = 1, 2, \ldots, 9.$$

• If $a_i \equiv 1 \pmod{3}$, then $a_i = 3\kappa + 1$, $\kappa \in \mathbb{Z}$.

Therefore,

$$a_i^2 = 9\,\kappa^2 + 6\,\kappa + 1 = 3(3\kappa^2 + 2\kappa) + 1 = 3\mu + 1,$$

where $\mu = 3\kappa^2 + 2\kappa \in \mathbb{Z}$.

• If $a_i \equiv 2 \pmod{3}$, then $a_i = 3\kappa + 2$, $\kappa \in \mathbb{Z}$.

Therefore,

$$a_i^2 = 9\kappa^2 + 12\kappa + 4 = 3(3\kappa^2 + 4\kappa + 1) + 1 = 3\lambda + 1,$$

where $\lambda = 3\,\kappa^2 + 4\,\kappa + 1 \in \mathbb{Z}$.

Thus, in every case the integer a_i^2 can be expressed in the form $3\,\rho_i + 1$, where $\rho_i \in \mathbb{Z}$ and $i = 1, 2, \ldots, 9$.

Hence,

$$a_1^2 + a_2^2 + \cdots + a_9^2 = (3\rho_1 + 1)^2 + (3\rho_2 + 1)^2 + \cdots + (3\rho_9 + 1)^2$$

$$= 9\,(\rho_1^2 + \rho_2^2 + \cdots + \rho_9^2) + 6\,(\rho_1 + \rho_2 + \cdots + \rho_9) + 9$$

$$\equiv 0 \pmod{3}.$$

Consequently,

$$a_1^2 + a_2^2 + \cdots + a_9^2 \equiv 0 \pmod{3}.$$ □

8) Let a be an integer. Prove that there are no integers b, c with $c > 1$, such that

$$(a+1)^2 + (a+2)^2 + \cdots + (a+99)^2 = b^c. \qquad \text{(a)}$$

(1998 Hungarian Mathematical Olympiad)

Proof. Assume that there exist integers a, b, c with $c > 1$, such that

$$(a+1)^2 + (a+2)^2 + \cdots + (a+99)^2 = b^c.$$

But

$$(a+1)^2 + (a+2)^2 + \cdots + (a+99)^2$$
$$= (a^2 + 2 \cdot a \cdot 1 + 1^2) + (a^2 + 2 \cdot a \cdot 2 + 2^2) \cdots + (a^2 + 2 \cdot a \cdot 99 + 99^2)$$
$$= 99a^2 + 2 \cdot \frac{99 \cdot 100}{2} \cdot a + \frac{99 \cdot 100 \cdot 199}{6}$$
$$= 33(3a^2 + 300a + 50 \cdot 199).$$

Therefore,

$$33(3a^2 + 300a + 50 \cdot 199) = b^c \qquad \text{(b)}$$

for $a, b, c \in \mathbb{Z}$ with $c > 1$. From (b) it follows that

$$3 \mid b.$$

Since $c = 2, 3, \ldots$ it follows that

$$3^2 \mid b^c.$$

Then from (b) it should hold

$$3^2 \mid 33(3a^2 + 300a + 50 \cdot 199),$$

which is not possible. Thus, (a) cannot be satisfied. $\qquad \square$

9) If m, n, m_1, n_1 are positive integers such that

$$(m+n)(m+n-1) + 2m = (m_1+n_1)(m_1+n_1-1) + 2m_1, \qquad \text{(a)}$$

prove that $m = m_1$ and $n = n_1$.

Proof. From (a) we obtain

$$(m+n)^2 + (m-n) = (m_1+n_1)^2 + (m_1-n_1)$$

or equivalently

$$(m+n+m_1+n_1)(m+n-m_1-n_1) = m_1 - n_1 - m + n. \qquad \text{(b)}$$

Case 1. Suppose that

$$m_1 - n_1 - m + n = 0. \tag{c}$$

Then because of the fact that m, n, m_1, n_1 are positive integers and thus $m + n + m_1 + n_1 \neq 0$ from (b) and (c) it follows that

$$m + n - m_1 - n_1 = 0. \tag{d}$$

From (c) and (d) by adding and subtracting, respectively, we get

$$n = n_1 \quad \text{and} \quad m = m_1.$$

Case 2. Suppose that

$$m_1 - n_1 - m + n \neq 0. \tag{e}$$

From (b) and (e) it follows that

$$m + n - m_1 - n_1 \neq 0,$$

as well as that

$$|m_1 - n_1 - m + n| = |m + n + m_1 + n_1||m + n - m_1 - n_1|$$
$$= (m + n + m_1 + n_1)|m + n - m_1 - n_1|,$$

that is,

$$|m_1 - n_1 - m + n| \geq m + n + m_1 + n_1. \tag{f}$$

The relation (f) is not possible because from the triangle inequality one has

$$|m_1 - n_1 - m + n| = |(m_1 + n) - (n_1 + m)|$$
$$< |m_1 + n| + |n_1 + m|$$
$$= (m_1 + n) + (n_1 + m),$$

that is,

$$|m_1 - n_1 - m + n| < m + n + m_1 + n_1,$$

which contradicts (f).

Hence, if (a) holds, then $m = m_1$ and $n = n_1$. $\qquad\square$

10) Prove that if m, n are integers, then the expression

$$E = m^5 + 3m^4 n - 5m^3 n^2 - 15m^2 n^3 + 4mn^4 + 12n^5$$

cannot take the value 33.

Solution. The expression E can be written as

$$
\begin{aligned}
E &= (m^5 - m^3 n^2) - 4(m^3 n^2 - mn^4) + 3(m^4 n - m^2 n^3) - 12(m^2 n^3 - n^5) \\
&= m^3(m^2 - n^2) - 4mn^2(m^2 - n^2) + 3m^2 n(m^2 - n^2) - 12n^3(m^2 - n^2) \\
&= (m^2 - n^2)(m^3 - 4mn^2 + 3m^2 n - 12n^3) \\
&= (m^2 - n^2)[(m^3 - 4mn^2) + (3m^2 n - 12n^3)] \\
&= (m^2 - n^2)[m(m^2 - 4n^2) + 3n(m^2 - 4n^2)] \\
&= (m^2 - n^2)(m + 3n)(m^2 - 4n^2) \\
&= (m - n)(m + n)(m + 3n)(m - 2n)(m + 2n).
\end{aligned}
$$

That is,
$$
E = (m - 2n)(m - n)(m + n)(m + 2n)(m + 3n).
$$

It is evident that for $n \neq 0$, $n \in \mathbb{Z}$, the expression E has been factored in five pairwise distinct factors.

Thus, the integer E has at least five pairwise different divisors. But the number 33 cannot be expressed as a product of five pairwise different factors. One can write

$$
33 = (-3) \cdot 11 \cdot (-1) \cdot 1 \quad \text{or} \quad 33 = 3 \cdot (-11) \cdot (-1) \cdot 1. \qquad \square
$$

11) Prove that
$$
(2m + 1)^{2^n} = 2^{n+2}\lambda_n + 1,
$$

for every positive integer n, with $\lambda_n \in \mathbb{Z}$.

(Elias Karakitsos, Sparta, Greece)

Proof. Let $P(n)$ denote the equality that we want to prove. By the Mathematical Induction Principle we have

- For $n = 1$, we get

$$
\begin{aligned}
(2m + 1)^2 &= 2^2 m^2 + 2^2 m + 1 = 2^2 m(m + 1) + 1 \\
&= 2^2 \cdot 2\lambda_1 + 1 = 2^3 \lambda_1 + 1 \\
&= 2^{1+2}\lambda_1 + 1 \,, \text{ where } \lambda_1 \in \mathbb{Z},
\end{aligned}
$$

which means that $P(1)$ holds true.
- We shall now prove that if $P(n)$ holds true, then $P(n + 1)$ also holds true. In other words, we shall prove that if

$$
(2m + 1)^{2^n} = 2^{n+2}\lambda_n + 1,
$$

then
$$(2m + 1)^{2^{n+1}} = 2^{n+3}\lambda_{n+1} + 1.$$

We have

$$(2m + 1)^{2^{n+1}} = (2m + 1)^{2^n \cdot 2} = [(2m + 1)^{2^n}]^2$$
$$= (2^{n+2}\lambda_n + 1)^2 = (2^{n+2}\lambda_n)^2 + 2 \cdot 2^{n+2}\lambda_n + 1$$
$$= 2^{2(n+2)}\lambda_n^2 + 2^{n+3}\lambda_n + 1 = 2^{2n+4}\lambda_n^2 + 2^{n+3}\lambda_n + 1$$
$$= 2^{n+3}(2^{n+1}\lambda_n^2 + \lambda_n) + 1 = 2^{n+3}\lambda_{n+1} + 1.$$

Hence, with the assumption that $P(n)$ holds true, it follows that $P(n + 1)$ also holds true. Thus, $P(n)$ holds true for every positive integer n.   □

12) Consider m, $n \in \mathbb{N}$, such that $m + n$ is odd. Prove that there is no $A \subseteq \mathbb{N}$ such that for all x, $y \in \mathbb{N}$, if $|x - y| = m$ then $x \in A$ or $y \in A$, and if $|x - y| = n$ then $x \notin A$ or $y \notin A$.

(Dimiter Skordev, Problem No. 11074, Amer. Math. Monthly, 2004. Solution by Gerry Myerson, Amer. Math. Monthly, 113(2006), p. 367)

Solution. If we assume that such a set A exists, since $y \notin A$ and $y + m \in A$, it follows that $A \neq \emptyset$.

Suppose that x is an element in A. Now, $x + n \notin A$, thus $x + n + m \in A$ and therefore $x + 2n + m \notin A$, that is, $x + 2n \in A$. If one repeats the same argument, it follows that $x + mn \in A$ if and only if m is an even integer.

If, however, $x + m + n \in A$, it follows that $x + m \notin A$, and thus $x + 2m \in A$, starting from $x \in A$. If one repeats the same argument, one obtains that $x + mn \in A$ if and only if n is even. So, both m and n have to be even, which is impossible, since by the hypothesis $m + n$ is an odd integer.   □

13) Let p be an odd prime number. If r_j is the remainder when the integer $j^{p-1} - 1/p$ is divided by p, where $j = 1, 2, \ldots, p - 1$, prove that

$$r_1 + 2r_2 + \cdots + (p - 1)r_{p-1} \equiv \frac{p + 1}{2} \pmod{p}.$$

(Dorin Andrica, "Babes-Bolyai" University, Cluj-Napoca, Romania)

Proof. For $j = 1, 2, \ldots, p - 1$, we have

$$\frac{j^{p-1} - 1}{p} = a_j p + r_j,$$

for some integer a_j. It follows that

$$\frac{j^p - j}{p} = j a_j p + j r_j,$$

hence

$$\frac{j^p - j + (p-j)^p - (p-j)}{p} = ja_j p + j r_j + (p-j)a_{p-j}p + (p-j)r_{p-j}.$$

We obtain

$$\frac{j^p + (p-j)^p}{p} = ja_j p + j r_j + (p-j)a_{p-j}p + (p-j)r_{p-j} + 1.$$

Because

$$j^p + (p-j)^p = \binom{p}{0}p^p - \binom{p}{1}p^{p-1}j + \cdots + \binom{p}{p-1}pj^{p-1},$$

it is evident that $p^2 \mid j^p + (p-j)^p$. Therefore, we get

$$jr_j + (p-j)r_{p-j} + 1 \equiv \pmod{p},$$

for all $j = 1, 2, \ldots, p-1$.

Adding up all these relations it follows that

$$2(r_1 + 2r_2 + \cdots + (p-1)r_{p-1}) \equiv -(p-1) \pmod{p},$$

hence

$$r_1 + 2r_2 + \cdots + (p-1)r_{p-1} \equiv \frac{p+1}{2} \pmod{p}. \qquad \square$$

14) Find all possible decimal digits a such that, for a given n, the decimal expansions of 2^n and 5^n both begin by a, and give a necessary and sufficient condition to determine all such integers n.

(Konstantinos Drakakis, University College Dublin, Ireland; Newsletter of the European Mathematical Society, Issue 73, 2009, Problem 48, p. 54)

Solution. Set

$$2^n = 10^{m(n)}(a(n) + a'(n)), \quad 5^n = 10^{l(n)}(b(n) + b'(n)),$$

so that $m, l \in \mathbb{N} \cup \{0\}$, $a, b \in \{1, \ldots, 9\}$, and $a', b' \in [0, 1)$. It follows that

$$10^n = 5^n 2^n \Leftrightarrow (a + a')(b + b') = 10^{n-m-l}.$$

However, $1 \le (a+a')(b+b') < 100$, forcing $n - m - l$ to be either 0 or 1 only. The former case leads to

$$(a+a')(b+b') = 1 \Leftrightarrow a = b = 1, \ a' = b' = 0 \Leftrightarrow 2^n = 5^n = 1 \Leftrightarrow n = 0,$$

while the latter case, setting $b = a$, leads to

$$a^2 \le (a+a')(a+b') = 10 < (a+1)^2 \Leftrightarrow a = 3.$$

This is indeed a possibility, as, for example, $2^5 = 32$, $5^5 = 3125$. But which values of n lead to such a coincidence?

For such an n, it helps to rewrite

$$2^n = 10^{m(n)}(3 + a'(n)) = 10^{m(n)}(\sqrt{10} + 3 + a'(n) - \sqrt{10})$$

$$= 10^{m(n)+\frac{1}{2}}(1 + u(n)), \quad u(n) := \frac{3 + a'(n)}{\sqrt{10}} - 1,$$

and similarly

$$5^n = 10^{l(n)+\frac{1}{2}}(1 + v(n)), \quad v(n) := \frac{3 + b'(n)}{\sqrt{10}} - 1.$$

Of course, u and v are not independent:

$$10^n = 5^n 2^n = 10^{m+l+1}(1 + u)(1 + v) \Leftrightarrow (1 + u)(1 + v) = 1$$

$$\Leftrightarrow v = -\frac{u}{1 + u},$$

and, furthermore,

$$\frac{3}{\sqrt{10}} - 1 \le u, v < \frac{4}{\sqrt{10}} - 1 \Leftrightarrow \frac{3}{\sqrt{10}} - 1 \le u, -\frac{u}{1 + u} < \frac{4}{\sqrt{10}} - 1.$$

Two inequalities for u have thus been obtained. It follows from the first that

$$2^n = 10^{m+\frac{1}{2}}(1 + u) \Leftrightarrow n \log 2 = m + \frac{1}{2} + \log(1 + u)$$

$$\Leftrightarrow \log 3 - \frac{1}{2} \le n \log 2 - m - \frac{1}{2} < \log 4 - \frac{1}{2},$$

which implies that

$$\left[n \log 2 - m - \frac{1}{2} \right] = 0 \Leftrightarrow m = \left[n \log 2 - \frac{1}{2} \right] = \lfloor n \log 2 \rfloor.$$

Here, the square brackets denote rounding to the nearest integer, while the L-shaped brackets denote the floor function, i.e., truncating to the largest integer not exceeding the given number. This result is obtained because

$$-1 < \log 3 - \frac{1}{2} < 0 < \log 4 - \frac{1}{2} < 1,$$

and it implies that

$$\log 3 \le n \log 2 - \left[n \log 2 - \frac{1}{2} \right] < \log 4.$$

It follows from the second inequality that

$$\frac{3}{\sqrt{10}} - 1 \leq -\frac{u}{1+u} < \frac{4}{\sqrt{10}} - 1$$

$$\Leftrightarrow \frac{3}{\sqrt{10}} \leq \frac{1}{1+u} < \frac{4}{\sqrt{10}} \Leftrightarrow \frac{\sqrt{10}}{4} - 1 < u \leq \frac{\sqrt{10}}{3} - 1,$$

and, combining the two inequalities,

$$\frac{3}{\sqrt{10}} \leq 1 + u = 2^n 10^{-\frac{1}{2} - \left[n \log 2 - \frac{1}{2}\right]} \leq \frac{\sqrt{10}}{3},$$

which, by taking logarithms, yields

$$\log 3 - \frac{1}{2} \leq n \log 2 - \left[n \log 2 - \frac{1}{2}\right] - \frac{1}{2} < \frac{1}{2} - \log 3$$

$$\Leftrightarrow \log 3 \leq n \log 2 - \left[n \log 2 - \frac{1}{2}\right] < 1 - \log 3.$$

To summarize,

$$|s - [s]| < \frac{1}{2} - \log 3, \quad s := n \log 2 - \frac{1}{2} \approx 0.0228787453,$$

is a necessary and sufficient condition for 2^n and 5^n to have the same first decimal digit in their decimal expansions. The first few such values of n can be found using a computer:

$$n = 5, 15, 78, 88, 98, 108, 118, 181, 191, 201, \ldots . \qquad \Box$$

15) Let a, n be positive integers such that a^n is a perfect number. Prove that

$$a^{n/\mu} > \frac{\mu}{2},$$

where μ denotes the number of distinct prime divisors of a^n.

(M.Th. Rassias, Proposed problem W. 27, Octogon Mathematical Magazine, 17(1)(2009), p. 311)

Proof. It is a known fact that for every perfect number m it holds

$$\sum_{d \mid m} \frac{1}{d} = 2.$$

Therefore,

$$\sum_{d \mid a^n} \frac{1}{d} = 2.$$

Assume that

$$a^n = p_1^{k_1} p_2^{k_2} \cdots p_\mu^{k_\mu}$$

is the unique prime factorization of the number a^n. It is obvious that

$$\frac{1}{p_1^{k_1}} + \frac{1}{p_2^{k_2}} + \cdots + \frac{1}{p_\mu^{k_\mu}} < 2. \tag{1}$$

From Cauchy's arithmetic-geometric mean inequality it follows

$$\sqrt[\mu]{p_1^{k_1} p_2^{k_2} \cdots p_\mu^{k_\mu}} \geq \frac{\mu}{\frac{1}{p_1^{k_1}} + \frac{1}{p_2^{k_2}} + \cdots + \frac{1}{p_\mu^{k_\mu}}},$$

that is,

$$\frac{1}{p_1^{k_1}} + \frac{1}{p_2^{k_2}} + \cdots + \frac{1}{p_\mu^{k_\mu}} \geq \frac{\mu}{\sqrt[\mu]{p_1^{k_1} p_2^{k_2} \cdots p_\mu^{k_\mu}}}.$$

Therefore,

$$\frac{1}{p_1^{k_1}} + \frac{1}{p_2^{k_2}} + \cdots + \frac{1}{p_\mu^{k_\mu}} \geq \frac{\mu}{a^{n/\mu}}. \tag{2}$$

From (1) and (2) it follows that

$$2 > \frac{\mu}{a^{n/\mu}} \Leftrightarrow a^{n/\mu} > \frac{\mu}{2}. \qquad \square$$

16) Prove that the sum

$$S = \frac{1}{2} + \frac{1}{3} + \cdots + \frac{1}{n},$$

where $n > 1$ cannot be an integer.

Proof. If we consider an integer A and we prove that the product $A \cdot S$ is not an integer, then we will have proved that S is also not an integer. Therefore, our purpose is to consider the suitable integer A.

Let $k \in \mathbb{Z}$ be the greatest integer, such that $2^k \leq n$. Consider, also, the integer B which represents the product of all odd numbers which do not exceed n.

Obviously $2^{k-1} B \in \mathbb{Z}$. This is exactly the integer A we are looking for.

$$A \cdot Ss = 2^{k-1} B \cdot S = \frac{2^{k-1}(3 \cdot 5 \cdots \lambda)}{2} + \frac{2^{k-1}(3 \cdot 5 \cdots \lambda)}{3} + \cdots$$
$$+ \frac{2^{k-1}(3 \cdot 5 \cdots \lambda)}{2^k} + \cdots + \frac{2^{k-1}(3 \cdot 5 \cdots \lambda)}{n},$$

where λ is the greatest odd integer, which does not exceed n.

All the terms in the above summation are integers with the exception of the number

$$\frac{2^{k-1}(3 \cdot 5 \cdots \lambda)}{2^k}.$$

Therefore, $A \cdot S \notin \mathbb{Z}$. Hence, $S \notin \mathbb{Z}$. $\qquad \square$

17) Let $n \geq 3$ be an odd positive integer. Prove that the set

$$A = \left\{ \binom{n}{1}, \binom{n}{2}, \ldots, \binom{n}{\frac{n-1}{2}} \right\}$$

contains an odd number of odd integers.

(Revista Matematică Timisoara, No.2 (1984), Problem 5346)

Proof. In order to prove that the set A contains an odd number of odd integers, it is enough to prove that the sum of its elements is an odd integer, because in every other case the sum of the elements of A is an even number. Let n be an odd positive integer with $n \geq 3$.

Set

$$S_n = \binom{n}{1} + \binom{n}{2} + \cdots + \binom{n}{(n-1)/2}.$$

It is enough to prove that S_n is an odd integer. We know that

$$\binom{n}{k} = \binom{n}{n-k}$$

for $k, n \in \mathbb{N}$ with $k \leq n$.

Therefore, it holds

$$\binom{n}{1} + \binom{n}{2} + \cdots + \binom{n}{\frac{n-1}{2}} = \binom{n}{n-1} + \binom{n}{n-2} + \cdots + \binom{n}{\frac{n+1}{2}}.$$

That is,

$$2S_n = \binom{n}{1} + \binom{n}{2} + \cdots + \binom{n}{n-1}$$

$$= \left[\binom{n}{1} + \binom{n}{2} + \cdots + \binom{n}{n} \right] - \left[\binom{n}{0} + \binom{n}{n} \right]$$

$$= 2^n - 2.$$

Therefore,

$$S_n = 2^{n-1} - 1.$$

It is clear that the number S_n is odd.   □

18) Prove that every positive rational number can be expressed in the form

$$\frac{a^3 + b^3}{c^3 + d^3},$$

where a, b, c, d are positive integers.

Proof. Firstly, we will prove that every rational number in the open interval $(1, 2)$ can be expressed in the form

$$\frac{a^3 + b^3}{a^3 + d^3}.$$

Let $\frac{\kappa}{\lambda} \in (1, 2)$, where κ, λ are positive integers with $\gcd(\kappa, \lambda) = 1$. Consider a, b, d such that

$$b \neq d \text{ and } a^2 - ab + b^2 = a^2 - ad + d^2, \text{ then } b + d = a.$$

Thus, we obtain

$$\frac{a^3 + b^3}{a^3 + d^3} = \frac{a + b}{a + d} = \frac{a + b}{2a - b}.$$

Therefore, for $a + b = 3\kappa$ and $2a - b = 3\lambda$, we obtain

$$\left. \begin{array}{r} a + b = 3\kappa \\ 2a - b = 3\lambda \end{array} \right\} \Leftrightarrow \left. \begin{array}{l} a = \kappa + \lambda \\ b = 2\kappa - \lambda. \end{array} \right\}$$

Thus, for $a = \kappa + \lambda$, $b = 2\kappa - \lambda$ the claim is proved.

Suppose now that r is any positive rational number, $r > 0$, and q_1, q_2 two positive integers such that

$$1 < \frac{q_1^3}{q_2^3} r < 2.$$

Therefore, there exist positive integers a, b, d such that

$$\frac{q_1^3}{q_2^3} r = \frac{a^3 + b^3}{a^3 + d^3}$$

and thus

$$r = \frac{(aq_2)^3 + (bq_2)^3}{(aq_1)^3 + (dq_1)^3}. \qquad \square$$

19) Prove that every composite positive integer can be represented in the form

$$xy + xz + yz + 1,$$

where x, y, z are positive integers.

(Problem 1, The Forty-Ninth William Lowell Putnam Mathematical Competition, 1988)

Proof. It is evident that every composite positive integer can be expressed in the form $a \cdot b$, where a and b are positive integers with $a, b \geq 2$. Therefore, if c is a composite positive integer, then

$$c = a \cdot b.$$

Set $z = 1$ in the expression $xy + xz + yz + 1$. Then

$$xy + x + y + 1 = (x+1)(y+1).$$

However,

$$(x+1)(y+1) = a \cdot b$$

for $x = a - 1$ and $y = b - 1$. Therefore, every composite positive integer can be represented in the form

$$xy + xz + yz + 1,$$

where $x = a - 1$, $y = b - 1$ and $z = 1$.   □

20) If for the rational number x the value of the expression

$$2x^4 + 3x + 1$$

is an integer, prove that x is also an integer.

(School of Aviation Engineers of Greece, Entrance Examinations, 1968)

Proof. Since x is a rational number, it follows that $x = \frac{a}{b}$ where a, b are integers, such that $b \neq 0$ and $\gcd(a, b) = 1$.

In addition, by hypothesis, it is true that

$$2x^4 + 3x + 1 = \lambda, \quad \lambda \in \mathbb{Z}. \tag{a}$$

Thus, (a) can be written in the form

$$2\left(\frac{a}{b}\right)^4 + 3 \cdot \frac{a}{b} + 1 = \lambda,$$

that is,

$$2a^4 + 3ab^3 + b^4 = \lambda b^4.$$

Therefore,

$$(\lambda b - b - 3a)b^3 = 2a^4. \tag{b}$$

However, $\gcd(a, b) = 1$, thus $\gcd(a^4, b^3) = 1$.

But $\lambda b - b - 3a \in \mathbb{Z}$ and

$$b^3 \mid (\lambda b - b - 3a)b^3,$$

thus

$$b^3 \mid 2a^4. \tag{c}$$

From (c) and the fact that $\gcd(a^4, b^3) = 1$, it follows that

$$b^3 \mid 2.$$

This implies that

$$b^3 = \pm 1 \quad \text{or} \quad b^3 = \pm 2.$$

The case $b^3 = \pm 2$ is impossible in the set of integers.
Thus, $b^3 = \pm 1$, that is, $b = \pm 1$. Hence, we obtain

$$x = \frac{a}{b} = \pm a \in \mathbb{Z},$$

that is, x is an integer.   □

21) Consider the sequence (x_n) of real numbers, which is defined by the recursive formula

$$x_1 = 0, \ x_{n+1} = 5x_n + \sqrt{24x_n^2 + 1},$$

where $n = 1, 2, 3, \ldots$.
Prove that all the terms of the sequence are integers.

Proof. It is true that

$$(x_{n+1} - 5x_n)^2 = (\sqrt{24x_n^2 + 1})^2$$
$$\Leftrightarrow x_{n+1}^2 - 10x_n x_{n+1} + 25x_n^2 = 24x_n^2 + 1$$
$$\Leftrightarrow x_{n+1}^2 + x_n^2 - 10x_n x_{n+1} - 1 = 0$$
$$\Leftrightarrow x_{n+1}^2 - 10x_n x_{n+1} + x_n^2 - 1 = 0. \tag{a}$$

Therefore, for $n \geq 2$ we obtain that

$$x_n^2 - 10x_n x_{n-1} + x_{n-1}^2 - 1 = 0. \tag{b}$$

Consider the equation

$$y^2 - 10x_n y + x_n^2 - 1 = 0. \tag{c}$$

From (a), (b) it follows that the numbers x_{n+1}, x_{n-1} are roots of the equation (c). It is obvious that $x_{n+1} > x_n > x_{n-1}$. Thus, the roots x_{n+1}, x_{n-1} are distinct.

Therefore, for $n \geq 2$ it follows from Viète's formulae that

$$x_{n+1} + x_{n-1} = 10x_n. \tag{d}$$

For $n = 2$ we obtain

$$x_3 + x_1 = 10x_2. \tag{e}$$

But, we know that $x_1 = 0$, that is,

$$x_2 = 5x_1 + \sqrt{24x_1^2 + 1} = 1.$$

Thus, from (e), one has $x_3 = 10$. Therefore, since $x_1 = 0$, $x_2 = 1$, $x_3 = 10$, that is, x_1, x_2, $x_3 \in \mathbb{Z}$ and $x_{n+1} + x_{n-1} = 10x_n$, it follows that all the terms of the sequence (x_n) are integers.   □

22) Let $a \in \mathbb{Z}$, $n, k \in \mathbb{N}$ with $k \equiv -a^2 \pmod{2n}$. Prove that

$$\sqrt{2}\sqrt{k+a^2} \leq \frac{k+a^2}{2n} + n \leq \frac{1}{2}(k+a^2) + 1.$$

Proof. It is evident that $n \geq 1$ and $k + a^2 \geq 2n$. Therefore, it holds

$$\frac{1}{2}(k+a^2) \geq n \geq 1. \qquad \text{(a)}$$

Consider the function

$$f : \left[1, \frac{1}{2}(k+a^2)\right] \to \mathbb{R}, \text{ defined by } f(x) = \frac{k+a^2}{2x} + x.$$

We will examine the monotonicity of the function f. We have

$$f'(x) = -\frac{k+a^2}{2x^2} + 1.$$

However, the following implications hold:

$$1 \leq x \leq \frac{1}{2}(k+a^2)$$

$$\frac{2}{k+a^2} \leq \frac{1}{x} \leq 1$$

$$\frac{4}{(k+a^2)^2} \leq \frac{1}{x^2} \leq 1$$

$$-1 \leq -\frac{1}{x^2} \leq -\frac{4}{(k+a^2)^2}$$

$$-\frac{k+a^2}{2} \leq -\frac{k+a^2}{2x^2} \leq -\frac{2(k+a^2)}{(k+a^2)^2}$$

$$-\frac{k+a^2}{2} + 1 \leq -\frac{k+a^2}{2x^2} + 1 \leq -\frac{2}{k+a^2} + 1,$$

that is,

$$-\frac{k+a^2}{2} + 1 \leq f'(x) \leq -\frac{2}{k+a^2} + 1.$$

Therefore, the derivative function f' can take both positive and negative values (because of (a)).
    For $f'(x) = 0$ we obtain

$$\frac{k+a^2}{2x^2} = 1 \Leftrightarrow x = \pm\sqrt{\frac{k+a^2}{2}}.$$

From the definition of f it must be $x \geq 1 > 0$ and hence for $f'(x) = 0$ one has

$$x = \sqrt{\frac{k + a^2}{2}} = \frac{\sqrt{2}\sqrt{k + a^2}}{2}.$$

Thus, for

$$x \leq \frac{\sqrt{2}\sqrt{k + a^2}}{2}$$

$$\Rightarrow x^2 \leq \frac{k + a^2}{2}$$

$$\Rightarrow \frac{1}{x^2} \geq \frac{2}{k + a^2}$$

$$\Rightarrow -\frac{1}{x^2} \leq -\frac{2}{k + a^2}$$

$$\Rightarrow -\frac{k + a^2}{2x^2} \leq -1$$

$$\Rightarrow -\frac{k + a^2}{2x^2} + 1 \leq 0,$$

that is,

$$f'(x) \leq 0.$$

Therefore, for

$$1 \leq x \leq \sqrt{\frac{k + a^2}{2}}$$

the function f is decreasing. Thus, it holds

$$f(1) \geq f(x) \geq f\left(\sqrt{\frac{k + a^2}{2}}\right).$$

For

$$x \geq \frac{\sqrt{2}\sqrt{k + a^2}}{2} \Rightarrow x^2 \geq \frac{k + a^2}{2}$$

$$\Rightarrow -\frac{1}{x^2} \geq -\frac{2}{k + a^2}$$

$$\Rightarrow -\frac{k + a^2}{2x^2} \geq -1$$

$$\Rightarrow -\frac{k + a^2}{2x^2} + 1 \geq 0,$$

that is,

$$f'(x) \geq 0.$$

Thus, for

$$\sqrt{\frac{k+a^2}{2}} \leq x \leq \frac{1}{2}(k+a^2)$$

the function f is increasing.

Therefore, the function f has total (global) maximum for

$$x = 1, \quad x = \frac{1}{2}(k+a^2)$$

and total (global) minimum for

$$x = \sqrt{\frac{k+a^2}{2}}.$$

Hence,

$$f\left(\sqrt{\frac{k+a^2}{2}}\right) \leq f(x) \leq f(1) = f\left(\frac{1}{2}(k+a^2)\right)$$

$$\Leftrightarrow \frac{k+a^2}{2\sqrt{\frac{k+a^2}{2}}} + \sqrt{\frac{k+a^2}{2}} \leq \frac{k+a^2}{2x} + x \leq \frac{k+a^2}{2} + 1$$

$$\Leftrightarrow \frac{k+a^2}{\sqrt{2}\sqrt{k+a^2}} + \frac{\sqrt{2}\sqrt{k+a^2}}{2} \leq \frac{k+a^2}{2x} + x \leq \frac{k+a^2}{2} + 1$$

$$\Leftrightarrow \sqrt{2}\sqrt{k+a^2} \leq \frac{k+a^2}{2x} + x \leq \frac{k+a^2}{2} + 1.$$

Therefore, the inequality we want to prove is valid for every real value of x, such that $x \in [1, \frac{1}{2}(k+a^2)]$ and thus it holds for every $n \in \mathbb{N}$ for which $1 \leq n \leq \frac{1}{2}(k+a^2)$. Thus,

$$\sqrt{2}\sqrt{k+a^2} \leq \frac{k+a^2}{2n} + n \leq \frac{1}{2(k+a^2)+1}. \qquad \square$$

23) Let $n_1 = \overline{abcabc}$ and $n_2 = \overline{d00d}$ be positive integers represented in the decimal system, where $a, b, c, d \in \{0, 1, 2, \ldots, 9\}$ with $a \neq 0$ and $d \neq 0$.

i) Prove that $\sqrt{n_1}$ cannot be an integer.
ii) Find all positive integers n_1 and n_2 such that $\sqrt{n_1 + n_2}$ is an integer.
iii) From all the pairs (n_1, n_2) such that $\sqrt{n_1 n_2}$ is an integer find those for which $\sqrt{n_1 n_2}$ has the greatest possible value.

(48th National Mathematical Olympiad, Suceava, 1997)

Solution.

i) It is clear that
$$n_1 = 1,001 \cdot \overline{abc}.$$
However, $1,001 = 7 \cdot 11 \cdot 13$, thus
$$n_1 = 7 \cdot 11 \cdot 13 \cdot \overline{abc}.$$

In case $\sqrt{n_1}$ is an integer, that is, n_1 is a square of a positive integer, then $\overline{abc}$ must necessarily be divisible by the numbers

$$7, 11, 13.$$

In that case $\overline{abc}$ must be divisible by $7 \cdot 11 \cdot 13$, namely, it must be divisible by $1,001$, which is impossible.

ii) We have
$$n_1 + n_2 = \overline{abcabc} + \overline{d00d}$$
$$= 1,001 \cdot \overline{abc} + 1,001 \cdot d,$$
that is,
$$n_1 + n_2 = 1,001 \cdot (\overline{abc} + d).$$

For $\sqrt{n_1 + n_2}$ to be an integer, it means that $n_1 + n_2$ is a square of a positive integer. This happens if and only if

$$\overline{abc} + d = 1,001.$$

This is valid if and only if $a = 9$, $b = 9$ and the numbers c, d are digits with sum $c + d = 11$, that is, $c = 2, d = 9$ or $c = 3, d = 8$ or $c = 4, d = 7$ or $c = 5, d = 6$ or $c = 6, d = 5$ or $c = 7, d = 4$ or $c = 8, d = 3$ or $c = 9$, $d = 2$.

iii) We know that
$$n_1 = 1,001 \cdot \overline{abc}$$
and
$$n_2 = 1,001 \cdot d.$$
Thus,
$$n_1 \cdot n_2 = 1,001^2 \cdot \overline{abc} \cdot d.$$

For $\sqrt{n_1 n_2}$ to be an integer, namely, $n_1 n_2$ to be a square of a positive integer, it means that
$$\overline{abc} \cdot d$$
is a square of a positive integer, which takes value not greater than

$$999 \cdot 9 = 8,991.$$

We can easily see that the greatest square of a positive integer less than $8,991$ is $94^2 = 2^2 \cdot 47^2$, which cannot be expressed in the representation

$$\overline{abc} \cdot d,$$

because $47^2 = 2,209$ is a four-digit integer and thus does not have the decimal form $\overline{abc}$.

However, $93^2 = 9 \cdot 961 = 3^2 \cdot 31^2$. Hence, the pair (n_1, n_2) satisfying the required property is

$$(961, 9). \qquad \square$$

24) Determine the number of real solutions a of the equation

$$\left\lfloor \frac{a}{2} \right\rfloor + \left\lfloor \frac{a}{3} \right\rfloor + \left\lfloor \frac{a}{5} \right\rfloor = a. \tag{a}$$

(Canadian Mathematical Olympiad, 1998)

Solution. The number a is an integer since it is the sum of three integers

$$\left\lfloor \frac{a}{2} \right\rfloor, \left\lfloor \frac{a}{3} \right\rfloor \text{ and } \left\lfloor \frac{a}{5} \right\rfloor.$$

Set

$$a = 30k + r,$$

where k, r are integers and $0 \leq r < 30$.

Then (a) can be written as follows

$$\left\lfloor 15k + \frac{r}{2} \right\rfloor + \left\lfloor 10k + \frac{r}{3} \right\rfloor + \left\lfloor 6k + \frac{r}{5} \right\rfloor = 30k + r,$$

that is,

$$31k + \left\lfloor \frac{r}{2} \right\rfloor + \left\lfloor \frac{r}{3} \right\rfloor + \left\lfloor \frac{r}{5} \right\rfloor = 30k + r$$

or

$$k = r - \left\lfloor \frac{r}{2} \right\rfloor - \left\lfloor \frac{r}{3} \right\rfloor - \left\lfloor \frac{r}{5} \right\rfloor.$$

To every value of r, $0 \leq r < 30$, corresponds a unique value of k and thus a unique value of a. However, r can be any one of the numbers $0, 1, 2, 3, \ldots, 29$. Hence, there exist exactly 30 real solutions a of the given equation (a). $\square$

25) Let $a, b \in \mathbb{N}$. Prove that

$$\left\lfloor \frac{2a}{b} \right\rfloor - 2 \left\lfloor \frac{a}{b} \right\rfloor = \begin{cases} 0, & \text{if } \lfloor \frac{2a}{b} \rfloor \text{ is an even integer} \\ 1, & \text{if } \lfloor \frac{2a}{b} \rfloor \text{ is an odd integer.} \end{cases}$$

Proof. Let $\left\lfloor \frac{2a}{b} \right\rfloor = k$. Then, it is clear that

$$k \leq \frac{2a}{b} < k+1.$$

Therefore, it is evident that there exists a real number r, $0 \leq r < 1$, such that

$$\frac{2a}{b} = k+r$$

or

$$\frac{a}{b} = \frac{k+r}{2}. \tag{1}$$

• If $\left\lfloor \frac{2a}{b} \right\rfloor$ is an even integer, then there exists a positive integer λ, such that $k = 2\lambda$. Thus, by (1) we have

$$\frac{a}{b} = \lambda + \frac{r}{2}. \tag{2}$$

However, since $0 \leq r/2 < 1$, by (2) it follows that

$$\left\lfloor \frac{a}{b} \right\rfloor = \lambda.$$

Thus,

$$\left\lfloor \frac{2a}{b} \right\rfloor - 2\left\lfloor \frac{a}{b} \right\rfloor = k - 2\lambda$$

$$= 2\lambda - 2\lambda = 0.$$

• If $\left\lfloor \frac{2a}{b} \right\rfloor$ is an odd integer, then similarly there exists a positive integer λ, such that $k = 2\lambda + 1$. Hence, we obtain

$$\frac{a}{b} = \lambda + \frac{r+1}{2}. \tag{3}$$

But, the fact that $0 \leq r < 1$, it follows that

$$\frac{1}{2} < \frac{r+1}{2} < 1.$$

Therefore, by (3) we get

$$\left\lfloor \frac{a}{b} \right\rfloor = \lambda.$$

Thus,

$$\left\lfloor \frac{2a}{b} \right\rfloor - 2\left\lfloor \frac{a}{b} \right\rfloor = k - 2\lambda = 2\lambda + 1 - 2\lambda = 1. \qquad \square$$

26) Prove that

$$\lfloor \sqrt{n} + \sqrt{n+1} + \sqrt{n+2} \rfloor = \lfloor \sqrt{9n+8} \rfloor \tag{1}$$

for $n = 0, 1, 2, \ldots$.

(Crux Mathematicorum 28(1)(2002). See also Amer. Math. Monthly, 1988, pp. 133–134)

Proof.
- If $n = 0, 1, 2$ the relation (1) obviously holds.
- If $n \geq 3$, $n \in \mathbb{N}$, then from Cauchy's inequality (arithmetic-geometric-harmonic mean inequality) one has

$$\frac{\sqrt{n} + \sqrt{n+1} + \sqrt{n+2}}{3} > \sqrt[3]{\sqrt{n}\sqrt{n+1}\sqrt{n+2}}$$

$$= \sqrt{\sqrt[3]{n(n+1)(n+2)}} \qquad (*)$$

$$> \sqrt{n + \frac{8}{9}}$$

$$= \frac{\sqrt{9n + 8}}{3}.$$

Therefore,
$$\sqrt{n} + \sqrt{n+1} + \sqrt{n+2} > \sqrt{9n + 8}. \qquad (2)$$

Furthermore,

$$\frac{\sqrt{n} + \sqrt{n+1} + \sqrt{n+2}}{3} < \sqrt{\frac{(\sqrt{n})^2 + (\sqrt{n+1})^2 + (\sqrt{n+2})^2}{3}}$$

$$= \sqrt{\frac{3n + 3}{3}} = \sqrt{n + 1}, \qquad (**)$$

that is,
$$\sqrt{n} + \sqrt{n+1} + \sqrt{n+2} < \sqrt{9n + 9}. \qquad (3)$$

From (2) and (3) it follows that

$$\sqrt{9n + 8} < \sqrt{n} + \sqrt{n+1} + \sqrt{n+2} < \sqrt{9n + 9}.$$

Hence,
$$\lfloor \sqrt{n} + \sqrt{n+1} + \sqrt{n+2} \rfloor = \lfloor \sqrt{9n + 8} \rfloor. \qquad \square$$

Remark 11.1. In the proof above, we used the following inequalities:

$(*)$ If $n \geq 3$, $n \in \mathbb{N}$, then

$$n(n+1)(n+2) > \left(n + \frac{8}{9}\right)^3.$$

For the proof of $(**)$ we consider the function $f : \mathbb{R} \to \mathbb{R}$ defined by

$$f(x) = (x-1)x(x+1) - \left(x - \frac{1}{9}\right)^3.$$

(**) If α, β, γ are distinct positive real numbers, then

$$\sqrt[3]{\alpha\beta\gamma} < \frac{\alpha + \beta + \gamma}{3} < \sqrt{\frac{\alpha^2 + \beta^2 + \gamma^2}{3}}.$$

27) Solve the equation

$$\lfloor 3x - 2 \rfloor - \lfloor 2x - 1 \rfloor = 2x - 6, \ x \in \mathbb{R}.$$

(Elias Karakitsos, Sparta, Greece)

Solution. Set $\lfloor 3x - 2 \rfloor = a$, where $a \in \mathbb{Z}$. Then, it follows that there exists ϑ_1, with $0 \le \vartheta_1 < 1$, such that

$$3x - 2 - \vartheta_1 = a$$

or

$$x = \frac{a + \vartheta_1 + 2}{3}.$$

Set $\lfloor 2x - 1 \rfloor = b$, where $b \in \mathbb{Z}$. Then, it follows that there exists ϑ_2, with $0 \le \vartheta_2 < 1$, such that

$$2x - 1 - \vartheta_2 = b$$

or

$$x = \frac{b + \vartheta_2 + 1}{2}.$$

Therefore,

$$a - b = 2x - 6$$

or

$$2x = a - b + 6$$

or

$$x = \frac{a - b + 6}{2}.$$

Thus,

$$\frac{a + \vartheta_1 + 2}{3} = \frac{b + \vartheta_2 + 1}{2} = \frac{a - b + 6}{2}.$$

Hence,

$$2a + 2\vartheta_1 + 4 = 3a - 3b + 18$$

or

$$\vartheta_1 = \frac{a - 3b + 14}{2}.$$

Similarly

$$\frac{b + \vartheta_2 + 1}{2} = \frac{a - b + 6}{2}$$

or

$$b + \vartheta_2 + 1 = a - b + 6$$

or

$$\vartheta_2 = a - 2b + 5.$$

Since $0 \leq \vartheta_1 < 1$, we obtain

$$0 \leq \frac{a - 3b + 14}{2} < 1$$

or

$$-14 \leq a - 3b < -12.$$

Thus,

$$a - 3b = -14 \quad \text{or} \quad a - 3b = -13.$$

Furthermore, since $0 \leq \vartheta_2 < 1$, we obtain

$$0 \leq a - 2b + 5 < 1$$

or

$$-5 \leq a - 2b < -4$$

or

$$a - 2b = -5.$$

By solving the systems of equations

$$a - 3b = -14$$
$$a - 2b = -5$$

and

$$a - 3b = -13$$
$$a - 2b = -5$$

it follows that $a = 13$, $b = 9$ and $a = 11$, $b = 8$, respectively.

The above solutions are acceptable, since a and b are integer numbers.

- For $a = 13$, $b = 9$ the solution of the initial equation is

$$x = \frac{a - b + 6}{2} = \frac{13 - 9 + 6}{2} = \frac{10}{2} = 5.$$

- For $a = 11$, $b = 8$ the solution of the initial equation is

$$x = \frac{a - b + 6}{2} = \frac{11 - 8 + 6}{2} = \frac{9}{2} = 4.5.$$

Therefore, the real solutions of the equation are the numbers 4.5 and 5.   □

28) Solve the equation

$$\lfloor x \rfloor^2 = \lfloor 3x - 2 \rfloor, x \in \mathbb{R}. \tag{1}$$

(Elias Karakitsos, Sparta, Greece)

Solution. Set $\lfloor x \rfloor = a$, where $a \in \mathbb{Z}$. Then it is evident that there exists θ_1, such that

$$x = a + \theta_1, \text{ with } 0 \le \theta_1 < 1.$$

In addition, by (1) we have

$$a^2 = \lfloor 3x - 2 \rfloor.$$

Therefore, there exists θ_2, such that

$$a^2 = 3x - 2 - \theta_2.$$

Thus,

$$x = \frac{a^2 + \theta_2 + 2}{3}.$$

By the above equalities, it follows that

$$\frac{a^2 + \theta_2 + 2}{3} = a + \theta_1$$

or

$$3\theta_1 - \theta_2 = a^2 - 3a + 2.$$

However, since $0 \le \theta_1 < 1$ and $0 \le \theta_2 < 1$, we obtain that

$$-1 < 3\theta_1 - \theta_2 < 3.$$

Hence,

$$-1 < a^2 - 3a + 2 < 3$$

and thus

$$a^2 - 3a + 2 = 0 \text{ or } 1 \text{ or } 2.$$

If $a^2 - 3a + 2 = 0$, then $a = 2$ or $a = 1$ which are both acceptable values since $a \in \mathbb{Z}$.

If $a^2 - 3a + 2 = 1$, then $a = (3 \pm \sqrt{5})/2 \notin \mathbb{Z}$, which is not acceptable.

If $a^2 - 3a + 2 = 2$, then $a = 0$ or $a = 3$, which are both acceptable values.

Therefore, we shall examine the cases when $a = 0$ or 1 or 2 or 3.

• For $a = 0$, we have $x = a + \theta_1 = \theta_1$ and

$$3\theta_1 - \theta_2 = 2$$

or

$$\theta_2 = 3\theta_1 - 2.$$

However, since $0 \leq \theta_2 < 1$, we get

$$0 \leq 3\theta_1 - 2 < 1$$

or

$$\frac{2}{3} \leq \theta_1 < 1.$$

Hence, the real numbers $x = \theta_1$, with $2/3 \leq \theta_1 < 1$, are solutions of (1).

• For $a = 1$, we have $x = a + \theta_1 = 1 + \theta_1$ and

$$3\theta_1 - \theta_2 = 1^2 - 3 \cdot 1 + 2$$

or

$$\theta_2 = 3\theta_1.$$

However, since $0 \leq \theta_2 < 1$, we get

$$0 \leq \theta_1 < \frac{1}{3}.$$

Hence, the real numbers $x = 1 + \theta_1$, with $0 \leq \theta_1 < \frac{1}{3}$, are solutions of (1).

• For $a = 2$, we have $x = a + \theta_1 = 2 + \theta_1$ and

$$3\theta_1 - \theta_2 = 2^2 - 3 \cdot 2 + 2$$

or

$$\theta_2 = 3\theta_1.$$

However, since $0 \leq \theta_2 < 1$, we get

$$0 \leq \theta_1 < \frac{1}{3}.$$

Hence, the real numbers $x = 2 + \theta_1$, with $0 \leq \theta_1 < 1/3$, are solutions of (1)

• For $a = 3$, we have $x = a + \theta_1 = 3 + \theta_1$ and

$$3\theta_1 - \theta_2 = 3^2 - 3 \cdot 3 + 2$$

or

$$\theta_2 = 3\theta_1 - 2.$$

However, since $0 \leq \theta_2 < 1$, we get

$$\frac{2}{3} \leq \theta_1 < 1.$$

Hence, the real numbers $x = 3 + \theta_1$, with $2/3 \leq \theta_1 < 1$, are solutions of (1).

Therefore, the real numbers

$$x = \theta_1, \ \frac{2}{3} \leq \theta_1 < 1$$

$$x = 1 + \theta_1, \ 0 \le \theta_1 < \frac{1}{3}$$

$$x = 2 + \theta_1, \ 0 \le \theta_1 < \frac{1}{3}$$

$$x = 3 + \theta_1, \ \frac{2}{3} \le \theta_1 < 1$$

are all the solutions of (1).   □

29) Solve the equation

$$\lfloor x^2 - 3x + 2 \rfloor = 3x - 7, \quad x \in \mathbb{R}.$$

(Elias Karakitsos, Sparta, Greece)

Solution. Set $\lfloor x^2 - 3x + 2 \rfloor = a$, where $a \in \mathbb{Z}$. Then, it follows that there exists ϑ, with $0 \le \vartheta < 1$, such that

$$x^2 - 3x + 2 - \vartheta = a.$$

Thus,

$$\vartheta = x^2 - 3x + 2 - a. \tag{1}$$

Furthermore,

$$3x - 7 = a$$

or

$$x = \frac{a + 7}{3}. \tag{2}$$

By (1) and (2), it follows that

$$\vartheta = \left(\frac{a+7}{3}\right)^2 - 3\left(\frac{a+7}{3}\right) + 2 - a$$

or

$$\vartheta = \frac{a^2 - 4a + 4}{9}.$$

However, since $0 \le \vartheta < 1$, we get

$$0 \le \frac{a^2 - 4a + 4}{9} < 1.$$

Thus, it is evident that

$$0 \le a^2 - 4a + 4$$

and

$$a^2 - 4a - 5 < 0.$$

Therefore, in order to determine the integer values of a, it suffices to solve the following system:

$$a^2 - 4a + 4 \geq 0$$

$$a^2 - 4a - 5 < 0.$$

However, $a^2 - 4a + 4 \geq 0$ always holds true, since $(a - 2)^2 \geq 0$, for every real value of a. Furthermore, by $a^2 - 4a - 5 < 0$ it follows that

$$-1 < a < 5,$$

and hence $a = 0$ or 1 or 2 or 3 or 4. Therefore, the solutions of the equation can be derived by calculating the value of $x = (a + 7)/3$, for $a = 0, 1, 2, 3, 4$.

- For $a = 0$, we have $x = \frac{0+7}{3} = \frac{7}{3}$.
- For $a = 1$, we have $x = \frac{1+7}{3} = \frac{8}{3}$.
- For $a = 2$, we have $x = \frac{2+7}{3} = \frac{9}{3} = 3$.
- For $a = 3$, we have $x = \frac{3+7}{3} = \frac{10}{3}$.
- For $a = 4$, we have $x = \frac{4+7}{3} = \frac{11}{3}$.

Therefore, the real solutions of the equation are the numbers

$$\frac{7}{3}, \frac{8}{3}, 3, \frac{10}{3} \text{ and } \frac{11}{3}. \qquad \square$$

30) Prove that for every real number x and a given positive integer n it holds

$$\lfloor x \rfloor + \left\lfloor x + \frac{1}{n} \right\rfloor + \left\lfloor x + \frac{2}{n} \right\rfloor + \cdots + \left\lfloor x + \frac{n-1}{n} \right\rfloor = \lfloor nx \rfloor.$$

(Charles Hermite, 1822–1901)

Proof. Consider the real-valued function f of the real variable x defined by

$$f(x) = \lfloor x \rfloor + \left\lfloor x + \frac{1}{n} \right\rfloor + \left\lfloor x + \frac{2}{n} \right\rfloor + \cdots + \left\lfloor x + \frac{n-1}{n} \right\rfloor - \lfloor nx \rfloor. \qquad (1)$$

It follows that

$$f\left(x + \frac{1}{n}\right) = \left\lfloor x + \frac{1}{n} \right\rfloor + \left\lfloor x + \frac{1}{n} + \frac{1}{n} \right\rfloor + \left\lfloor x + \frac{1}{n} + \frac{2}{n} \right\rfloor$$

$$+ \cdots + \left\lfloor x + \frac{1}{n} + \frac{n-2}{n} \right\rfloor + \left\lfloor x + \frac{1}{n} + \frac{n-1}{n} \right\rfloor - \left\lfloor n\left(x + \frac{1}{n}\right) \right\rfloor$$

$$= \left\lfloor x + \frac{1}{n} \right\rfloor + \left\lfloor x + \frac{2}{n} \right\rfloor + \left\lfloor x + \frac{3}{n} \right\rfloor$$

$$+ \cdots + \left\lfloor x + \frac{n-1}{n} \right\rfloor + \lfloor x + 1 \rfloor - \lfloor nx + 1 \rfloor$$

$$= \left\lfloor x + \frac{1}{n} \right\rfloor + \left\lfloor x + \frac{2}{n} \right\rfloor + \cdots + \left\lfloor x + \frac{n-1}{n} \right\rfloor + \lfloor x \rfloor + 1 - \lfloor nx \rfloor - 1$$

$$= f(x).$$

Therefore, for a given positive integer n, it follows that

$$f\left(x + \frac{1}{n}\right) = f(x)$$

for all real values of x.

This implies that f is a periodic function with period $1/n$.

However, for every real value of x such that $x \in [0, 1/n)$ the function f vanishes, because every term in the right-hand side of (1) equals zero. Since $f(x) = 0$ for every $x \in [0, 1/n)$ and f is periodic with period $1/n$, it follows that $f(x) = 0$ for every

$$x \in \left[\frac{k}{n}, \frac{k+1}{n}\right) \quad \text{for } k \in \mathbb{Z}.$$

Thus,

$$f(x) = 0 \tag{2}$$

for every real number x.

From (1) and (2) it follows that

$$\lfloor x \rfloor + \left\lfloor x + \frac{1}{n} \right\rfloor + \left\lfloor x + \frac{2}{n} \right\rfloor + \cdots + \left\lfloor x + \frac{n-1}{n} \right\rfloor = \lfloor nx \rfloor. \qquad \square$$

31) Let k be a positive integer. Prove that there exist polynomials $P_0(n), P_1(n), \ldots, P_{k-1}(n)$ (which may depend on k) such that for any integer n,

$$\left\lfloor \frac{n}{k} \right\rfloor^k = P_0(n) + P_1(n) \left\lfloor \frac{n}{k} \right\rfloor + \cdots + P_{k-1}(n) \left\lfloor \frac{n}{k} \right\rfloor^{k-1}.$$

(Problem B5, The Sixty-Eighth William Lowell Putnam Mathematical Competition, 2007. Amer. Math. Monthly, 115(2008), pp. 732, 737)

Proof. Assume that

$$x = \left\lfloor \frac{n}{k} \right\rfloor.$$

It follows that x must be equal to exactly one of the numbers

$$\frac{n}{k}, \frac{n-1}{k}, \frac{n-2}{k}, \ldots, \frac{n-k+1}{k}.$$

Thus,

$$\left(x - \frac{n}{k}\right)\left(x - \frac{n-1}{k}\right)\left(x - \frac{n-2}{k}\right)\cdots\left(x - \frac{n-k+1}{k}\right) = 0.$$

If we expand the above product and bring x^k to one side of the equation, we obtain an equation of the form

$$x^k = P_0(n) + P_1(n)x + P_2(n)x^2 + \cdots + P_{k-1}(n)x^{k-1},$$

namely,

$$\left\lfloor\frac{n}{k}\right\rfloor^k = P_0(n) + P_1(n)\left\lfloor\frac{n}{k}\right\rfloor + \cdots + P_{k-1}(n)\left\lfloor\frac{n}{k}\right\rfloor^{k-1}. \qquad \square$$

32) A rational number $r = a/b$, where a, b are coprime positive integers, is called *good* if and only if $r > 1$ and there exist integers N, c, such that for every positive integer $n \geq N$, it holds

$$|\{r^n\} - c| \leq \frac{1}{2(a+b)},$$

where $\{r\} = r - \lfloor r \rfloor$.
Prove that every *good* rational number is an integer.

(Chinese National Team Selection Contest, 2007)

Proof. Let us suppose that the rational number r is *good* and set

$$A_n = \lfloor r^{n+1} \rfloor - \lfloor r^n \rfloor.$$

Therefore, we get

$$\begin{aligned}
|(r-1)r^n - A_n| &= |r^{n+1} - \lfloor r^{n+1} \rfloor - (r^n - \lfloor r^n \rfloor)| \\
&= |\{r^{n+1}\} - \{r^n\}| \\
&= |\{r^{n+1}\} - c - (\{r^n\} - c)| \\
&\leq |\{r^{n+1}\} - c| + |\{r^n\} - c| \\
&\leq \frac{1}{2(a+b)} + \frac{1}{2(a+b)} = \frac{1}{a+b}.
\end{aligned}$$

However, since $r^n = a^n/b^n$, it is evident that the denominator of the rational number $|(r-1)r^n - A_n|$ is a power of the positive integer b. Thus,

$$|(r-1)r^n - A_n| < \frac{1}{a+b}. \qquad (1)$$

In addition, it holds

$$(r-1)r^{n+1} = (r-1)r^n \frac{a}{b}$$

or

$$b(r-1)r^{n+1} = a(r-1)r^n.$$

Therefore,

$$
\begin{aligned}
|bA_{n+1} - aA_n| &= |bA_{n+1} - b(r-1)r^{n+1} + a(r-1)r^n - aA_n| \\
&= |b(A_{n+1} - (r-1)r^{n+1}) - a(A_n - (r-1)r^n)| \\
&\leq |b(A_{n+1} - (r-1)r^{n+1})| + |a(A_n - (r-1)r^n)|.
\end{aligned}
$$

Hence, by (1) it follows that

$$|bA_{n+1} - aA_n| < \frac{b}{a+b} + \frac{a}{a+b} = 1.$$

Thus, evidently

$$|bA_{n+1} - aA_n| = 0$$

or

$$A_{n+1} = \frac{a}{b} A_n.$$

By the above relation, it is clear that

$$A_{n+m} = \frac{a^m}{b^m} A_n,$$

for every positive integer m.

If $b > 1$, for m sufficiently large, b^m fails to divide A_n, contradicting the fact that both A_{m+n} and A_n are integers, so that necessarily $b = 1$. □

33) Determine the integer part of

$$\sum_{n=1}^{10^9} \frac{1}{\sqrt[3]{n^2}},$$

where $n \in \mathbb{N}$.

Solution. Since

$$\left(n + \frac{1}{3}\right)^3 = n^3 + n^2 + \frac{n}{3} + \frac{1}{27}$$

and

$$\left(n - \frac{1}{3}\right)^3 = n^3 - n^2 + \frac{n}{3} - \frac{1}{27},$$

for $n \in \mathbb{N}$, we obtain

$$\left(n + \frac{1}{3}\right)^3 > n^3 + n^2 \quad \text{and} \quad \left(n - \frac{1}{3}\right)^3 > n^3 - n^2.$$

Therefore, if we divide by n^3, we get

$$\left(1 + \frac{1}{3n}\right)^3 > 1 + \frac{1}{n} \quad \text{and} \quad \left(1 - \frac{1}{3n}\right)^3 > 1 - \frac{1}{n},$$

and thus

$$1 + \frac{1}{3n} > \left(1 + \frac{1}{n}\right)^{1/3} \quad \text{and} \quad 1 - \frac{1}{3n} > \left(1 - \frac{1}{n}\right)^{1/3}.$$

Hence,

$$\left(1 + \frac{1}{n}\right)^{1/3} - 1 < \frac{1}{3n} < 1 - \left(1 - \frac{1}{n}\right)^{1/3},$$

or

$$3\left(\left(1 + \frac{1}{n}\right)^{1/3} - 1\right) < \frac{1}{n} < 3\left(1 - \left(1 - \frac{1}{n}\right)^{1/3}\right).$$

Thus,

$$3\left((n+1)^{1/3} - n^{1/3}\right) < \frac{1}{n^{2/3}} < 3(n^{1/3} - (n-1)^{1/3}).$$

By the above inequality, for $n = 2, 3, \ldots, 10^9$, we obtain

$$3(3^{1/3} - 2^{1/3}) < \frac{1}{2^{2/3}} < 3(2^{1/3} - 1)$$

$$3(4^{1/3} - 3^{1/3}) < \frac{1}{3^{2/3}} < 3(3^{1/3} - 2^{1/3})$$

$$\vdots$$

$$3\left((10^9 + 1)^{1/3} - 10^3\right) < \frac{1}{(10^9)^{2/3}} < 3(10^3 - (10^9 - 1)^{1/3}).$$

Therefore, by adding the above inequalities by parts, we get

$$3\left((10^9 + 1)^{1/3} - 2^{1/3}\right) < \sum_{n=2}^{10^9} \frac{1}{n^{2/3}} < 3(10^3 - 1).$$

Hence,

$$3\left((10^9 + 1)^{1/3} - 2^{1/3}\right) + 1 < \sum_{n=1}^{10^9} \frac{1}{\sqrt[3]{n^2}} < 3(10^3 - 1) + 1 = 3 \cdot 10^3 - 2.$$

However,

$$3 \cdot 10^3 - 2 - 3((10^9 + 1)^{1/3} - 2^{1/3}) - 1$$
$$= 3 \cdot 10^3 - 3 - 3(10^9 + 1)^{1/3} + 3 \cdot 2^{1/3}$$
$$= 3 \cdot 2^{1/3} - 3 + (3 \cdot 10^3 - 3(10^9 + 1)^{1/3})$$
$$< 3 \cdot 2^{1/3} - 3 < 1.$$

Hence,

$$\left\lfloor \sum_{n=2}^{10^9} \frac{1}{\sqrt[3]{n^2}} \right\rfloor = 3 \cdot 10^3 - 2 - 1 = 2997. \qquad \square$$

34) Calculate the integer part of

$$\sum_{n=1}^{+\infty} \sum_{m=1}^{+\infty} \frac{1}{m^2 n + mn^2 + 2mn}.$$

Solution.

$$\sum_{n=1}^{+\infty} \sum_{m=1}^{+\infty} \frac{1}{m^2 n + mn^2 + 2mn}$$

$$= \sum_{n=1}^{+\infty} \frac{1}{n} \sum_{m=1}^{+\infty} \frac{1}{m(m+n+2)}$$

$$= \sum_{n=1}^{+\infty} \frac{1}{n} \sum_{m=1}^{+\infty} \left(\frac{1}{(n+2)m} - \frac{1}{(n+2)(m+n+2)} \right)$$

$$= \sum_{n=1}^{+\infty} \frac{1}{n(n+2)} \sum_{m=1}^{+\infty} \left(\frac{1}{m} - \frac{1}{m+n+2} \right)$$

$$= \sum_{n=1}^{+\infty} \frac{1}{n(n+2)} \left(1 + \frac{1}{2} + \frac{1}{3} + \cdots + \frac{1}{n+2} \right)$$

$$= \sum_{n=1}^{+\infty} \left(\frac{1}{2n} - \frac{1}{2n+4} \right) \left(1 + \frac{1}{2} + \frac{1}{3} + \cdots + \frac{1}{n+2} \right)$$

$$= \frac{1}{2} \sum_{n=1}^{+\infty} \left(\frac{1}{n} - \frac{1}{n+2} \right) \left(1 + \frac{1}{2} + \frac{1}{3} + \cdots + \frac{1}{n+2} \right)$$

$$= \frac{1}{2} \left[\left(1 - \frac{1}{3} \right) \left(1 + \frac{1}{2} + \frac{1}{3} \right) \right]$$

$$+ \frac{1}{2} \left[\left(\frac{1}{2} - \frac{1}{4} \right) \left(1 + \frac{1}{2} + \frac{1}{3} + \frac{1}{4} \right) \right]$$

$$+ \frac{1}{2} \left[\left(\frac{1}{3} - \frac{1}{5} \right) \left(1 + \frac{1}{2} + \frac{1}{3} + \frac{1}{4} + \frac{1}{5} \right) \right] + \cdots$$

$$= \cdots = \frac{7}{4}.$$

Therefore, it is clear that the integer part of

$$\sum_{n=1}^{+\infty} \sum_{m=1}^{+\infty} \frac{1}{m^2 n + mn^2 + 2mn}$$

is equal to 1.   □

35) Find all positive integers a, b such that

$$a^4 + 4b^4$$

is a prime number.

Solution. We have

$$a^4 + 4b^4 = (a^4 + 4a^2 b^2 + 4b^4) - 4a^2 b^2$$

$$= (a^2 + 2b^2)^2 - (2ab)^2$$

$$= (a^2 + 2b^2 + 2ab)(a^2 + 2b^2 - 2ab)$$

$$= [(a + b)^2 + b^2][(a - b)^2 + b^2].$$

However, for all positive integers a, b it holds

$$(a + b)^2 + b^2 > 1,$$

therefore for the number

$$a^4 + 4b^4$$

to be prime it must be true that

$$(a - b)^2 + b^2 = 1.$$

This happens if

$$a - b = 0 \quad \text{and} \quad b = 1,$$

namely, when $a = b = 1$.   □

36) Let n and $8n^2+1$ be two prime numbers. Prove that the number $8n^2 - 1$ is also a prime number.

Proof.
- For $n = 2$, it follows that $8n^2 + 1 = 33$, which is a composite number. (This case is not acceptable by the hypothesis.)
- For $n = 3$, one has $8n^2 + 1 = 73$ and $8n^2 - 1 = 71$ which is a prime number.
- For $n > 3$, $n \in \mathbb{N}$, the positive integer n can be expressed in one of the following forms:

$$n = 3m + 1 \quad \text{or} \quad n = 3m + 2,$$

where n is a prime number by the hypothesis.
    If $n = 3m + 1$, then

$$n^2 = 9m^2 + 6m + 1 = 3k + 1,$$

where $k = 3m^2 + 2m \in \mathbb{N}$.
    Therefore, $8n^2 + 1 = 8(3k + 1) + 1 = 3\lambda$, where $\lambda = 8k + 3 \in \mathbb{N}$.
    Thus, the integer $8n^2 + 1$ is composite. (This case is not acceptable by the hypothesis.)
    If $n = 3m+2$, then similarly we obtain that the integer $8n^2+1$ is composite. (This case is not acceptable by the hypothesis.)

Note. For $n > 3$, we could also deal with the problem in the following way:
    It is evident that $3 \mid (8n^2 - 1)8n^2(8n^2 + 1)$. However, since n and $8n^2 + 1$ are prime numbers, we obtain

$$3 \mid 8n^2 - 1$$

or

$$3 \mid 9n^2 - n^2 - 1$$

or

$$3 \mid n^2 + 1, \tag{1}$$

which is impossible, since

- If $n = 3m+1$, then by (1) it would follow that $3 \mid 2$, which is a contradiction.
- If $n = 3m+2$, then by (1) it would follow that $3 \mid 5$, which is a contradiction.

In conclusion, the only case where both n and $8n^2 + 1$ are prime numbers is when $n = 3$. In this case $8n^2 - 1$ is also a prime number. $\square$

37) Prove that there does not exist a nonconstant polynomial $p(n)$ with integer coefficients, such that for every natural number n, the number $p(n)$ is prime.

Proof. We consider a polynomial $p(x)$ in its general form

$$p(x) = a_k x^k + a_{k-1} x^{k-1} + \cdots + a_1 x + a_0, \quad x \in \mathbb{N}.$$

Set $p(x) = y$. Then,

$$p(by+x) = a_k(by+x)^k + a_{k-1}(by+x)^{k-1} + \cdots + a_1(by+x) + a_0, \quad b \in \mathbb{Z}. \quad (1)$$

However, we know that

$$(x+y)^n = x^n + \binom{n}{1} x^{n-1} y + \cdots + \binom{n}{n-1} x\, y^{n-1} + y^n.$$

Thus, from (1) we obtain

$$
\begin{aligned}
p(by + x) &= a_k(\lambda_k y + x^k) + a_{k-1}(\lambda_{k-1} y + x^{k-1}) + \cdots + a_1(\lambda_1 y + x) + a_0 \\
&= (a_k \lambda_k + a_{k-1}\lambda_{k-1} + \cdots + a_1 \lambda_1)\, y \\
&\quad + (a_k x^k + a_{k-1} x^{k-1} + \cdots + a_1 x + a_0) \\
&= \lambda y + y,
\end{aligned}
$$

where $\lambda_1, \lambda_2, \ldots \lambda_k, \lambda \in \mathbb{Z}$.

Obviously it holds that $y \mid p(by + x)$ for every integer b and every natural number x. $\qquad \square$

38) Let n be an odd integer greater than or equal to 5. Prove that

$$\binom{n}{1} - 5\binom{n}{2} + 5^2 \binom{n}{3} - \cdots + 5^{n-1} \binom{n}{n}$$

is not a prime number.

(Titu Andreescu, Korean Mathematical Competition, 2001)

Proof. Set

$$A = \binom{n}{1} - 5\binom{n}{2} + 5^2 \binom{n}{3} - \cdots + 5^{n-1}\binom{n}{n}.$$

Then

$$
\begin{aligned}
5A &= 5\binom{n}{1} - 5^2\binom{n}{2} + 5^3\binom{n}{3} - \cdots + 5^n\binom{n}{n} \\
&= 1 - 1 + 5\binom{n}{1} - 5^2\binom{n}{2} + 5^3\binom{n}{3} - \cdots + 5^n\binom{n}{n} \\
&= 1 + (-1 + 5)^n \\
&= 1 + 4^n.
\end{aligned}
$$

Thus,

$$A = \frac{1}{5}(4^n + 1) = \frac{1}{5}(4^n + 2 \cdot 2^n + 1 - 2^{n+1})$$

$$= \frac{1}{5}[(2^n + 1)^2 - (2^{(n+1)/2})^2]$$

$$= \frac{1}{5}(2^n + 2^{(n+1)/2} + 1)(2^n - 2^{(n+1)/2} + 1)$$

$$= \frac{1}{5}[(2^{(n-1)/2} + 1)^2 + 2^{n-1}][(2^{(n-1)/2} - 1)^2 + 2^{n-1}].$$

It is evident that A is a composite integer, since one of the factors within brackets is divisible by 5 and n is greater than or equal to 5. □

39) Prove that there are infinitely many prime numbers of the form $4n + 3$, where $n \in \mathbb{N}$.

Proof. We will apply a similar argument with the one followed by Euclid to prove that prime numbers are infinitely many. Suppose that the set of prime numbers of the form $4n + 3$ is finite and that p is the last and largest among all such primes. Consider the integer q where

$$q = 2^2 \cdot 3 \cdot 5 \cdots p - 1.$$

The integer q is of the form $4n + 3$, $n \in \mathbb{N}$, because

$$q = 2^2 \cdot 3 \cdot 5 \cdots p - 1 = 4k - 1,$$

where $k = 3 \cdot 5 \cdots p \in \mathbb{Z}$.

Thus,

$$q = 4k + 3 - 4 = 4(k - 1) + 3, k \in \mathbb{N}.$$

The integer q is a prime or it can be represented as a product of powers of prime numbers.

• If q is a prime number, then there exists a prime number of the form $4n + 3$ which is greater than p. That is a contradiction. (The fact that $q > p$ can be easily proved by applying Mathematical Induction.) Therefore the set of prime numbers of the form $4n + 3$ is infinite.

• If q is not a prime number, then q can be represented as a product of powers of primes. These prime factors can be represented in the form $4n + 1$ or $4n + 3$ (since $4n$, $4n + 2$ are not prime numbers).

However, not all prime factors can be expressed in the form $4n + 1$, because in that case q should be expressed in the form $4n + 1$. This happens since the product of two integers of the form $4n + 1$ is also an integer of the form $4n + 1$.

Therefore, at least one of the prime factors of the integer q will be of the form $4n + 3$.

But, none of the prime numbers which are less than or at most equal to p, divide q. Thus, every prime factor of q of the form $4n + 3$ will be greater than p. That is a contradiction. Therefore, in this case too, the prime numbers of the form $4n + 3$ are infinitely many.   □

Remark 11.2. We should mention here that this is an elementary proof of a general fact, which was proved by Dirichlet using L-series and which asserts that there are infinitely many primes of the form $a + km$, where $k \in \mathbb{N}$ and $\gcd(a, m) = 1$.

40) Let $y \in \mathbb{Z}^* = \mathbb{Z} - \{0\}$. If $x_1, x_2, \ldots, x_n \in \mathbb{Z}^* - \{1\}$ with $n \in \mathbb{N}$ and

$$(x_1 x_2 \cdots x_n)^2 y \leq 2^{2(n+1)},$$

as well as

$$x_1 x_2 \cdots x_n y = z + 1 \quad z \in \mathbb{N},$$

prove that at least one of the integers $x_1, x_2, \ldots, x_n, z$ is a prime number.

(M. Th. Rassias, Proposed problem W.3, Octogon Mathematical Magazine 15 (1) (2007), page 291. See also Proposed problem No. 109, Euclid Mathematical Magazine B', Greek Math. Soc. 66 (2007), p. 71)

Proof. Set $x_1 x_2 \cdots x_n = x$. Then it follows

$$xy = z + 1.$$

Assume that the integers $x_1, x_2, \ldots, x_n, z$ are not all prime. In this case each of the integers $x_1, x_2, \ldots, x_n, z$ can be expressed as a product of at least two prime factors. Thus, finally, it holds

$$xz = p_{\lambda_1} p_{\lambda_2} p_{\lambda_3} \cdots p_{\lambda_m} \quad \text{with} \quad m \geq 2(n+1), \quad \text{where} \quad \lambda, m \in \mathbb{N}$$

(such that the primes $p_{\lambda_1}, p_{\lambda_2}, p_{\lambda_3}, \ldots, p_{\lambda_m}$ are not necessarily pairwise distinct).

The inequality $m \geq 2(n+1)$ holds, because each of the integers $x_1, x_2, \ldots, x_n, z$ can be expressed as a product of at least two primes, since none of these is a prime number.

Thus, we obtain

$$xz \geq p_{\lambda_1} p_{\lambda_2} p_{\lambda_3} \cdots p_{\lambda_{2(n+1)}}.$$

However,

$$p_{\lambda_i} \geq 2, \quad \text{where} \quad i = 1, 2, \ldots, 2(n+1).$$

Thus,

$$xz \geq 2^{2(n+1)}.$$

Let $xz = a$, $a \in \mathbb{Z}$. Then,

$$\left.\begin{array}{c} xy = z + 1 \\ xz = a \end{array}\right\} \Rightarrow x^2 yz = a(z+1) \Leftrightarrow x^2 y = a\left(1 + \frac{1}{z}\right) > a.$$

But

$$a \geq 2^{2(n+1)},$$

therefore

$$x^2 y > 2^{2(n+1)},$$

that is,

$$(x_1 x_2 \cdots x_n)^2 y > 2^{2(n+1)},$$

which is a contradiction, because of hypothesis. Therefore, at least one of the integers $x_1, x_2, \ldots, x_n, z$ is a prime number. $\qquad \square$

Remark 11.3. If we assume that

$$x_1 = x_2 = \cdots = x_n = \beta,$$

where β is not a prime number, then z must necessarily be a prime number. This means that if $y \in \mathbb{Z}^*$, $\beta \in \mathbb{Z}^* - \{1\}$ and $z \in \mathbb{N}$ such that

$$\beta^{2n} y \leq 2^{2(n+1)} \quad \text{and} \quad \beta^n y = z + 1,$$

where β is not a prime number, then z must be a prime number.

41) Let p be a prime number. Let $h(x)$ be a polynomial with integer coefficients such that $h(0), h(1), \ldots, h(p^2 - 1)$ are distinct modulo p^2. Prove that $h(0), h(1), \ldots, h(p^3 - 1)$ are distinct modulo p^3.

(Problem B4, The Sixty-Ninth William Lowell Putnam Mathematical Competition, 2008, Amer. Math. Monthly 116(2009), pp. 722, 725)

Proof. Assume that this is not the case. Then there exist a and b such that $0 \leq a < b < p^3$ with

$$h(a) \equiv h(b) \pmod{p^3}.$$

By the hypothesis it follows that the mapping h induces an injection

$$\mathbb{Z}/p^2\mathbb{Z} \longrightarrow \mathbb{Z}/p^2\mathbb{Z},$$

and thus

$$a \equiv b \pmod{p^2}.$$

It follows that

$$h(a + x) = h(a) + h'(a)x + x^2 d(x)$$

for certain $d(x) \in \mathbb{Z}[x]$. For $x = p$, applying the property that

$$h(a + p) \not\equiv h(a) \, (\mathrm{mod} \, p^2),$$

it yields

$$p \nmid h'(a).$$

Set

$$x = b - a,$$

then

$$x = p^2 c$$

for a certain value of $c \in \{1, 2, \ldots, p - 1\}$.

Therefore,

$$0 \equiv h(b) - h(a) \equiv h'(a)p^2 c + p^4 c^2 d(p^2 c)$$
$$\equiv h'(a)p^2 c \, (\mathrm{mod} \, p^3),$$

which contradicts the fact that both $h'(a)$ and c are not multiples of p.

Hence, $h(0)$, $h(1),\ldots$, $h(p^3 - 1)$ are distinct modulo p^3. $\qquad \square$

42) Prove that every odd perfect number has at least three distinct prime factors.

Proof. Let us suppose that there exists an odd perfect number n, which does not have at least three distinct prime factors. In that case n would either be of the form

$$n = p^k,$$

where p is a prime number with $p \geq 3$ and k is a positive integer, or of the form

$$n = p_1^{k_1} p_2^{k_2},$$

where p_1, p_2 are prime numbers with $3 \leq p_1 < p_2$ and k_1, k_2 are positive integers.

However, if $n = p^k$, then

$$\sigma_1(n) = 1 + p + p^2 + \cdots + p^k.$$

By induction, we can easily prove that

$$1 + p + p^2 + \cdots + p^k < 2p^k = 2n.$$

Thus, it follows that

$$\sigma_1(n) < 2n,$$

which is impossible, since by the hypothesis n is a perfect number.

Furthermore, if $n = p_1^{k_1} p_2^{k_2}$, we have

$$\sigma_1(n) = \sigma_1(p_1^{k_1} p_2^{k_2}) = \sigma_1(p_1^{k_1})\sigma_1(p_2^{k_2}),$$

since $p_1^{k_1}$, $p_2^{k_2}$ are obviously positive coprime integers. Therefore,

$$\sigma_1(n) = (1 + p_1 + p_1^2 + \cdots + p_1^{k_1})(1 + p_2 + p_2^2 + \cdots p_2^{k_2})$$

$$= p_1^{k_1}\left(1 + \frac{1}{p_1} + \frac{1}{p_1^2} + \cdots + \frac{1}{p_1^{k_1}}\right) p_2^{k_2}\left(1 + \frac{1}{p_2} + \frac{1}{p_2^2} + \cdots + \frac{1}{p_2^{k_2}}\right)$$

$$< p_1^{k_1} p_2^{k_2} \sum_{m=0}^{+\infty} \frac{1}{p_1^m} \sum_{m=0}^{+\infty} \frac{1}{p_2^m} = p_1^{k_1} p_2^{k_2} \frac{1}{1 - 1/p_1} \cdot \frac{1}{1 - 1/p_2}$$

$$\leq p_1^{k_1} p_2^{k_2} \frac{1}{1 - 1/3} \cdot \frac{1}{1 - 1/5} = \frac{15}{8} p_1^{k_1} p_2^{k_2}$$

$$< 2 p_1^{k_1} p_2^{k_2} = 2n.$$

Hence,

$$\sigma_1(n) < 2n,$$

which is impossible, since by the hypothesis n is a perfect number.   □

43) Let (a_n) be a sequence of positive integers, such that $(a_i, a_j) = 1$ for every $i \neq j$. If

$$\sum_{n=0}^{+\infty} \frac{1}{a_n} = +\infty,$$

prove that the sequence (a_n) contains infinitely many prime numbers.

(K. Gaitanas, student of the School of Applied Mathematics and Physical Sciences, NTUA, Greece, 2005)

Proof. Suppose that there exists $n_0 \in \mathbb{N}$, such that every term of the sequence with $n \geq n_0$ is a composite integer. Let p_n be the smallest prime divisor of a_n (for $n \geq n_0$). Then

$$a_n = p_n \cdot q \quad \text{where} \quad q \geq p_n.$$

Thus, $a_n \geq p_n^2$, that is, $1/a_n \leq 1/p_n^2$. Therefore,

$$\sum_{n=n_0}^{+\infty} \frac{1}{a_n} \leq \sum_{n=n_0}^{+\infty} \frac{1}{p_n^2} \leq \sum_{n=1}^{+\infty} \frac{1}{n^2} < +\infty.$$

Hence, the series

$$\sum_{n=n_0}^{+\infty} \frac{1}{a_n}$$

converges in $\mathbb{R}$.

Because of the fact that

$$\sum_{n=0}^{+\infty} \frac{1}{a_n} = \sum_{n=0}^{n_0-1} \frac{1}{a_n} + \sum_{n=n_0}^{+\infty} \frac{1}{a_n}$$

and the series $\sum_{n=n_0}^{+\infty} \frac{1}{a_n}$ converges, it follows that the series $\sum_{n=0}^{+\infty} \frac{1}{a_n}$ converges as well. That is a contradiction.

Therefore, for every term a_m, $m \in \mathbb{N}$, we can find at least one term a_{m+k}, $k \in \mathbb{N}$ which is a prime number. There are infinitely many choices for a_m, $m \in \mathbb{N}$ and thus one can find infinitely many primes in the sequence (a_n). $\square$

44) Let p_i denote the ith prime number. Prove that

$$p_1^k + p_2^k + \cdots + p_n^k > n^{k+1},$$

for every pair of positive integers n, k.

(Dorin Andrica, Revista Matematică Timisoara, No. 2(1978), p. 45, Problem 3483)

Proof. It is evident that
$$p_{i+1} - p_i \geq 2,$$

for every positive integer i. Thus, by adding the inequalities that occur for $i = 1, 2, \ldots, n-1$, it follows that

$$p_n - 2 \geq 2(n-1)$$

or

$$p_n \geq 2n. \tag{1}$$

Generally, by the Power Mean Inequality we know that for real numbers r_1, r_2, with $r_1 \leq r_2$, $r_1, r_2 \neq 0$ and positive real numbers a_i, $i = 1, 2, \ldots, n$, it holds

$$\left(\frac{1}{n} \sum_{i=1}^{n} a_i^{r_1} \right)^{1/r_1} \leq \left(\frac{1}{n} \sum_{i=1}^{n} a_i^{r_2} \right)^{1/r_2}.$$

Thus, for $r_1 = 1$ and $r_2 = k$, we obtain

$$\frac{1}{n} \sum_{i=1}^{n} p_i \leq \left(\frac{1}{n} \sum_{i=1}^{n} p_i^k \right)^{1/k}$$

or

$$\sum_{i=1}^{n} p_i^k \geq n \left(\frac{1}{n} \sum_{i=1}^{n} p_i \right)^k. \tag{2}$$

However, by (1) it follows that

$$\frac{1}{n} \sum_{i=1}^{n} p_i \geq \frac{1}{n} \sum_{i=1}^{n} 2i > \frac{1}{n} \sum_{i=1}^{n} (2i - 1)$$

$$= \frac{1}{n} \left(2 \cdot \frac{n(n+1)}{2} - n \right) = \frac{n^2}{n}$$

$$= n.$$

Therefore, by (2) we get

$$\sum_{i=1}^{n} p_i^k > n \cdot n^k = n^{k+1}. \qquad \square$$

45) Prove that 7 divides the number

$$1^{47} + 2^{47} + 3^{47} + 4^{47} + 5^{47} + 6^{47}.$$

Proof. It is a standard fact that if a, b are positive integers and n is an odd positive integer, then

$$a + b \mid a^n + b^n.$$

Thus,

$$1^{47} + 6^{47} = (1 + 6) \cdot k \quad \text{where} \quad k \in \mathbb{N}, \tag{1}$$

$$2^{47} + 5^{47} = (2 + 5) \cdot l \quad \text{where} \quad l \in \mathbb{N} \tag{2}$$

and

$$3^{47} + 4^{47} = (3 + 4) \cdot m \quad \text{where} \quad m \in \mathbb{N}. \tag{3}$$

From (1), (2) and (3) it follows that

$$(1^{47} + 6^{47}) + (2^{47} + 5^{47}) + (3^{47} + 4^{47}) = 7 \cdot (k + l + m),$$

that is,

$$1^{47} + 2^{47} + 3^{47} + 4^{47} + 5^{47} + 6^{47} = 7 \cdot r,$$

where

$$r = k + l + m \in \mathbb{N}.$$

Hence, 7 divides the number

$$1^{47} + 2^{47} + 3^{47} + 4^{47} + 5^{47} + 6^{47}. \qquad \square$$

46) Prove that if $3 \nmid n$, then

$$13 \mid 3^{2n} + 3^n + 1,$$

where $n \in \mathbb{N}$.

Proof. Since $3 \nmid n$, $n \in \mathbb{N}$, it follows that

$$n = 3k + 1 \quad \text{or} \quad n = 3k + 2, \quad \text{where} \quad k \in \mathbb{N}.$$

• If $n = 3k + 1$, then

$$
\begin{aligned}
3^{2n} + 3^n + 1 &= (3^{3k+1})^2 + 3^{3k+1} + 1 \\
&= 9 \cdot 27^{2k} + 3 \cdot 27^k + 1 \\
&= 9 \cdot (2 \cdot 13 + 1)^{2k} + 3 \cdot (2 \cdot 13 + 1)^k + 1 \\
&= 9 \cdot (\text{mult. } 13 + 1) + 3 \cdot (\text{mult. } 13 + 1) + 1 \\
&= 12 \cdot \text{mult. } 13 + 13.
\end{aligned}
$$

Therefore,

$$3^{2n} + 3^n + 1 = 13l, \quad \text{where} \quad l \in \mathbb{N},$$

that is,

$$13 \mid 3^{2n} + 3^n + 1, \quad \text{where} \quad n \in \mathbb{N}.$$

• If $n = 3k + 2$, then

$$
\begin{aligned}
3^{2n} + 3^n + 1 &= (3^{3k+2})^2 + 3^{3k+2} + 1 \\
&= 9^2 \cdot 27^{2k} + 9 \cdot 27^k + 1 \\
&= 9^2 \cdot (2 \cdot 13 + 1)^{2k} + 9 \cdot (2 \cdot 13 + 1)^k + 1 \\
&= 9^2 \cdot (\text{mult. } 13 + 1) + 9 \cdot (\text{mult. } 13 + 1) + 1 \\
&= 90 \cdot \text{mult. } 13 + 91 \\
&= 90 \cdot \text{mult. } 13 + 13 \cdot 7.
\end{aligned}
$$

Thus,

$$3^{2n} + 3^n + 1 = 13m, \quad \text{where} \quad m \in \mathbb{N},$$

that is,

$$13 \mid 3^{2n} + 3^n + 1.$$

Hence, in every case if $3 \nmid n$, it follows that

$$13 \mid 3^{2n} + 3^n + 1. \qquad \square$$

47) Prove that for every positive integer n **the value of the expression**

$$2^{4n+1} - 2^{2n} - 1$$

is divisible by 9.

Proof. We have

$$2^{4n+1} - 2^{2n} - 1 = 2^{4n} \cdot 2 - 2^{2n} - 1 = (2^{4n} - 2^{2n}) + (2^{4n} - 1)$$
$$= 2^{2n}(2^{2n} - 1) + (2^{2n} - 1)(2^{2n} + 1)$$
$$= (2^{2n} - 1)(2^{2n} + 2^{2n} + 1)$$
$$= (2^{2n} - 1)(2^{2n+1} + 1).$$

That is,

$$2^{4n+1} - 2^{2n} - 1 = (4^n - 1)(2^{2n+1} + 1).$$

However, for every positive integer n it is clear that

$$4^n - 1 = \text{mult. } 3 \quad \text{and} \quad 2^{2n+1} + 1 = \text{mult. } 3,$$

since

$$(4 - 1) \mid (4^n - 1) \quad \text{and} \quad (2 + 1) \mid (2^{2n+1} + 1)$$

for every positive integer n.

Therefore,

$$2^{4n+1} - 2^{2n} - 1 = \text{mult. } 9.$$

Hence, the value of the expression

$$2^{4n+1} - 2^{2n} - 1$$

is divisible by 9 for every positive integer n. $\qquad\square$

48) Prove that 7 divides the number

$$2222^{5555} + 5555^{2222}.$$

Solution. We will prove that

$$2222^{5555} + 5555^{2222} = \text{mult. } 7.$$

We have

$$2222^{5555} + 5555^{2222} = (22 \cdot 100 + 22)^{5555} + (55 \cdot 100 + 55)^{2222}$$
$$= (22 \cdot 101)^{5555} + (55 \cdot 101)^{2222}$$

$$= (\text{mult. } 7 + 1)^{5555}(\text{mult. } 7 + 3)^{5555}$$
$$+ (\text{mult. } 7 - 1)^{2222}(\text{mult. } 7 + 3)^{2222}$$
$$= (\text{mult. } 7 + 1)(\text{mult. } 7 + 3)^{5555}$$
$$+ (\text{mult. } 7 + 1)(\text{mult. } 7 + 3)^{2222}.$$

That is,

$$2222^{5555} + 5555^{2222} = \text{mult. } 7 + 3^{5555} + \text{mult. } 7 + 3^{2222}.$$

However,

$$3^{5555} = (3^5)^{1111} = 243^{1111} = (245 - 2)^{1111}$$
$$= (\text{mult. } 7 - 2)^{1111} = \text{mult. } 7 - 2^{1111}$$

as well as

$$3^{2222} = (3^2)^{1111} = 9^{1111} = (7 + 2)^{1111}$$
$$= (\text{mult. } 7 + 2)^{1111} = \text{mult. } 7 + 2^{1111}.$$

Therefore,

$$3^{5555} + 3^{2222} = (\text{mult. } 7 - 2^{1111}) + (\text{mult. } 7 + 2^{1111})$$
$$= \text{mult. } 7.$$

Hence,

$$7 \mid 2222^{5555} + 5555^{2222}. \qquad \qquad \square$$

49) If p is a prime number and a, λ are two positive integers such that $p^\lambda \mid (a - 1)$, prove that

$$p^{n+\lambda} \mid (a^{p^n} - 1)$$

for every $n \in \mathbb{N} \cup \{0\}$.

(*Crux Mathematicorum, 1992, p. 84, Problem 1617. Proposed by Stanley Rabinowitz, Westford, Massachusetts. The proof is due to Ian Goldberg, University of Toronto Schools*)

Proof. We will apply the Principle of Mathematical Induction on the natural number n. Trivially for $n = 0$ one has $p^\lambda \mid (a - 1)$, which is true because of the hypothesis.
Suppose that

$$p^{n+\lambda} \mid (a^{p^n} - 1).$$

We will prove that

$$p^{n+\lambda+1} \mid (a^{p^{n+1}} - 1).$$

Set $x = a^{p^n}$. It is enough to prove that

$$p^{n+\lambda+1} \mid (x^p - 1).$$

However, $p^{n+\lambda} \mid (x - 1)$. From this relation it follows

$$x \equiv 1 \,(\text{mod}, p)$$
$$x^2 \equiv 1 \,(\text{mod}, p)$$

$$\vdots$$

$$x^{p-1} \equiv 1 \,(\text{mod}, p).$$

It is also true that
$$1 \equiv 1 \,(\text{mod}, p).$$

If we add by parts the above congruences we obtain

$$x^{p-1} + \cdots + x^2 + x + 1 \equiv p \,(\text{mod}, p)$$

or
$$p \mid (x^{p-1} + \cdots + x + 1). \tag{1}$$

However, we have assumed that

$$p^{n+\lambda} \mid (x - 1). \tag{2}$$

Multiplying (1) and (2) by parts we obtain

$$p^{n+\lambda+1} \mid (x - 1)(x^{p-1} + \cdots + x + 1)$$

or
$$p^{n+\lambda+1} \mid (x^p - 1). \qquad \square$$

50) Prove that for any prime number p greater than 3, the number

$$\frac{2^p + 1}{3}$$

is not divisible by 3.

Proof. It is evident that if n is an even integer, one has

$$2^n \equiv 1 \,(\text{mod } 3),$$

but if n is an odd integer,
$$2^n \equiv 2 \,(\text{mod } 3).$$

Using the hypothesis that p is a prime number and $p > 3$, we obtain

$$\frac{2^p + 1}{3} = \frac{2^p + 1}{2 + 1} = 2^{p-1} - 2^{p-2} + 2^{p-3} - 2^{p-4} + \cdots - 2 + 1$$

$$\equiv (1 - 2) + (1 - 2) + \cdots + (1 - 2) + 1 \,(\mathrm{mod}\,3)$$

$$\equiv \left(\frac{p - 1}{2}\right)(-1) + 1 \,(\mathrm{mod}\,3)$$

$$\equiv \left(\frac{p - 1}{2}\right)2 + 1 \,(\mathrm{mod}\,3)$$

$$\equiv p - 1 + 1 \equiv p \,(\mathrm{mod}\,3),$$

which is nonzero for $p > 3$.   □

51) Determine all positive integers n for which the number $n^8 - n^2$ is not divisible by 72.

(38th National Mathematical Olympiad, Slovenia, 1997)

Solution. We have

$$n^8 - n^2 = n^2(n^6 - 1) = n^2(n - 1)(n + 1)(n^2 - n + 1)(n^2 + n + 1)$$

and

$$72 = 8 \cdot 9.$$

We claim that the number $n^8 - n^2$ is a multiple of 9.

We examine the following cases:

If $n \equiv 0 \,(\mathrm{mod}\,3)$, then $n^2 \equiv 0 \,(\mathrm{mod}\,9)$.

If $n \equiv 1 \,(\mathrm{mod}\,3)$, then $n - 1 \equiv 0 \,(\mathrm{mod}\,3)$ as well as $n^2 + n + 1 \equiv 0 \,(\mathrm{mod}\,3)$.

If $n \equiv -1 \,(\mathrm{mod}\,3)$, then $n + 1 \equiv 0 \,(\mathrm{mod}\,3)$ as well as $n^2 - n + 1 \equiv 0 \,(\mathrm{mod}\,3)$.

Therefore, the number $n^8 - n^2$ is a multiple of 9 for any positive integer n.

We will examine whether $n^8 - n^2$ is divisible by 8.

In case n is an odd integer, then the number $(n - 1)(n + 1)$ is a multiple of 8.

If n is divisible by 4, then $n^2 \equiv 0 \,(\mathrm{mod}\,8)$.

If $n \equiv 2 \,(\mathrm{mod}\,4)$, then $n^8 - n^2$ is a multiple of 4, but is not a multiple of 8. Hence, the number $n^8 - n^2$ is not divisible by 72 if and only if $n \equiv 2 \,(\mathrm{mod}\,4)$.   □

52) Prove that for every positive integer n the number $3^n + n^3$ is a multiple of 7 if and only if the number $3^n \cdot n^3 + 1$ is a multiple of 7.

(Bulgarian Mathematical Competition, 1995)

Proof.

- In case 7 divides n, then 7 is neither a divisor of $3^n + n^2$ nor a divisor of

$$3^n \cdot n^3 + 1.$$

- If 7 does not divide n, then 7 divides

$$n^6 - 1 = (n^3 - 1)(n^3 + 1)$$

and because of the fact that 7 is a prime number, it follows that 7 must divide either $n^3 - 1$ or $n^3 + 1$. However,

$$3^n \cdot n^3 + 1 = 3^n \cdot n^3 - 3^n - n^3 + n^3 + 3^n + 1$$
$$= (n^3 - 1)(3^n - 1) + (n^3 + 3^n)$$

as well as

$$3^n \cdot n^3 + 1 = (n^3 + 1)(3^n + 1) - (n^3 + 3^n).$$

Therefore, the number 7 is a divisor of $3^n + n^3$ if and only if 7 is a divisor of $3^n \cdot n^3 + 1$.   □

53) Find the sum of all positive integers that are less than 10,000 whose squares divided by 17 leave remainder 9.

Solution. Let x be any one of these integers.
Then

$$x^2 = 17k + 9, \quad \text{where} \quad k \in \mathbb{N}. \tag{1}$$

Then

$$k = \frac{x^2 - 9}{17},$$

that is,

$$k = \frac{(x - 3)(x + 3)}{17}. \tag{2}$$

Because of the fact that $k \in \mathbb{N}$ and 17 is a prime number, it follows that

$$\frac{(x - 3)(x + 3)}{17} \in \mathbb{N}$$

in the following cases:

$$x - 3 = 17\lambda, \quad \lambda \in \mathbb{N} \tag{3}$$

or

$$x + 3 = 17\lambda, \quad \lambda \in \mathbb{N}. \tag{4}$$

- For case (3), since x depends upon λ, we will denote the solution of (3) by x_λ and thus we get

$$x_\lambda = 17\lambda + 3, \quad \lambda \in \mathbb{N}.$$

Since

$$x_\lambda < 10,000$$

it follows that

$$17\lambda + 3 < 10,000$$

which implies

$$\lambda < 588\frac{1}{7}.$$

Therefore, $\lambda = 1, 2, 3, \ldots, 588$.

For these values of λ, from the formula

$$x_\lambda = 17\lambda + 3,$$

we obtain

$$\begin{cases} x_1 = 17 \cdot 1 + 3 \\ x_2 = 17 \cdot 2 + 3 \\ \vdots \\ x_{588} = 17 \cdot 588 + 3. \end{cases}$$

Adding by parts the above equalities, we get

$$x_1 + x_2 + \cdots + x_{588} = 17(1 + 2 + \cdots + 588) + 588 \cdot 3,$$

that is,

$$x_1 + x_2 + \cdots + x_{588} = 17 \cdot \frac{(588 + 1) \cdot 588}{2} + 588 \cdot 3. \tag{5}$$

- For case (4), since x depends upon λ, we will denote the solution of (4) by $\bar{x}_\lambda$ and we have

$$\bar{x}_\lambda = 17\lambda - 3, \quad \lambda \in \mathbb{N}.$$

Since

$$\bar{x}_\lambda < 10,000$$

it follows that

$$17\lambda - 3 < 10,000$$

which implies

$$\lambda < 588\frac{7}{17}.$$

Thus,

$$\lambda = 1, 2, \ldots, 588.$$

For these values of λ from the formula

$$\bar{x}_\lambda = 17\lambda - 3,$$

we obtain

$$\begin{cases} \bar{x}_1 = 17 \cdot 1 - 3 \\ \bar{x}_2 = 17 \cdot 2 - 3 \\ \quad \vdots \\ \bar{x}_{588} = 17 \cdot 588 - 3. \end{cases}$$

Adding by parts the above equalities, we have

$$\bar{x}_1 + \bar{x}_2 + \cdots + \bar{x}_{588} = 17 \cdot (1 + 2 + \cdots + 588) - 588 \cdot 3,$$

that is,

$$\bar{x}_1 + \bar{x}_2 + \cdots + \bar{x}_{588} = 17 \cdot \frac{(588 + 1) \cdot 588}{2} - 588 \cdot 3. \tag{6}$$

Hence, the total sum S of

$$x_1, x_2, \ldots, x_{588}, \bar{x}_1, \bar{x}_2, \ldots, \bar{x}_{588}$$

is

$$S = \left(\frac{17}{2} \cdot (588 + 1) \cdot 588 + 588 \cdot 3\right) + \left(\frac{17}{2} \cdot (588 + 1) \cdot 588 - 588 \cdot 3\right)$$

$$= 17 \cdot 589 \cdot 588$$

$$= 5,887,644.$$

Therefore,

$$S = 5,889,644. \qquad \square$$

54) What is the largest positive integer m with the property that, for any positive integer n, m divides $n^{241} - n$? What is the new value of m if n is restricted to be odd?

(Konstantinos Drakakis, University College Dublin, Ireland; Newsletter of the European Mathematical Society, Issue 77, 2010, Problem 69)

Solution. To begin with, observe that $n^{241} - n = n(n^{240} - 1)$, and that $240 = 2^4 \cdot 3 \cdot 5$. Fermat's Little Theorem guarantees that $p \mid n^s - n$ for all positive integers n, where p is a prime number, as long as $p - 1 \mid s - 1$; furthermore, assuming that this condition holds, in general $p^2 \nmid n^s - n$ (although this may be true for some n it is certainly not true for all n: a counterexample is $n = p$ for $s \geq 2$). In this particular case, which prime numbers p have the property that $p-1 \mid 240$? Testing exhaustively all integers of the form $2^i \cdot 3^j \cdot 5^k + 1$ for primality, where $0 \le i \le 4$, $0 \le j, k \le 1$, the following list of such primes is obtained: 2, 3, 5, 7, 11, 13, 17, 31, 41, 61, 241, so that m is their least common multiple, namely, their product: $m = 2 \cdot 3 \cdot 5 \cdot 7 \cdot 11 \cdot 13 \cdot 17 \cdot 31 \cdot 41 \cdot 61 \cdot 241 = 9,538,864,545,210$.

Restrict now n to be an odd positive integer. Repeated use of the identities

$$x^2 - 1 = (x-1)(x+1) \quad \text{and} \quad x^{2k+1} + 1 = (x+1)\sum_{i=0}^{2k}(-x)^i$$

eventually factorizes $n^{241} - n$ into irreducible polynomials of the form $\sum_{i=0}^{l} a_i n^i$ where $a_i \in \{0, 1, -1\}$. Any such polynomial with an odd number of nonzero coefficients produces odd values for odd n: to see this, use

$$n \equiv 1 \,(\text{mod}\, 2) \quad \text{and} \quad -1 \equiv 1 \,(\text{mod}\, 2)$$

to reduce such a polynomial into the sum of an odd number of 1s, which is equivalent to 1 modulo 2. Thus,

$$n^{241} - n = (n-1)(n+1)(n^2+1)(n^4+1)(n^8+1)F(n),$$

where $F(n)$ is a polynomial producing odd values for odd n.

Since $n \equiv 1 \,(\text{mod}\, 2)$ implies both that $n^2 \equiv 1 \,(\text{mod}\, 4)$ and that $8 \parallel n^2 - 1$[1] (these facts are easy to verify), it follows that $(n-1)(n+1)$ has three factors of 2, and the remaining three binomials one factor of 2 each, hence

$$2^6 = 64 \parallel n^{241} - n.$$

Therefore, the new value of m is

$$\text{lcm}(64; 9,538,864,545,210)$$
$$= 9,538,864,545,210 \cdot 32 = 305,243,665,446,720. \qquad \square$$

55) Let f be a nonconstant polynomial with positive integer coefficients. Prove that if n is a positive integer, then $f(n)$ divides $f(f(n)+1)$ if and only if $n = 1$.

(Problem B1, The Sixty-Eighth William Lowell Putnam Mathematical Competition, 2007. Amer. Math. Monthly, 115(2008), pp. 731, 735)

Proof. Set

$$f(x) = \sum_{i=0}^{m} a_i x^i.$$

---

[1] The notation $m^k \parallel n$ will be used to denote the fact that $m^k \mid n$ but $m^{k+1} \nmid n$.

Then, we obtain

$$f(f(n) + 1) = \sum_{i=0}^{m} a_i (f(n) + 1)^i$$

$$\equiv \sum_{i=0}^{m} a_i \,(\mathrm{mod}\ f(n)),$$

that is,

$$f(f(n) + 1) \equiv f(1)\,(\mathrm{mod}\ f(n)). \tag{1}$$

If $n = 1$, (1) yields

$$f(f(1) + 1) \equiv 0\,(\mathrm{mod}\ f(1)),$$

and therefore

$$f(1) \mid f(f(1) + 1).$$

If

$$f(n) \mid f(f(n) + 1),$$

for some $n > 1$ it follows from (1) that

$$f(n) \mid f(1).$$

But f is a nonconstant polynomial with positive integer coefficients, thus

$$f(n) > f(1) > 0,$$

a contradiction. □

56) Let N_n and D_n be two relatively prime positive integers. If

$$1 + \frac{1}{2} + \frac{1}{3} + \cdots + \frac{1}{n} = \frac{N_n}{D_n},$$

find all prime numbers p with $p \geq 5$, such that

$$p \mid N_{p-4}.$$

(Crux Mathematicorum, 1989, p. 62, Problem 1310. Proposed by Robert E. Shafer, Berkeley, California. The solution was given by Colin Springer, University of Waterloo, Canada)

Solution.

• If $p = 5$, then

$$1 = \frac{N_{5-4}}{D_{5-4}},$$

which implies $N_{5-4} = D_{5-4}$, that is, $N_1 = D_1$, which is impossible since by hypothesis N_n and D_n are relatively prime positive integers.

• If $p = 7$, then

$$1 + \frac{1}{2} + \frac{1}{3} = \frac{N_{7-4}}{D_{7-4}}.$$

This implies that for $p = 7$ the number p does not divide the number N_{p-4} since

$$\frac{11}{6} = \frac{N_3}{D_3}$$

with $\gcd(11, 6) = \gcd(N_3, D_3) = 1$ yields $N_3 = 11$ with $7 \nmid 11$.

• Therefore, let us assume that $p \geq 11$. Then,

$$\frac{N_{p-4}}{D_{p-4}} = 1 + \frac{1}{2} + \frac{1}{3} + \cdots + \frac{1}{p-4}$$

$$= 1 + \frac{1}{2} + \frac{1}{3} + \left(\frac{1}{4} + \frac{1}{p-4}\right) + \left(\frac{1}{5} + \frac{1}{p-5}\right) + \cdots$$

$$+ \left[\frac{1}{(p-1)/2} + \frac{1}{(p+1)/2}\right]$$

after a simple rearrangement of the terms of the summation.

But, in general it holds

$$\frac{1}{k} + \frac{1}{p-k} = p \cdot \frac{1}{k(p-k)}, \quad \text{where} \quad p \neq k.$$

Thus, we obtain that

$$\frac{N_{p-4}}{D_{p-4}} = \frac{11}{6} + \left[\frac{1}{4(p-4)} + \frac{1}{5(p-5)} + \cdots + \frac{1}{(p^2-1)/4}\right].$$

From the above relation it follows that for the prime number p to divide N_{p-4} it must hold $p \mid 11$, but we have assumed that $p \geq 11$, thus

$$p = 11.$$

Hence, the only prime number p which solves the problem is $p = 11$.   □

57) Given the positive integer n and the prime number p such that $p^p \mid n!$, prove that

$$p^{p+1} \mid n!.$$

(Proposed by D. Beckwith, Amer. Math. Monthly, Problem No. 11158, 2005)

Proof.

First way. It is given that $p^p \mid n!$, that is,

$$p^p \mid 1 \cdot 2 \cdot 3 \cdots p(p+1) \cdots [p+(p-1)]$$
$$\cdot 2p(2p+1) \cdots [2p+(p-1)]$$
$$\cdot 3p(3p+1) \cdots [3p+(p-1)]$$

$$\cdot \; \cdots$$

$$\cdot [(p-1)p][(p-1)p+1] \cdots [(p-1)p+(p-1)]$$
$$\cdot pp(pp+1) \cdots n. \tag{1}$$

However, it is clear that

$$p \mid p, 2p, 3p, \ldots, (p-1)p, pp$$

and

$$p \nmid (kp+\lambda), \quad \text{for all} \;\; k, \lambda \in \mathbb{Z} \;\; \text{with} \;\; 1 \leq \lambda \leq p-1.$$

Thus, by several eliminations in relation (1) we get

$$p \mid pp(pp+1) \cdots n,$$

that is, the inequality

$$n \geq pp.$$

must hold.

In every other case we have

$$p^p \nmid n!.$$

Therefore,

$$p^2 \mid pp(pp+1) \cdots n, \quad \text{with} \;\; n \geq pp.$$

Hence,

$$p^3 \mid (p-1)p \cdot [(p-1)p+1] \cdots [(p-1)p+(p-1)] \cdot pp(pp+1) \cdot n.$$

Thus,

$$p^4 \mid (p-2)p \cdot [(p-2)p+1] \cdots [(p-2)p+(p-1)]$$
$$(p-1)p \cdot [(p-1)p+1] \cdots [(p-1)p+(p-1)]$$
$$\cdot pp(pp+1) \cdots n.$$

Following the same argument we get

$$p^{p+1} \mid 1 \cdot p(1 \cdot p+1) \cdots (1 \cdot p+p-1) \cdots$$
$$\cdots (p-1)p \cdot [(p-1)p+1] \cdots [(p-1)p+(p-1)]$$
$$\cdot pp(pp+1) \cdots n,$$

where all the terms of the above product are consecutive integers.

Hence,

$$p^{p+1} \mid n!$$

Second way. (Harris Kwong, SUNY at Fredonia, New York)
We will make use of Legendre's theorem. According to this theorem the
greatest power of p which can divide the number $n!$ is given by the formula

$$\sum_{k=1}^{+\infty} \left\lfloor \frac{n}{p^k} \right\rfloor.$$

We will prove that

$$n \geq pp.$$

If $n < pp$, then

$$\sum_{k=1}^{+\infty} \left\lfloor \frac{n}{p^k} \right\rfloor = \left\lfloor \frac{n}{p} \right\rfloor < p.$$

This means that the integer p^p cannot divide the number $n!$, which is impossible because of the hypothesis.

Consequently, if

$$p^p \mid n!, \quad \text{then} \quad n \geq pp.$$

Therefore,

$$\sum_{k=1}^{+\infty} \left\lfloor \frac{n}{p^k} \right\rfloor = \left\lfloor \frac{n}{p} \right\rfloor + \left\lfloor \frac{n}{p^2} \right\rfloor + \left\lfloor \frac{n}{p^3} \right\rfloor + \cdots$$

$$\geq \left\lfloor \frac{n}{p} \right\rfloor + \left\lfloor \frac{n}{p^2} \right\rfloor$$

$$\geq p + 1.$$

Hence, the prime number p raised to a power which is at least equal to $p+1$, divides $n!$. $\qquad\square$

**58) Prove that there are no integer values of x, y, z, where x is of
the form $4k + 3 \in \mathbb{Z}$, such that**

$$x^n = y^n + z^n,$$

where $n \in \mathbb{N} - \{1\}$.

Proof. According to Fermat's Last Theorem[2] the diophantine equation $x^n = y^n + z^n$ does not accept nonzero integer solutions for x, y, z, where $n \geq 3$.

---

[2] See Appendix for more details on Fermat's Last Theorem.

Therefore, it suffices to prove that for $n = 2$, the equation $x^n = y^n + z^n$ does not accept any nonzero integer solutions, where x is of the form $4k + 3$, $k \in \mathbb{Z}$.

However, if the equation $x^2 = y^2 + z^2$ has a solution, then the equation $x = y^2 + z^2$ has also a solution. This is clear, since if (x_0, y_0, z_0) is a solution of the equation $x^2 = y^2 + z^2$, then (x_0^2, y_0, z_0) is a solution of the equation $x = y^2 + z^2$. Therefore, it suffices to prove that the diophantine equation

$$x = y^2 + z^2,$$

where x is of the form $4k + 3$, does not accept any nonzero integer solutions.

But, no integer of the form $4k + 3$ can be represented as the sum of two squares of integers. We shall now prove that fact.

Let us suppose that $x = y^2 + z^2$, where $x = 4k + 3$. Then, we obtain

$$y^2 + z^2 \equiv 3 \,(\mathrm{mod}\, 4).$$

However, generally it holds

$$a^2 \equiv 0 \ \text{ or } \ 1 \,(\mathrm{mod}\, 4).$$

This is true, because of the following reasoning.

- If a is an even integer, then $a^2 = 4\lambda$, $\lambda \in \mathbb{Z}$, and thus

$$a^2 \equiv 0 \,(\mathrm{mod}\, 4).$$

- If a is an odd integer, then $a^2 = 8\lambda + 1$, $\lambda \in \mathbb{Z}$, and thus

$$a^2 \equiv 1 \,(\mathrm{mod}\, 4).$$

Hence, for the sum of two squares of integers, we have

$$a^2 + b^2 \equiv 0 \ \text{ or } \ 1 \ \text{ or } \ 2 \,(\mathrm{mod}\, 4)$$

and

$$a^2 + b^2 \not\equiv 3 \,(\mathrm{mod}\, 4).$$

Therefore, the claim that if $x = y^2 + z^2$ then $y^2 + z^2 \equiv 3 \,(\mathrm{mod}\, 4)$, leads to a contradiction.

Thus, the diophantine equation

$$x = y^2 + z^2$$

does not accept any nonzero integer solutions when x is of the form $4k + 3$ and hence the diophantine equation

$$x^2 = y^2 + z^2$$

does not accept any nonzero integer solutions when x is of the form $4k + 3$. Therefore, the diophantine equation

$$x^n = y^n + z^n$$

does not accept any nonzero integer solutions when x is of the form $4k + 3$.

$\square$

59) Let P_n denote the product of all distinct prime numbers p_1, $p_2, \ldots, p_k$, which are less than or equal to n (where $k < n$). Prove that P_n divides the integer

$$n^k \sum_{\lambda=0}^{p_1-1} (-1)^\lambda \binom{n}{\lambda} \cdot \sum_{\lambda=0}^{p_2-1} (-1)^\lambda \binom{n}{\lambda} \cdots \sum_{\lambda=0}^{p_k-1} (-1)^\lambda \binom{n}{\lambda}.$$

Proof. For the purposes of this proof, we will follow the convention that, for $n > 0$, $\binom{n}{-1} = 0$. Generally, for every integer i, with $1 \leq i \leq k$, we have

$$n \sum_{\lambda=0}^{p_i-1} (-1)^\lambda \binom{n}{\lambda} = n \sum_{\lambda=0}^{p_i-1} (-1)^\lambda \left(\binom{n-1}{\lambda} + \binom{n-1}{\lambda-1} \right)$$

$$= n \left(\sum_{\lambda=0}^{p_i-1} (-1)^\lambda \binom{n-1}{\lambda} + \sum_{\lambda=0}^{p_i-1} (-1)^\lambda \binom{n-1}{\lambda-1} \right)$$

$$= n \left(\sum_{\lambda=0}^{p_i-1} (-1)^\lambda \binom{n-1}{\lambda} - \sum_{\lambda=0}^{p_i-2} (-1)^\lambda \binom{n-1}{\lambda} \right)$$

$$= (-1)^{p_i-1} n \binom{n-1}{p_i-1}$$

$$= (-1)^{p_i-1} \frac{n!}{(p_i-1)!(n-p_i)!}$$

$$= p_i (-1)^{p_i-1} \binom{n}{p_i}.$$

Therefore,

$$n \sum_{\lambda=0}^{p_i-1} (-1)^\lambda \binom{n}{\lambda} \equiv 0 \,(\mathrm{mod}\, p_i).$$

However, since $p_1, p_2, \ldots, p_k$ are relatively prime integers, it follows that

$$n^k \sum_{\lambda=0}^{p_1-1} (-1)^\lambda \binom{n}{\lambda} \cdot \sum_{\lambda=0}^{p_2-1} (-1)^\lambda \binom{n}{\lambda} \cdots \sum_{\lambda=0}^{p_k-1} (-1)^\lambda \binom{n}{\lambda}$$

$$\equiv 0 \,(\mathrm{mod}\, p_1 p_2 \cdots p_k)$$

$$\equiv 0 \,(\mathrm{mod}\, P_n).$$

$\square$

60) Determine all pairs of positive integers (a, b), such that the number

$$\frac{a^2}{2ab^2 - b^3 + 1}$$

is a positive integer.

(44th IMO, Tokyo, Japan)

Solution. Let (a, b) be a pair of positive integers, such that

$$\frac{a^2}{2ab^2 - b^3 + 1} = m,$$

where $m \in \mathbb{N}$.

Then, clearly

$$a^2 - 2ab^2 m + b^4 m^2 = b^4 m^2 - b^3 m + m.$$

Thus,

$$(a - b^2 m)^2 = b^4 m^2 - b^3 m + m,$$

which is equivalent to

$$(a - b^2 m)^2 = \left((b^2 m)^2 - 2b^2 m \frac{b}{2} + \frac{b^2}{4} \right) - \frac{b^2}{4} + m$$

or

$$(a - b^2 m)^2 = \left(b^2 m - \frac{b}{2} \right)^2 + \frac{4m - b^2}{4}$$

$$= \frac{(2b^2 m - b)^2 + (4m - b^2)}{4}.$$

Therefore, we obtain

$$(2(a - b^2 m))^2 = (2b^2 m - b)^2 + 4m - b^2. \tag{1}$$

Now, we shall distinguish three cases, concerning $4m - b^2$.

- If $4m - b^2 > 0$, then by (1) we obtain

$$4m - b^2 = (2(a - b^2 m))^2 - (2b^2 m - b)^2 \geq (2b^2 m - b + 1)^2 - (2b^2 m - b)^2$$

$$= 2(2b^2 m - b) + 1.$$

This holds true, since if we set

$$M = 2(a - b^2 m)$$

and
$$N = 2b^2 m - b,$$

then due to the fact that $M^2 - N^2 > 0$ with $N > 0$, where $M, N \in \mathbb{Z}$, we obtain
$$|M| \geq N + 1.$$

Therefore,
$$M^2 - N^2 \geq (N+1)^2 - N^2$$
$$= 2N + 1.$$

Hence,
$$2(2b^2 m - b) + 1 + b^2 - 4m \leq 0$$

or
$$4m(b^2 - 1) + (b - 1)^2 \leq 0.$$

But, this is possible only in the case when $b = 1$. Therefore, it follows that
$$(a, b) = (2m, 1), \quad m \in \mathbb{N},$$

is a pair of positive integers which satisfies the desired property.
• If $4m - b^2 < 0$, then by (1) we obtain
$$4m - b^2 = (2(a - b^2 m))^2 - (2b^2 m - b)^2 \leq (2b^2 m - b - 1)^2 - (2b^2 m - b)^2$$
$$= -2(2b^2 m - b) + 1.$$

Thus, equivalently, we have
$$(4m - 1)b^2 - 2b + (4m - 1) \leq 0. \tag{2}$$

However,
$$(4m - 1)b^2 - 2b + (4m - 1)$$

is a quadratic polynomial with respect to b, with discriminant
$$D = 4(1 - (4m - 1)^2) < 0,$$

for every $m \in \mathbb{N}$. Thus, since $4m - 1 > 0$ it follows that (2) can never hold true. Therefore, in this case there does not exist a pair of positive integers (a, b) with the desired property.
• If $4m - b^2 = 0$, then
$$(2(a - b^2 m))^2 = (2b^2 m - b)^2$$

and thus
$$2a - 2b^2 m = \pm(2b^2 k - b).$$

Therefore,

- If $2a - 2b^2m = -2b^2m + b$, it follows that $b = 2a$. Thus, the pair of positive integers $(a, b) = (k, 2k)$, $k \in \mathbb{N}$, satisfies the desired property.
- If $2a - 2b^2m = 2b^2m - b$, it follows that

$$4b^2m - b = 2a. \tag{3}$$

However, since $4m - b^2 = 0$ it is evident that b is an even integer. Therefore, there exists a positive integer c, such that $b = 2c$ and $m = c^2$. Hence, by (3) we obtain that

$$a = 8c^4 - c.$$

Thus, the pair of positive integers $(a, b) = (8c^4 - c, 2c)$ satisfies the desired property.

To sum up, the pairs of positive integers (a, b) such that

$$\frac{a^2}{2ab^2 - b^3 + 1}$$

is a positive integer are:

$$(a, b) = (2m, 1), \quad (a, b) = (m, 2m) \quad \text{and} \quad (a, b) = (8m^4 - m, 2m),$$

where $m \in \mathbb{N}$. $\qquad\qquad\qquad\square$

61) Prove that for every integer $m \geq 2$ we have

$$F_m^{(F_{m+1}-1)} \equiv 1 \, (\mathrm{mod}\, F_{m+1}),$$

where F_m denotes the mth Fermat number.

Proof. We will construct step by step the first term of the congruence. For $r \geq 3$ we have

$$F_m^{2^r} = (F_m^{2^2})^{2^{r-2}}. \tag{1}$$

But

$$F_m^2 = (2^{2^m} + 1)^2 = 2^{2^{m+1}} + 2 \cdot 2^{2^m} + 1 = F_{m+1} + 2 \cdot 2^{2^m}$$
$$\equiv 2 \cdot 2^{2^m} \, (\mathrm{mod}\, F_{m+1}).$$

Hence, by the above relation we get

$$F_m^{2^2} \equiv (2 \cdot 2^{2^m})^2 \equiv 4 \cdot 2^{2^{m+1}} \equiv 4(F_{m+1} - 1)$$
$$\equiv -2^2 \, (\mathrm{mod}\, F_{m+1}).$$

Therefore, by (1) we obtain

$$F_m^{2^r} \equiv (-2^2)^{2^{r-2}} \equiv 2^{2^{r-1}} \pmod{F_{m+1}}.$$

We set $r = 2^{m+1} - 1$ in the above relation and we get

$$F_m^{2^{2^{m+1}-1}} = F_m^{2^{2^{m+1}}/2} = F_m^{(F_{m+1}-1)/2} \equiv 2^{2^{2^{m+1}-2}} \pmod{F_{m+1}}. \tag{2}$$

Also

$$\frac{2^{2^{m+1}} - 2}{2^{m+2}} = 2^{2^{m+1}-m-4}.$$

So, if

$$2^{m+1} \geq m + 4, \tag{3}$$

this means that 2^{m+2} divides $2^{2^{m+1}} - 2$. But (3) holds for every $m \geq 2$. Therefore, there exists a positive integer c such that

$$2^{2^{m+1}} - 2 = c \cdot 2^{m+2}.$$

Then

$$2^{2^{2^{m+1}}-2} = 2^{2c2^{m+1}}$$

and for that reason, (2) can take the form

$$F_m^{(F_{m+1}-1)/2} \equiv (2^{2^{m+1}})^{2c} \equiv (-1)^{2c} \equiv 1 \pmod{F_{m+1}}. \qquad \square$$

62) Prove that

$$\phi(n) \geq \frac{\sqrt{n}}{2}$$

for every positive integer n.

Proof. Let

$$n = 2^{a_0} p_1^{a_1} p_2^{a_2} \cdots p_k^{a_k}$$

be the canonical form of n, with $a_0 > 0$ and $p_i \neq 2$, for every $i = 1, 2, \ldots, k$. Then, we have

$$\phi(n) = n \left(1 - \frac{1}{2}\right)\left(1 - \frac{1}{p_1}\right)\left(1 - \frac{1}{p_2}\right) \cdots \left(1 - \frac{1}{p_k}\right)$$

$$= 2^{a_0-1} p_1^{a_1-1} p_2^{a_2-1} \cdots p_k^{a_k-1}(p_1 - 1)(p_2 - 1) \cdots (p_k - 1).$$

However, for every prime number $p \geq 3$ it holds

$$p - 1 > \sqrt{p}.$$

Hence,

$$\phi(n) > 2^{a_0-1}p_1^{a_1-1}p_2^{a_2-1}\cdots p_k^{a_k-1}\sqrt{p_1}\sqrt{p_2}\cdots\sqrt{p_k}r$$

$$= \frac{2^{a_0}}{2}\cdot p_1^{a_1-1/2}p_2^{a_2-1/2}\cdots p_k^{a_k-1/2}.$$

But, generally, it holds

$$a - \frac{1}{2} \geq \frac{a}{2},$$

for every positive integer a. Therefore, we obtain

$$\phi(n) \geq \frac{2^{a_0/2}p_1^{a_1/2}p_2^{a_2/2}\cdots p_k^{a_k/2}}{2}$$

$$= \frac{\sqrt{2^{a_0}p_1^{a_1}p_2^{a_2}\cdots p_k^{a_k}}}{2}$$

or

$$\phi(n) \geq \frac{\sqrt{n}}{2}. \qquad\qquad \square$$

Remark 11.4. If $a_0 = 0$, that means 2 is not a prime factor of n, similarly we obtain that

$$\phi(n) > \sqrt{n} > \frac{\sqrt{n}}{2}.$$

63) Let n be a perfect even number. Prove that the integer

$$n - \phi(n)$$

is a square of an integer and determine an infinity of integer values of k, such that the integer

$$k - \phi(k)$$

is a square of an integer.

(Crux Mathematicorum, 1988, p. 93, Problem 1204. Proposed by Thomas E. Moore, Bridgewater State College, Bridgewater, Massachusetts. Solved by Bob Prielipp, University of Wisconsin–Oshkosh)

Solution. According to Euler's Theorem 3.1.3 for perfect numbers, we obtain

$$n = 2^{\lambda-1}(2^\lambda - 1),$$

where the integer $2^\lambda - 1$ is a prime number. Therefore,

$$n - \phi(n) = 2^{\lambda-1}(2^\lambda - 1) - \phi(2^{\lambda-1}(2^\lambda - 1)). \qquad (1)$$

However, the integers $2^{\lambda-1}$ and $2^\lambda - 1$ are relatively prime. Hence,

$$\phi(2^{\lambda-1}(2^\lambda - 1)) = \phi(2^{\lambda-1})\phi(2^\lambda - 1) = (2^{\lambda-1} - 2^{\lambda-2})(2^\lambda - 1 - 1), \qquad (2)$$

since the integers 2 and $2^\lambda - 1$ are prime numbers.

By (1) and (2), it follows

$$n - \phi(n) = 2^{\lambda-1}(2^\lambda - 1) - 2^{\lambda-1}(2^{\lambda-1} - 1)$$
$$= (2^{\lambda-1})^2,$$

which is a square of an integer.

Now, it remains to determine an infinity of integer values of k, such that the difference

$$k - \phi(k)$$

is a square of an integer.

Let $k = p^{2m-1}$, where p is a prime number and $m \in \mathbb{N}$. The possible values of k are infinite. In this case, we have

$$k - \phi(k) = p^{2m-1} - \phi(p^{2m-1})$$
$$= p^{2m-1} - p^{2m-1} + p^{2m-2}$$
$$= (p^{m-1})^2,$$

which is a square of an integer. □

64) Let n be an integer greater than one. If $n = p_1^{k_1} p_2^{k_2} \cdots p_r^{k_r}$ is the canonical form of n, then prove that

$$\sum_{d|n} d\phi(d) = \frac{p_1^{2k_1+1} + 1}{p_1 + 1} \cdot \frac{p_2^{2k_2+1} + 1}{p_2 + 1} \cdots \frac{p_r^{2k_r+1} + 1}{p_r + 1}.$$

Proof. Set

$$F(n) = \sum_{d|n} d\phi(d).$$

Because of the fact that the function $F(n)$, as defined above, is multiplicative, for a prime number p one can write

$$F(p^k) = \sum_{d|p^k} d\phi(d)$$

$$= 1\phi(1) + p\phi(p) + p^2\phi(p^2) + \cdots + p^k\phi(p^k)$$

$$= 1 + p(p-1) + p^2(p^2 - p) + \cdots + p^k(p^k - p^{k-1})$$

$$= 1 + (p^2 - p) + (p^4 - p^3) + \cdots + (p^{2k} - p^{2k-1})$$

$$= 1 - p + p^2 - p^3 + p^4 - \cdots - p^{2k-1} + p^{2k}$$

$$= \frac{(-p)^{2k+1} - 1}{-p - 1}$$

$$= \frac{p^{2k+1} + 1}{p + 1}.$$

Therefore,

$$F(p^k) = \frac{p^{2k+1} + 1}{p + 1}.$$

Thus,

$$F(n) = F(p_1^{k_1} p_2^{k_2} \cdots p_r^{k_r})$$

$$= F(p_1^{k_1}) F(p_2^{k_2}) \cdots F(p_r^{k_r}).$$

Hence,

$$F(n) = \frac{p_1^{2k_1+1} + 1}{p_1 + 1} \cdot \frac{p_2^{2k_2+1} + 1}{p_2 + 1} \cdots \frac{p_r^{2k_r+1} + 1}{p_r + 1}. \qquad \square$$

65) Let n be an integer greater than one. If $n = p_1^{k_1} p_2^{k_2} \cdots p_r^{k_r}$ is the canonical form of n, then prove that

$$\sum_{d|n} \mu(d)\phi(d) = (2 - p_1)(2 - p_2) \cdots (2 - p_r).$$

Proof. If n is an integer greater than one with

$$n = p_1^{k_1} p_2^{k_2} \cdots p_r^{k_r},$$

then by Problem 2.2.5 it follows that

$$\sum_{d|n} \mu(d)\phi(d) = (1 - \phi(p_1))(1 - \phi(p_2)) \cdots (1 - \phi(p_r))$$

$$= (1 - (p_1 - 1))(1 - (p_2 - 1)) \cdots (1 - (p_r - 1)).$$

Hence,

$$\sum_{d|n} \mu(d)\phi(d) = (2 - p_1)(2 - p_2) \cdots (2 - p_r). \qquad \square$$

66) Let $n, \lambda \in \mathbb{N}$ with $\lambda > 1$ and $4 \mid n$. Solve the diophantine equation

$$\Phi(n)x + \phi(n)y = \phi(n)^\lambda, \tag{1}$$

where

$$\Phi(n) = \sum_{\substack{1 \le q < n \\ \gcd(n,q)=1}} q$$

and $\phi(n)$ is the Euler function.

Proof. We will simplify the expression (1) by finding the value of the function $\Phi(n)$.

The number of positive integers q, where $q < n$, which are relatively prime to the positive integer n is $\phi(n)$.

Consider the set

$$A = \{q_1, q_2, \ldots, q_{\phi(n)}\}.$$

For every $q_i \in A$, it follows that

$$n - q_i \in A, \quad \text{where} \quad i = 1, 2, \ldots, \phi(n).$$

This is true because if $(n, q_i) = 1$, then the $\gcd(n - q_i, n) = 1$.

Therefore, we obtain

$$\sum_{i=1}^{\phi(n)} [q_i + (n - q_i)] = \Phi(n) + \Phi(n) = 2\Phi(n). \tag{2}$$

However,

$$\sum_{i=1}^{\phi(n)} [q_i + (n - q_i)] = \sum_{i=1}^{\phi(n)} n = n\phi(n). \tag{3}$$

Therefore, by (1) and (2) we get

$$\Phi(n) = \frac{n\phi(n)}{2}.$$

Thus, the diophantine equation (1) can be written in the form

$$\frac{n\phi(n)}{2} x + \phi(n)y = \phi(n)^{\lambda}. \tag{4}$$

But, by the hypothesis it follows that $4 \mid n$, that is, $n \equiv 0 \pmod{2^2}$.

Set $n = 2^k m$, where $k \ge 2$. Then from (4), we obtain

$$2^{k-1} m \, \phi(n)x + \phi(n)y = \phi(n)^{\lambda},$$

that is,

$$2^{k-1} mx + y = \phi(n)^{\lambda-1}. \tag{5}$$

Set $a = 2^{k-1}m$, $b = 1$ and $\gcd(a, b) = d$. The diophantine equation (5) has infinitely many integer solutions since $\gcd(2^{k-1}m, 1) = 1$.

It is enough to find one solution (x_0, y_0) of (5).
One trivial solution is

$$x_0 = 0, \quad y_0 = \phi(n)^{\lambda-1}.$$

Therefore, an infinite family of solutions of (1) is defined by

$$(x, y) = \left(x_0 + \frac{b}{d}t, y_0 - \frac{a}{d}t \right) = (t, \phi(n)^{\lambda-1} - 2^{k-1}mt),$$

where $t \in \mathbb{Z}$.   □

67) Prove that

$$\sigma_1(n!) < \frac{(n+1)!}{2}$$

for all positive integers n, where $n \geq 8$.

(Crux Mathematicorum, 1990, Problem 1399, p. 58. Proposed by Sydney Bulman-Fleming and Edward T.H. Wang, Wilfried Laurier, University of Waterloo, Ontario. It was proved by Robert E. Shafer, Berkeley, California)

Proof. Set $n! = 2^{a_1} \cdot 3^{a_2} \cdot 5^{a_3} \cdots p$ where p is the greatest prime number such that $p \leq n$. Then according to Theorem 2.5.4 we get

$$
\begin{aligned}
\sigma_1(n!) &= \frac{2^{a_1+1} - 1}{1} \cdot \frac{3^{a_2+1} - 1}{2} \cdots \frac{p^2 - 1}{p - 1} \\
&= n! \cdot \frac{2^{a_1+1} - 1}{2^{a_1}} \cdot \frac{3^{a_2+1} - 1}{2 \cdot 3^{a_2}} \cdots \frac{p^2 - 1}{(p - 1)p}.
\end{aligned}
\tag{1}
$$

However, in general, if $n, a \in \mathbb{N}$, $n \neq 1$, it is clear that

$$\frac{n^{a+1} - 1}{(n - 1)n^a} < \frac{n}{n - 1}.$$

If we apply this inequality in (1), we get

$$
\begin{aligned}
\sigma_1(n!) &< n! \cdot \frac{2}{1} \cdot \frac{3}{2} \cdot \frac{5}{4} \cdots \frac{p}{p - 1} \\
&\leq n! \cdot \frac{\frac{2}{1} \cdot \frac{3}{2} \cdot \frac{4}{3} \cdot \frac{5}{4} \cdots \frac{n}{n-1} \cdot \frac{n+1}{n}}{\frac{4}{3} \cdot \frac{6}{5} \cdot \frac{8}{7} \cdot \frac{9}{8}} \\
&= \frac{(n + 1)!}{72/35} \\
&< \frac{(n + 1)!}{2}.
\end{aligned}
$$

Therefore,

$$\sigma_1(n!) < \frac{(n+1)!}{2}$$

for all positive integers n, where $n \geq 8$. $\square$

Remark 11.5. The above inequality holds also for all positive integers n, such that $n < 8$. The equality is satisfied when n takes the values $1, 2, 3, 4, 5$. The proof of these claims is a matter of several straightforward calculations.

68) Prove that

$$\sum_{n=1}^{+\infty} \frac{\tau(n)}{2^n} = \sum_{n=1}^{+\infty} \frac{1}{\phi(2^{n+1}) - 1}.$$

Proof. We have

$$\sum_{n=1}^{+\infty} \frac{1}{\phi(2^{n+1}) - 1} = \sum_{n=1}^{+\infty} \frac{1}{2^{n+1} - 2^n - 1} = \sum_{n=1}^{+\infty} \frac{1}{2^n(2-1) - 1} = \sum_{n=1}^{+\infty} \frac{1}{2^n - 1}$$

$$= \sum_{n=1}^{+\infty} \left(\frac{1}{2^n} \frac{1}{1 - 1/2^n} \right) = \sum_{n=1}^{+\infty} \left[\frac{1}{2^n} \cdot \left(1 + \frac{1}{2^n} + \frac{1}{2^{2n}} + \cdots \right) \right]$$

$$= \sum_{n=1}^{+\infty} \left(\frac{1}{2^n} + \frac{1}{2^{2n}} + \cdots \right) = \sum_{n=1}^{+\infty} \sum_{k=1}^{+\infty} \frac{1}{2^{kn}}.$$

Thus,

$$\sum_{n=1}^{+\infty} \frac{1}{\phi(2^{n+1}) - 1} = \sum_{n=1}^{+\infty} \sum_{k=1}^{+\infty} \frac{1}{2^{kn}}. \tag{1}$$

But, in the sum

$$\sum_{n=1}^{+\infty} \sum_{k=1}^{+\infty} \frac{1}{2^{kn}},$$

the integer kn obtains certain values more than once. Therefore, we can write

$$\sum_{n=1}^{+\infty} \sum_{k=1}^{+\infty} \frac{1}{2^{kn}} = \sum_{\lambda=1}^{+\infty} \frac{1}{2^\lambda} \sum_{kn=\lambda} 1 = \sum_{\lambda=1}^{+\infty} \frac{\tau(\lambda)}{2^\lambda}. \tag{2}$$

By (1) and (2) it follows that

$$\sum_{n=1}^{+\infty} \frac{\tau(n)}{2^n} = \sum_{n=1}^{+\infty} \frac{1}{\phi(2^{n+1}) - 1}.$$ $\square$

69) If f is a multiplicative arithmetic function, then

(α) Prove that

$$\sum_{d|n} f(d) = \prod_{p^a \| n} (1 + f(p) + f(p^2) + \cdots + f(p^a)),$$

where $p^a \| n$ **denotes the greatest power of the prime number p which divides n.**

(β) Prove that the function

$$g(n) = \sum_{d|n} f(d)$$

is multiplicative.

(γ) Prove that

$$\left(\sum_{d|n} \tau(d)\right)^2 = \sum_{d|n} \tau^3(d).$$

Proof.

(α)

$$\prod_{p^a \| n} (1 + f(p) + f(p^2) + \cdots + f(p^a))$$

$$= (1 + f(p_1) + f(p_1^2) + \cdots + f(p_1^{a_1})) \cdot$$
$$\cdot (1 + f(p_2) + f(p_2^2) + \cdots + f(p_2^{a_2})) \cdots$$
$$\cdots (1 + f(p_k) + f(p_k^2) + \cdots + f(p_k^{a_k})), \tag{1}$$

where the integers $p_1, p_2, \ldots, p_k$ denote the prime divisors of n and $a_1, a_2, \ldots, a_k$ satisfy the property

$$p_1^{a_1} \| n, p_2^{a_2} \| n, \ldots, p_k^{a_k} \| n.$$

Then if we carry over the calculations in the right-hand side of (1), we will derive a summation of the form

$$A = 1 + \sum_{p_1^{m_1} p_2^{m_2} \cdots p_k^{m_k} \, | \, n} f(p_1^{m_1}) f(p_2^{m_2}) \cdots f(p_k^{m_k}),$$

where $1 \le m_i \le a_i$, $1 \le i \le k$ and therefore

$$A = f(1) + \sum_{p_1^{m_1} p_2^{m_2} \cdots p_k^{m_k} | n} f(p_1^{m_1} p_2^{m_2} \cdots p_k^{m_k})$$

$$= \sum_{d|n} f(d).$$

(β) Consider two integers m, n such that $m, n \geq 1$ and $\gcd(m, n) = 1$. Then

$$g(mn) = \sum_{d|mn} f(d) = \prod_{p^a \,\|\, mn} (1 + f(p) + f(p^2) + \cdots + f(p^a)).$$

But $\gcd(m, n) = 1$, thus $p^a \,\|\, m$ or $p^a \,\|\, n$, where p^a does not divide m and n, simultaneously.

Thus, if $m = p_1^{m_1} p_2^{m_2} \cdots p_k^{m_k}$ and $n = q_1^{n_1} q_2^{n_2} \cdots q_\lambda^{n_\lambda}$, where $k, \lambda \in \mathbb{N}$ with $p_i \neq q_j$ for all i, j such that $1 \leq i \leq k$, $1 \leq j \leq \lambda$, then

$$g(mn) = \prod_{p^a \,\|\, mn} (1 + f(p) + f(p^2) + \cdots + f(p^a)) \cdot$$

$$\cdot \prod_{q^a \,\|\, mn} (1 + f(q) + f(q^2) + \cdots + f(q^a))$$

$$= g(m) \cdot g(n).$$

Therefore,

$$g(mn) = g(m) \cdot g(n),$$

where $m, n \geq 1$ and $\gcd(m, n) = 1$.

(γ) Since the function $\tau(n)$ is multiplicative, it follows that

$$\left(\sum_{d|n} \tau(d) \right)^2 = \left(\prod_{p^a \,\|\, n} (1 + \tau(p) + \tau(p^2) + \cdots + \tau(p^a)) \right)^2$$

$$= \left(\prod_{p^a \,\|\, n} (1 + 2 + 3 + \cdots + (a + 1)) \right)^2$$

$$= \prod_{p^a \,\|\, n} (1 + 2 + 3 + \cdots + (a + 1))^2$$

$$= \prod_{p^a \,\|\, n} (1^3 + 2^3 + 3^3 + \cdots + (a + 1)^3)$$

$$= \prod_{p^a \,\|\, n} (1^3 + \tau^3(p) + \tau^3(p^2) + \cdots + \tau^3(p^a))$$

$$= \sum_{d|n} \tau^3(d).$$

Thus,

$$\left(\sum_{d|n} \tau(d) \right)^2 = \sum_{d|n} \tau^3(d). \qquad \square$$

70) Consider two arithmetic functions f, g, such that

$$A(n) = \sum_{d|n} f(d)g\left(\frac{n}{d}\right)$$

and g are multiplicative.
Prove that f must also be multiplicative.

Proof. Let us suppose that f is not a multiplicative function. Therefore, there exists at least one pair of coprime positive integers a, b, such that

$$f(ab) \neq f(a)f(b).$$

Let a_0, b_0 be the pair with the above property for which the product $a_0 b_0$ has the least possible value.

It is evident that $a_0 b_0 \neq 1$, because if $a_0 b_0$ was equal to 1, then we would have $a_0 = 1$ and $b_0 = 1$ and thus

$$f(a_0 b_0) = f(1)$$

and

$$f(a_0)f(b_0) = f(1)f(1).$$

Hence,

$$f(1) \neq f(1)f(1)$$

and thus $f(1) \neq 1$. But in that case, we would have

$$A(1) = f(1)g(1) = f(1) \cdot 1 \neq 1,$$

which is impossible. Therefore, $a_0 b_0 > 1$. However, due to the property of a_0, b_0 it follows that for every pair of coprime positive integers c, d, such that $cd < a_0 b_0$, we obtain $f(cd) = f(c)f(d)$. Hence,

$$A(a_0 b_0) = \sum_{c|a_0,\, d|b_0,\, cd \leq a_0 b_0} f(cd)g\left(\frac{a_0 b_0}{cd}\right)$$

$$= f(a_0 b_0)g(1) + \sum_{c|a_0,\, d|b_0,\, cd < a_0 b_0} f(c)f(d)g\left(\frac{a_0 b_0}{cd}\right).$$

But, since g is a multiplicative function, we get

$$A(a_0 b_0) = f(a_0 b_0) + \sum_{c|a_0,\, d|b_0,\, cd < a_0 b_0} f(c)f(d)g\left(\frac{a_0}{c}\right)g\left(\frac{b_0}{d}\right)$$

$$= f(a_0 b_0) + \sum_{c|a_0} f(c)g\left(\frac{a_0}{c}\right)\sum_{d|b_0} f(d)g\left(\frac{b_0}{d}\right) - f(a_0)f(b_0)$$

$$= A(a_0)A(b_0) + f(a_0 b_0) - f(a_0)f(b_0).$$

However, since A is a multiplicative function, it is evident that

$$f(a_0 b_0) = f(a_0) f(b_0),$$

which is a contradiction. $\square$

71) Prove that

$$\sum_{n=2}^{+\infty} f(\zeta(n)) = 1,$$

where $f(x) = x - \lfloor x \rfloor$ denotes the fractional part of $x \in \mathbb{R}$ and $\zeta(s)$ is the Riemann zeta function.

(H. M. Srivastava, University of Victoria, Canada)

Proof. For $n \geq 2$, by the definition of the Riemann zeta function, it is evident that

$$1 < \zeta(n) \leq \zeta(2) = \sum_{n=1}^{+\infty} \frac{1}{n^2} = \frac{\pi^2}{6}.^3$$

Therefore, for $n \geq 2$ we have

$$0 < \zeta(n) - 1 < 1$$

and hence, for $n \geq 2$

$$\zeta(n) - 1 = f(\zeta(n)). \tag{1}$$

Thus, by (1) we obtain

$$\sum_{n=2}^{+\infty} f(\zeta(n)) = \sum_{n=2}^{+\infty} (\zeta(n) - 1) = \sum_{n=2}^{+\infty} \sum_{\lambda=2}^{+\infty} \frac{1}{\lambda^n}$$

$$= \sum_{\lambda=2}^{+\infty} \sum_{n=2}^{+\infty} \left(\frac{1}{\lambda} \right)^n = \sum_{\lambda=2}^{+\infty} \left(\left(1 - \frac{1}{\lambda}\right)^{-1} - 1 - \frac{1}{\lambda} \right)$$

$$= \sum_{\lambda=2}^{+\infty} \left(\frac{1}{\lambda - 1} - \frac{1}{\lambda} \right) = 1. \qquad \square$$

72) Prove that

$$\pi(x) \geq \log \log x$$

where $x \geq 2$.
    (Hint: Prove first the inequality

$$p_n < 2^{2^n},$$

where p_n denotes the nth prime number.)

---

[3] For further details concerning the calculation of $\zeta(2)$, see Property 7.2.4.

Proof. Because of the fact that in the inequality that we wish to prove, the double logarithm $\log \log x$ appears, we will try to bound the number x between powers of the form e^{e^n}, where $n \in \mathbb{N}$.

Thus, we consider

$$e^{e^{n-1}} < x \leq e^{e^n}, \quad \text{where} \quad n \geq 4. \tag{1}$$

(We will examine the case $n < 4$ separately.)

- We will first prove the inequality

$$p_n < 2^{2^n}.$$

For $n = 1$ it is true that $2 < 2^2$. Assume that $p_n < 2^{2^n}$, it is enough to prove that

$$p_{n+1} < 2^{2^{n+1}}.$$

It is a known fact that

$$p_{n+1} \leq p_1 p_2 \cdots p_n + 1.$$

The proof of this inequality follows easily by the fact that none of the primes $p_1, p_2, \ldots, p_n$ divides $p_1 p_2 \cdots p_n + 1$. Thus, p_{n+1} is its smallest possible prime factor. Therefore,

$$p_{n+1} \leq p_1 p_2 \cdots p_n + 1 < 2^{2^1 + 2^2 + \cdots + 2^n} + 1$$
$$< 2^{2(2^n - 1)} + 2^2 < 2^{2(2^n - 1)} \cdot 2^2 = 2^{2^{n+1}},$$

that is,

$$p_{n+1} < 2^{2^{n+1}}.$$

Therefore, by applying mathematical induction, we have shown that

$$p_n < 2^{2^n}, \quad \text{for every} \quad n \in \mathbb{N}.$$

- We will prove by applying again mathematical induction that

$$e^{n-1} > 2^n, \quad \text{forevery} \quad n \in \mathbb{N} \text{ with } n \geq 4.$$

For $n = 4$ it is true that $e^3 > 2^4$. We assume that $e^{n-1} > 2^n$ holds true and we shall prove that $e^n > 2^{n+1}$.

One has

$$e^n = e^{n-1} e > 2^n e > 2^n 2 = 2^{n+1}.$$

Therefore,

$$e^{n-1} > 2^n, \quad \text{for every} \quad n \text{ with } n \geq 4.$$

Hence,

$$e^{e^{n-1}} > 2^{2^n}.$$

- From the above inequality and (1) we derive that

$$\pi(x) \geq \pi(e^{e^{n-1}}) \geq \pi(2^{2^n}).$$

But

$$p_n < 2^{2^n} \quad \text{and} \quad \pi(p_n) = n,$$

that is,

$$\pi(2^{2^n}) > \pi(p_n) = n. \tag{2}$$

Thus,

$$\pi(x) \geq n.$$

From (1) it follows that

$$\log\log x \leq n. \tag{3}$$

From (2) and (3) one has

$$\pi(x) \geq \log\log x, \quad \text{for} \quad x > e^{e^3}.$$

It is enough to prove that

$$\pi(x) \geq \log\log x, \quad \text{for} \quad 2 \leq x \leq e^{e^3} \quad \text{(that is, for } n < 4\text{)}.$$

If $5 \leq x \leq e^{e^3}$, the proof is obvious because

$$\log\log x \leq 3 \quad \text{and} \quad \pi(x) \geq \pi(5) = 3.$$

If $2 \leq x \leq 5$, one has $\log\log x \leq \log\log 5 < 0.48$ and $\pi(x) \geq \pi(2) = 1$.
    Therefore,

$$\pi(x) > \log\log x.$$

Hence, in general it is true that

$$\pi(x) \geq \log\log x, \quad \text{for} \quad x \geq 2. \qquad \square$$

Remark 11.6. The result in the above property is "very weak," since the values of the function $\pi(x)$ are increasing much faster than the function $\log\log x$.
    For example, if $x = 10^{12}$, the inequality gives

$$\pi(10^{12}) \geq 3.318\ldots,$$

however

$$\pi(10^{12}) = 37607912018.$$

73) Prove that any integer can be expressed as the sum of the cubes of five integers not necessarily distinct.

(T. Andreescu, D. Andrica and Z. Feng)

Proof. It is evident that for every integer m the following equality holds:

$$(m+1)^3 + (m-1)^3 - 2m^3 = 6m. \tag{1}$$

Set in (1) in place of m the number

$$\frac{n^3 - n}{6}, \tag{2}$$

which is an integer since

$$n^3 - n = (n-1)n(n+1)$$

is the product of three consequtive integers and thus

$$n^3 - n = \text{mult. } 6.$$

From (1) and (2), one obtains

$$n^3 - n = \left(\frac{n^3-n}{6} + 1\right)^3 + \left(\frac{n^3-n}{6} - 1\right)^3 - \left(\frac{n^3-n}{6}\right)^3 - \left(\frac{n^3-n}{6}\right)^3. \tag{3}$$

From (3) it follows

$$\left(\frac{n^3-n}{6}\right)^3 + \left(\frac{n^3-n}{6}\right)^3 + \left(-\frac{n^3-n}{6} - 1\right)^3 + \left(-\frac{n^3-n}{6} + 1\right)^3 + n^3 = n.$$

Hence,

$$n^3 + \left(\frac{n^3-n}{6}\right)^3 + \left(\frac{n^3-n}{6}\right)^3 + \left(\frac{n-n^3}{6} - 1\right)^3 + \left(\frac{n-n^3}{6} + 1\right)^3 = n.$$

This completes the proof of the property. □

74) Let n be an integer. An integer A is formed by $2n$ digits each of which is 4; however, another integer B is formed by n digits each of which is 8. Prove that the integer

$$A + 2B + 4$$

is a perfect square of an integer.

(7ᵗʰ Balcan Mathematical Olympiad, Kusadasi, Turkey)

Proof. One has

$$A = \underbrace{444\ldots4}_{2n} \quad \text{and} \quad B = \underbrace{888\ldots8}_{n}.$$

Thus,

$$A = \underbrace{444\ldots4}_{2n} = \underbrace{444\ldots4}_{n}\underbrace{000\ldots0}_{n} + \underbrace{444\ldots4}_{n}$$

$$= \underbrace{444\ldots4}_{n}\cdot(10^n - 1) + \underbrace{888\ldots8}_{n}$$

$$= \underbrace{444\ldots4}_{n}\cdot(10^n - 1) + B.$$

Therefore,

$$A = 4\cdot\underbrace{111\ldots1}_{n}\cdot\underbrace{999\ldots9}_{n} + B = 4\cdot\underbrace{111\ldots1}_{n}\cdot9\cdot\underbrace{111\ldots1}_{n} + B$$

$$= 6\cdot\underbrace{111\ldots1}_{n}\cdot6\cdot\underbrace{111\ldots1}_{n} + B = (\underbrace{666\ldots6}_{n})^2 + B$$

$$= (3\cdot\underbrace{222\ldots2}_{n})^2 + B = \left(\frac{3}{4}\cdot\underbrace{888\ldots8}_{n}\right)^2 + B = \left(\frac{3}{4}\cdot B\right)^2 + B.$$

Thus, we obtain

$$A + 2B + 4 = \left(\frac{3}{4}\cdot B\right)^2 + B + 2B + 4$$

$$= \left(\frac{3}{4}\cdot B\right)^2 + 2\cdot\frac{3}{4}B\cdot2 + 2^2$$

$$= \left(\frac{3}{4}\cdot B + 2\right)^2 = \left(\frac{3}{4}\cdot\underbrace{888\ldots8}_{n} + 2\right)^2$$

$$= \left(3\cdot\underbrace{222\ldots2}_{n} + 2\right)^2 = \left(\underbrace{666\ldots68}_{n-1}\right)^2,$$

which is a perfect square of an integer. □

75) Find the integer values of x for which the expression $x^2 + 6x$ is a square of an integer.

Solution. It is enough to determine the integer values of x for which

$$x^2 + 6x = y^2, \quad \text{where} \quad y \in \mathbb{Z}. \tag{1}$$

From (1), we obtain the quadratic equation for x:

$$x^2 + 6x - y^2 = 0.$$

Therefore,

$$x = -3 \pm \sqrt{y^2 + 9}, \quad \text{where} \quad y \in \mathbb{Z}. \tag{2}$$

For (2) to be an integer there must exist $u \in \mathbb{Z}$, such that

$$y^2 + 9 = u^2,$$

that is,

$$u^2 - y^2 = 9$$

which implies

$$(|u| - |y|)(|u| + |y|) = 9. \tag{3}$$

Equation (3) is equivalent to the following:

$$(|u| - |y|)(|u| + |y|) = 9 \cdot 1. \tag{4}$$

However,

$$|u| - |y| < |u| + |y|, \quad \text{where} \quad u, y \in \mathbb{Z},$$

thus (4) can be written in the form of a system

$$\begin{cases} |u| + |y| = 9 \\ |u| - |y| = 1. \end{cases} \tag{5}$$

From (5), we obtain

$$|u| = 5, |y| = 4. \tag{6}$$

Therefore, from (2) and (6) it follows

$$x = -8 \quad \text{or} \quad x = 2.$$

Hence, the integer values of x for which $x^2 + 6x$ is square of an integer are

$$x = -8, x = 2. \qquad \square$$

76) Express the integer 459 as the sum of four squares of integers.

Solution. It is a known fact that if the integers m and n are integers which can be expressed as the sum of four squares of integers, then the product mn can also be expressed as the sum of four squares of integers.

More explicitly, by Euler's identity one has the following:

If

$$m = a_1^2 + a_2^2 + a_3^2 + a_4^2$$

and

$$n = b_1^2 + b_2^2 + b_3^2 + b_4^2,$$

where $a_1, a_2, a_3, a_4, b_1, b_2, b_3, b_4$ are integers, then

$$
\begin{aligned}
mn &= (a_1^2 + a_2^2 + a_3^2 + a_4^2)(b_1^2 + b_2^2 + b_3^4 + b_4^2) \\
&= (a_1b_1 + a_2b_2 + a_3b_3 + a_4b_4)^2 \\
&\quad + (a_1b_2 - a_2b_1 + a_3b_4 - a_4b_3)^2 \\
&\quad + (a_1b_3 - a_2b_4 - a_3b_1 + a_4b_2)^2 \\
&\quad + (a_1b_4 + a_2b_3 - a_3b_2 - a_4b_1)^2.
\end{aligned}
$$

Thus,

$$
\begin{aligned}
459 = 3^3 \cdot 17 &= 3^2 \cdot 3 \cdot 17 \\
&= 3^2(1^2 + 1^2 + 1^2 + 0^2)(4^2 + 1^2 + 0^2 + 0^2) \\
&= 3^2[(4 + 1 + 0 + 0)^2 + (1 - 4 + 0 - 0)^2 \\
&\quad + (0 - 0 - 4 + 0)^2 + (0 + 0 - 1 - 0)^2] \\
&= 3^2(5^2 + 3^2 + 4^2 + 1^2).
\end{aligned}
$$

Hence,

$$459 = (3 \cdot 5)^2 + (3 \cdot 3)^2 + (3 \cdot 4)^2 + (3 \cdot 1)^2,$$

that is,

$$459 = 15^2 + 9^2 + 12^2 + 3^2. \qquad \square$$

77) Find the three smallest positive consecutive natural numbers, whose sum is a perfect square and a perfect cube of a natural number.

(M. Th. Rassias, Proposed problem No. 94, Euclid Mathematical Magazine B', Greek Math. Soc., 62(2006), p. 80)

Solution. Let $n-1, n, n+1$ be the three consecutive natural numbers, where $n \in \mathbb{N} - \{1\}$. Then, we obtain

$$(n-1) + n + (n+1) = 3n.$$

Thus, the natural number $3n$ must be a perfect square as well as a perfect cube of a natural number, that is

$$3n = a^2 \quad \text{and} \quad 3n = b^3, \quad \text{where} \quad a, b \in \mathbb{N}.$$

Thus,

$$3 \mid a^2 \quad \text{and} \quad 3 \mid b^3,$$

and therefore
$$3 \mid a \quad \text{and} \quad 3 \mid b$$
(since 3 is a prime number).

From the Fundamental Theorem of Arithmetic we can write
$$a = p_1^{a_1} 3^{a_2} \cdots p_k^{a_k} \quad \text{and} \quad b = p_1^{b_1} 3^{b_2} \cdots p_k^{b_k},$$
where $p_1, \ldots, p_k$ are prime numbers.

Therefore,
$$p_1^{2a_1} 3^{2a_2} \cdots p_k^{2a_k} = p_1^{3b_1} 3^{3b_2} \cdots p_k^{3b_k}.$$

Consequently,
$$2a_2 = 3b_2. \tag{1}$$

This implies that if the natural number $3n$ is expressed in canonical form, the exponent of 3 will be simultaneously a multiple of 2 as well as a multiple of 3. For the determination of the smallest natural numbers with the requested property, the number $3n$ must be the smallest possible.

This happens if
$$3n = 3^k,$$
where k is the smallest possible exponent of 3.

From (1) it follows that the number k is the smallest common multiple of 2 and 3.

Therefore,
$$3n = 3^6 \Leftrightarrow n = 3^5.$$

Hence, the requested natural numbers are the following:
$$3^5 - 1, \quad 3^5, \quad 3^5 + 1.$$

In fact, one has
$$(3^5 - 1) + 3^5 + (3^5 + 1) = 3 \cdot 3^5 = 3^6 = (3^3)^2 = (3^2)^3. \qquad \square$$

78) Find all prime numbers p such that the number
$$\frac{2^{p-1} - 1}{p}$$
is a square of an integer.

(S. E. Louridas, Athens, Greece)

Solution. If $p = 2$, then obviously
$$\frac{2^{p-1} - 1}{p} = \frac{1}{2}$$
is not a square of an integer.

For $p > 2$, it follows that

$$\frac{2^{p-1} - 1}{p} = \frac{(2^{\frac{p-1}{2}} + 1)(2^{\frac{p-1}{2}} - 1)}{p}.$$

Furthermore, we have

$$\gcd(2^{\frac{p-1}{2}} + 1, 2^{\frac{p-1}{2}} - 1) = 1.$$

That is true because if there existed a prime number q such that

$$q \mid 2^{\frac{p-1}{2}} + 1 \quad \text{and} \quad q \mid 2^{\frac{p-1}{2}} - 1,$$

then clearly we would have $q \mid 2$ and thus $q = 2$, which is impossible.

Let us consider that $2^{\frac{p-1}{2}} + 1$ is divisible by p. Then we obtain that

$$\frac{2^{\frac{p-1}{2}} + 1}{p} \cdot (2^{\frac{p-1}{2}} - 1) = a^2,$$

for some integer a. However, since

$$\gcd\left(\frac{2^{\frac{p-1}{2}} + 1}{p}, 2^{\frac{p-1}{2}} - 1\right) = 1,$$

it is evident that both

$$\frac{2^{\frac{p-1}{2}} + 1}{p} \quad \text{and} \quad 2^{\frac{p-1}{2}} - 1$$

must be squares of integers. Now, set $2^{\frac{p-1}{2}} - 1 = b^2$, where $b = 2k+1$, $k \in \mathbb{Z}$. Thus,

$$2^{\frac{p-1}{2}} - 1 = 4k^2 + 4k + 1 = 4k(k + 1) + 1.$$

But, by the above relation it follows that $(p-1)/2 < 2$, because if that was not the case, then $2^{\frac{p-1}{2}}$ would be divisible by 4, which is impossible. Therefore, $2 < p < 5$ or $p = 3$. In this case

$$\frac{2^{p-1} - 1}{p} = 1,$$

which is a square of an integer.

Let us now set

$$2^{\frac{p-1}{2}} + 1 = c^2, \quad \text{where} \quad c = 2m + 1, m \in \mathbb{Z}.$$

Then, we obtain that

$$2^{\frac{p-1}{2}} = 4m(m + 1).$$

If $m > 1$, then because of the fact that either m or $m + 1$ is odd, it follows that $4m(m + 1)$ has an odd divisor and therefore $2^{\frac{p-1}{2}}$ has an odd divisor, which is impossible. Thus, $m = 1$ and hence

$$2^{\frac{p-1}{2}} = 8$$

or

$$\frac{p - 1}{2} = 3$$

or

$$p = 7.$$

In this case

$$\frac{2^{p-1} - 1}{p} = 9 = 3^2.$$

Therefore,

$$p = 3 \quad \text{or} \quad p = 7. \qquad \qquad \square$$

79) Let n be a positive integer, such that the $\gcd(n, 6) = 1$. Prove that the sum of n squares of consecutive integers is a multiple of n.

Proof. Let

$$k = a^2 + (a + 1)^2 + (a + 2)^2 + \cdots + (a + n - 1)^2.$$

Then we obtain

$$k = a^2 + (a^2 + 2a + 1) + (a^2 + 2 \cdot 2a + 2^2) + \cdots$$
$$+ [a^2 + 2(n - 1)a + (n - 1)^2]$$
$$= na^2 + 2 \cdot \frac{(n - 1)n}{2} a + \frac{(n - 1)n(2n - 1)}{6}$$
$$= na^2 + an(n - 1) + n \cdot \frac{(n - 1)(2n - 1)}{6}.$$

To complete the proof it is enough to prove that the number

$$\frac{(n - 1)(2n - 1)}{6}$$

is an integer.

We know that the $\gcd(n, 6) = 1$, that is, the integer n is odd and therefore $n - 1$ is even. Since the $\gcd(n, 6) = 1$, it follows that

$$n = 3\lambda + 1 \quad \text{or} \quad n = 3\lambda - 1, \quad \text{where } \lambda \in \mathbb{Z},$$

because the integer n cannot be a multiple of 3.

- If $n = 3\lambda + 1$, then $n - 1 = 3\lambda$. But, the number $n - 1$ is also even. Thus,

$$n - 1 \equiv 0 \, (\text{mod } 6) \text{ and thus } (n-1)(2n-1) \equiv 0 \, (\text{mod } 6).$$

- If $n = 3\lambda - 1$, then $2n - 1 = 3(2\lambda - 1) \equiv 0 \, (\text{mod } 3)$.

    Since $2 \nmid n$ and $n = 3\lambda - 1$, it follows that $2 \nmid \lambda - 1$. Thus, $\lambda - 1 = 2\kappa - 1$, $\kappa \in \mathbb{Z}$. Therefore, $\lambda = 2\kappa$. Hence, $n = 3\lambda - 1 = 6\kappa - 1$ and thus

$$n - 1 = 6\kappa - 2 \equiv 0 \, (\text{mod } 2).$$

Thus,

$$(n - 1)(2n - 1) \equiv 0 \, (\text{mod } 6).$$

Therefore,

$$\frac{(n-1)(2n-1)}{6} \in \mathbb{Z}$$

and thus the integer k is a multiple of n, which completes the proof. □

80) Prove that for every $m \in \mathbb{N} - \{1, 2\}$, such that the integer $7 \cdot 4^m$ can be expressed as a sum of four squares of nonnegative integers a, b, c, d, each of the numbers a, b, c, d is at least equal to 2^{m-1}.

(W. Sierpiński, 250 Problèmes de Théorie Élémentaire des Nombres, P.W., Warsaw, 1970)

Proof. From Lagrange's theorem we know that every positive integer can be represented as a sum of four squares of positive integers. Therefore, the integer $7 \cdot 4^m$ can be expressed as a sum of four squares of positive integers.

In the special case of the given integer $7 \cdot 4^m$, we can prove the above claim without making use of Lagrange's theorem and this happens because

$$7 \cdot 4^m = 3 \cdot 4^m + 4 \cdot 4^m = 3 \cdot (2^m)^2 + 3 \cdot 4^m$$

$$= (2^m)^2 + (2^m)^2 + (2^m)^2 + (2^{m+1})^2.$$

Let

$$7 \cdot 4^m = a^2 + b^2 + c^2 + d^2$$

with one of the numbers a, b, c, d less than 2^{m-1}.

Without loss of generality we assume that

$$0 \le a \le 2^{m-2}.$$

However, we don't allow $a = 0$ because in that case the number $7 \cdot 4^m$ will be represented as a sum of three squares. To prove that

$$7 \cdot 4^m \ne k_1^2 + k_2^2 + k_3^2, \text{ for every } k_1, k_2, k_3 \in \mathbb{N}$$

let

$$7 \cdot 4^m = k_1^2 + k_2^2 + k_3^2, \text{ for some } k_1, k_2, k_3 \in \mathbb{N}.$$

Then

$$4 \mid 7 \cdot 4^m \Rightarrow 4 \mid (k_1^2 + k_2^2 + k_3^2).$$

Therefore, the integer number $k_1^2 + k_2^2 + k_3^2$ is even, which means either k_1, k_2, k_3 must all be even or two of the numbers k_1, k_2, k_3 must be odd and the third one must be even. We will prove that the second case leads to a contradiction.

- If k_1, k_2, k_3 are even numbers, then the claim is obviously true.
- If two of the numbers k_1, k_2, k_3 are odd but the third one is even, then without loss of generality we consider k_3 to be even and k_1, k_2 to be odd numbers.
  Then

$$\frac{k_1^2 + k_2^2 + k_3^2}{4} \in \mathbb{Z} \Rightarrow \left(\frac{k_1}{2}\right)^2 + \left(\frac{k_2}{2}\right)^2 + \left(\frac{k_3}{2}\right)^2 \in \mathbb{Z}.$$

However,

$$\left(\frac{k_3}{2}\right)^2 \in \mathbb{Z},$$

thus

$$\left(\frac{k_1}{2}\right)^2 + \left(\frac{k_2}{2}\right)^2 \in \mathbb{Z}.$$

Therefore,

$$\frac{k_1^2 + k_2^2}{4} \in \mathbb{Z}$$

or

$$\frac{(2\lambda_1 + 1)^2 + (2\lambda_2 + 1)^2}{4} \in \mathbb{Z}, \text{ where } \lambda_1, \lambda_2 \in \mathbb{N}$$

or

$$\frac{(8\mu_1 + 1)^2 + (8\mu_2 + 1)^2}{4} \in \mathbb{Z}, \text{ where } \mu_1, \mu_2 \in \mathbb{N}$$

or

$$2(\mu_1 + \mu_2) + \frac{1}{2} \in \mathbb{Z},$$

which is a contradiction.

Therefore, we have proved that all three integers k_1, k_2, k_3 are even.

Thus, we obtain that

$$7 \cdot 4^m = 4(\xi_1^2 + \xi_2^2 + \xi_3^2), \text{ where } \xi_1, \xi_2, \xi_3 \in \mathbb{N},$$

that is,

$$7 \cdot 4^{m-1} = \xi_1^2 + \xi_2^2 + \xi_3^2.$$

If $m - 1 \geq 1$, then, similarly, the numbers ξ_1, ξ_2, ξ_3 are also even numbers and therefore we continue in the same way until we come up with the result that

$$7 \cdot 4^0 = r_1^2 + r_2^2 + r_3^2, \text{ where } r_1, r_2, r_3 \in \mathbb{N}.$$

This means that the number 7 is written as a sum of three squares. This is obviously impossible.

Hence, the integer $7 \cdot 4^m$ cannot be written as a sum of three squares of integers. This implies that a can never take zero value.

Because of the fact that $0 < a \leq 2^{m-2}$, it follows that we can consider a to be written in the form

$$a = 2^r(2k - 1), \text{ where } r, k \in \mathbb{N} \text{ with } r \leq m - 2$$

for suitable values of r, k.

Thus, we obtain that

$$7 \cdot 4^m = a^2 + b^2 + c^2 + d^2 \Leftrightarrow 7 \cdot 4^m - a^2 = b^2 + c^2 + d^2$$

$$\Leftrightarrow 7 \cdot 4^m - [2^r(2k - 1)]^2 = b^2 + c^2 + d^2.$$

Therefore,

$$7 \cdot 4^m - [2^r(2k - 1)]^2 = 7 \cdot 4^m - 4^r(8\mu + 1), \ \mu \in \mathbb{N}$$

$$= 4^r(7 \cdot 4^{m-r} - (8\mu + 1)).$$

However,

$$r \leq m - 2 \Leftrightarrow m - r \geq 2.$$

Therefore, in the product form of 4^{m-r} the factor 8 appears at least once.[4] That is,

$$7 \cdot 4^m = 4^r(8\mu' - 8\mu - 1), \mu' \in \mathbb{N}.$$

Thus,

$$7 \cdot 4^m = 4^r(8z - 1), z \in \mathbb{N}$$

$$= 4^r[8(z - 1) + 7].$$

This implies that the integer $7 \cdot 4^m$ can be written in the form $4^r(8n + 7)$, $n \in \mathbb{N}$. Therefore,

$$4^r(8n + 7) = b^2 + c^2 + d^2.$$

We will now prove that a number of the form $4^r(8n+7)$ cannot be represented as a sum of three squares of integers. If we follow the same method with the one we used to prove that the integer $7 \cdot 4^m$ cannot be represented as a sum of three squares of integers, we conclude with the result that

$$8n + 7 = b_1^2 + c_1^2 + d_1^2,$$

which is not possible.

---

[4] $4^{m-r} \geq 4^2 = 4 \cdot 4 = 8 \cdot 2.$

This is the case because if

$$8n + 7 = b_1^2 + c_1^2 + d_1^2$$

for some positive integers b_1, c_1, d_1, then these numbers will all be odd or two of these will be even and the third one will be odd.

- If all integers b_1, c_1, d_1 are odd, then we have

$$b_1^2 + c_1^2 + d_1^2 = (8\mu_1 + 1) + (8\mu_2 + 1) + (8\mu_3 + 1), \quad \mu_1, \mu_2, \mu_3 \in \mathbb{N}$$

and therefore

$$b_1^2 + c_1^2 + d_1^2 \equiv 3 \,(\mathrm{mod}\,8) \Rightarrow 8n + 7 \equiv 3 \,(\mathrm{mod}\,8) \tag{1}$$

However,

$$8n + 7 \equiv 7 \,(\mathrm{mod}\,8). \tag{2}$$

From (1) and (2) it follows that

$$0 \equiv 4 \,(\mathrm{mod}\,8),$$

which is impossible.

- If two of the numbers b_1, c_1, d_1 are even and the third number is odd, then without loss of generality we can assume d_1 to be an odd number. Thus,

$$b_1^2 + c_1^2 + d_1^2 = 4\mu_1^2 + 4\mu_2^2 + 8\mu_3 + 1 \equiv 1 \,(\mathrm{mod}\,4)$$

$$\Leftrightarrow b_1^2 + c_1^2 + d_1^2 \equiv 1 \,(\mathrm{mod}\,4)$$

$$\Leftrightarrow 8n + 7 \equiv 1 \,(\mathrm{mod}\,4). \tag{3}$$

But, it is true that

$$8n + 7 \equiv 3 \,(\mathrm{mod}\,4). \tag{4}$$

From (3) and (4) it follows that

$$0 \equiv 2 \,(\mathrm{mod}\,4),$$

which is impossible.

Therefore, in all cases, the hypothesis that at least one of the numbers a, b, c, d is less than 2^{m-1}, leads to a contradiction.

Hence, each of the positive integers a, b, c, d must be at least equal to 2^{m-1}.

$\square$

81) Let n be a positive integer and d_1, d_2, d_3, d_4 the smallest positive integer divisors of n with $d_1 < d_2 < d_3 < d_4$. Find all integer values of n, such that

$$n = d_1^2 + d_2^2 + d_3^2 + d_4^2.$$

Proof. If x is an even positive integer, then $x^2 \equiv 0 \,(\mathrm{mod}\ 4)$ and if x is an odd positive integer, then $x^2 \equiv 1 \,(\mathrm{mod}\ 4)$.

In the case when n is an odd positive integer, then all four of its divisors d_1, d_2, d_3, d_4 are necessarily odd positive integers.

Then
$$n = d_1^2 + d_2^2 + d_3^2 + d_4^2 \equiv 0 \,(\mathrm{mod}\ 4),$$
which is not possible.

Therefore, the positive integer n is even, that is $2 \mid n$.

If $4 \mid n$, then $d_1 = 1$ and $d_2 = 2$ and thus
$$n = 1 + 4 + d_3^2 + d_4^2. \tag{1}$$

Because of the fact n is an even number, it follows that one of the numbers d_3, d_4 must be odd and the other one must be even.

From (1) we obtain
$$n = 1 + 4 + (8k + 1) + 4\lambda = 4(2k + \lambda + 1) + 2, \ \ \text{where } k, \lambda \in \mathbb{N}.$$

Thus,
$$4 \nmid n.$$

It is evident that d_3 must be a prime number, because otherwise there should exist a positive integer m such that $m \mid d_3$. Then we would have
$$1 < 2 < m < d_3 \text{ and } m \mid n,$$
which is not possible, because in that case d_3 would not be the third smallest divisor of n. However, this contradicts the hypothesis.

Since d_3 is an odd number, it follows that d_4 will be an even number.

For d_4 to obtain the smallest value, it must take the form 2μ, $\mu \in \mathbb{N}$, where μ is one of the numbers d_1, d_2, d_3. Thus, it must hold
$$d_4 = 2d_3.$$

Consequently, we obtain $(d_1, d_2, d_3, d_4) = (1, 2, p, 2p)$, where p is a prime number. Therefore,
$$n = 5(1 + p^2), \text{ that is, } 5 \mid n.$$

Thus, $p = d_3 = 5$ and $n = 130$. Hence, the only positive integer which satisfies the hypothesis of the problem is the number 130. $\square$

82) Let a, b be two positive integers, such that
$$ab + 1 \mid a^2 + b^2.$$

Prove that the integer
$$\frac{a^2 + b^2}{ab + 1}$$
is a perfect square of a positive integer.

(Shortlist, 29th International Mathematical Olympiad, 1988)

Proof.
- Set

$$\frac{a^2 + b^2}{a\,b + 1} = q \in \mathbb{N}.$$

Then

$$a^2 + b^2 = q\,a\,b + q. \tag{1}$$

Without loss of generality we consider $a \leq b$. We claim that

$$q\,a - a < b \leq q\,a. \tag{2}$$

Proof of (2). It is true that

$$q\,a\,b < q\,a\,b + q = a^2 + b^2 \leq a\,b + b^2$$

$$\Leftrightarrow q\,a\,b < a\,b + b^2$$

$$\Leftrightarrow q\,a < a + b$$

$$\Leftrightarrow q\,a - a < b,$$

which proves the left-hand side of inequality (2). Suppose that $b > qa$. Then $b = qa + c$, where $c \in \mathbb{N}$. Therefore,

$$(1) \Leftrightarrow a^2 + (q\,a + c)^2 = q\,a\,(q\,a + c) + q$$

$$\Leftrightarrow a^2 + q^2 a^2 + 2\,q\,a\,c + c^2 = q^2 a^2 + q\,a\,c + q$$

$$\Leftrightarrow a^2 + q\,a\,c + c^2 = q.$$

However, it holds

$$q \leq q\,a\,c < a^2 + q\,a\,c + c^2 = q \Leftrightarrow q < q,$$

which is not possible. Thus,

$$b \leq q\,a.$$

This completes the proof of inequality (2).
- From (2) it follows that

$$b = q\,a - m,$$

where $0 \leq m < a$ and $m \in \mathbb{N} \cup \{0\}$.

From (1) it follows that

$$a^2 + b^2 = q\,a\,b + q \Leftrightarrow a^2 + (q\,a - m)^2 = q\,a\,(q\,a - m) + q$$

$$\Leftrightarrow a^2 + q^2 a^2 - 2\,q\,a\,m + m^2 = q^2 a^2 - q\,a\,m + q$$

$$\Leftrightarrow a^2 + m^2 = q\,a\,m + q$$

$$\Leftrightarrow a^2 + m^2 = q\,(a\,m + 1)$$

$$\Leftrightarrow \frac{a^2 + m^2}{a\,m + 1} = q.$$

Thus, if there exists a pair (a, b) which satisfies the relation

$$\frac{a^2 + b^2}{a\,b + 1} = q,$$

then there exists a pair (m, a) which satisfies the property

$$\frac{m^2 + a^2}{m\,a + 1} = q.$$

Similarly, since there exists the pair (m, a) which satisfies

$$\frac{m^2 + a^2}{m\,a + 1} = q$$

there will also exist another pair (m_1, a) which will satisfy

$$\frac{m_1^2 + a^2}{m_1 a + 1} = q.$$

Continuing with this process and since $0 \leq m < a \leq b$ we get $m = 0$.
    Therefore,
$$\frac{a^2 + m^2}{a\,m + 1} = q \Leftrightarrow q = a^2 \quad (\text{for} \quad m = 0),$$

which is a perfect square of a positive integer. $\square$

83) Let k be an integer, which can be expressed as a sum of two squares of integers, that is,

$$k = a^2 + b^2 \quad \text{with} \quad a, b \in \mathbb{Z}.$$

If p is a prime number greater than 2, which can be expressed as a sum of two squares of integers c, d for which it holds

$$(c^2 + d^2) \mid (a^2 + b^2) \quad \text{and} \quad (c^2 + d^2) \nmid (a + b),$$

prove that the integer

$$\frac{a^2 + b^2}{c^2 + d^2} = \frac{k}{p}$$

can be expressed as a sum of two squares of integers.

Proof. We will make use of the following identity:

$$(x_1^2 + x_2^2)(x_3^2 + x_4^2) = (x_1 x_3 \pm x_2 x_4)^2 + (x_1 x_4 \mp x_2 x_3)^2. \tag{1}$$

• Set $k/p = m \in \mathbb{Z}$. Then,

$$m = \frac{a^2 + b^2}{c^2 + d^2} = \frac{(a^2 + b^2)(c^2 + d^2)}{(c^2 + d^2)^2}.$$

Therefore, by means of (1), we obtain

$$m = \frac{(a\,c \pm b\,d)^2 + (a\,d \mp b\,c)^2}{(c^2 + d^2)^2} = \frac{(a\,c \pm b\,d)^2}{(c^2 + d^2)^2} + \frac{(a\,d \mp b\,c)^2}{(c^2 + d^2)^2}. \qquad (2)$$

It is enough to prove that

$$(c^2 + d^2) \mid (a\,c + b\,d) \quad \text{and} \quad (c^2 + d^2) \mid (a\,d - b\,c)$$

or

$$(c^2 + d^2) \mid (a\,c - b\,d) \quad \text{and} \quad (c^2 + d^2) \mid (a\,d + b\,c).$$

• We will first prove that

$$(c^2 + d^2) \mid (a\,c + b\,d) \quad \text{or} \quad (c^2 + d^2) \mid (a\,c - b\,d).$$

In order to prove this, it is the same to show that

$$(c^2 + d^2) \mid (a\,c + b\,d)\,(a\,c - b\,d)$$

(since $c^2 + d^2$ is a prime number).

However,

$$(a\,c + b\,d)(a\,c - b\,d) = a^2c^2 - b^2d^2$$

$$= a^2c^2 + b^2c^2 - (b^2c^2 + b^2d^2)$$

$$= (a^2 + b^2)c^2 - b^2(c^2 + d^2)$$

$$= m(c^2 + d^2)c^2 - b^2(c^2 + d^2)$$

$$= (c^2m - b^2)(c^2 + d^2).$$

Thus,

$$(c^2 + d^2) \mid (a\,c + b\,d) \quad \text{or} \quad (c^2 + d^2) \mid (a\,c - b\,d).$$

• Similarly, we can prove that

$$(c^2 + d^2) \mid (a\,d - b\,c) \quad \text{or} \quad (c^2 + d^2) \mid (a\,d + b\,c)$$

because

$$(a\,d - b\,c)(a\,d + b\,c) = a^2d^2 - b^2c^2$$

$$= a^2d^2 + d^2b^2 - (d^2b^2 + b^2c^2)$$

$$= d^2(a^2 + b^2) - b^2(c^2 + d^2)$$

$$= d^2m(c^2 + d^2) - b^2(c^2 + d^2)$$

$$= (d^2m - b^2)(c^2 + d^2).$$

Thus,
$$(c^2 + d^2) \mid (a\,d - b\,c) \quad \text{or} \quad (c^2 + d^2) \mid (a\,d + b\,c).$$

- In the case where $(c^2 + d^2) \mid (a\,c + b\,d)$ we will prove that
$$(c^2 + d^2) \mid (a\,d - b\,c).$$

This is true since if $c^2 + d^2$ was a divisor of $a\,d + b\,c$, then we would have
$$(c^2 + d^2) \mid [(a\,c + b\,d) + (a\,d + b\,c)] \Rightarrow (c^2 + d^2) \mid [a(c + d) + b(c + d)]$$
$$\Rightarrow (c^2 + d^2) \mid (a + b)(c + d).$$

Therefore,
$$(c^2 + d^2) \mid (a + b) \quad \text{or} \quad (c^2 + d^2) \mid (c + d).$$

But $c^2 + d^2 > c + d$, thus
$$(c^2 + d^2) \mid (a + b),$$

which is not possible because of the hypothesis.

Thus in this case, because of (2), it follows
$$m = \frac{(a\,c + b\,d)^2}{(c^2 + d^2)^2} + \frac{(a\,d - b\,c)^2}{(c^2 + d^2)^2} = \frac{\lambda_1^2(c^2 + d^2)^2}{(c^2 + d^2)^2} + \frac{\lambda_2^2(c^2 + d^2)^2}{(c^2 + d^2)^2},$$

where
$$\lambda_1 = \frac{ac + bd}{c^2 + d^2} \quad \text{and} \quad \lambda_2 = \frac{ad - bc}{c^2 + d^2},$$

that is,
$$m = \lambda_1^2 + \lambda_2^2, \quad \text{where} \quad \lambda_1, \lambda_2 \in \mathbb{Z},$$

which proves the claim.

- In the case where $(c^2 + d^2) \mid (a\,c - b\,d)$ we will prove that
$$(c^2 + d^2) \mid (a\,d + b\,c).$$

Similarly, this holds because if $(c^2 + d^2) \mid (a\,d - b\,c)$, then
$$(c^2 + d^2) \mid ac - bd - (ad - bc) \Rightarrow (c^2 + d^2) \mid (a + b)c - (a + b)d$$
$$\Rightarrow (c^2 + d^2) \mid (a + b)(c - d).$$

Therefore,
$$(c^2 + d^2) \mid (a + b) \quad \text{or} \quad (c^2 + d^2) \mid (c - d).$$

But
$$c^2 + d^2 > |c - d| \quad \text{or} \quad c \neq d.$$

It follows that
$$(c^2 + d^2) \mid (a + b),$$

which is not possible. Therefore, in this case, because of (2) it follows that

$$m = \frac{(ac - bd)^2}{(c^2 + d^2)^2} + \frac{(ad + bc)^2}{(c^2 + d^2)^2}$$

$$= \frac{\lambda_3^2(c^2 + d^2)^2}{(c^2 + d^2)^2} + \frac{\lambda_4^2(c^2 + d^2)^2}{(c^2 + d^2)^2}$$

$$= \lambda_3^2 + \lambda_4^2,$$

where

$$\lambda_3 = \frac{ac - bd}{c^2 + d^2} \quad \text{and} \quad \lambda_4 = \frac{ad + bc}{c^2 + d^2}$$

and $\lambda_3, \lambda_4 \in \mathbb{Z}$. Hence, the integer m can be expressed as a sum of two squares of integers. $\qquad\square$

84) Prove that the integer

$$p_1 p_2 \cdots p_n - 1, \text{ where } n \in \mathbb{N} \text{ with } n > 1,$$

cannot be represented as a perfect power of an integer.

(By $p_1, p_2, \ldots, p_n$ we denote, respectively, the 1st, 2nd, 3rd, ..., nth prime number.)

(M. Le, The perfect powers in $\{p_1 p_2 \cdots p_n\}_{n=1}^{+\infty}$, Octogon Mathematical Magazine 13(2)(2005), pp. 1101–1102)

Proof. We assume that the integer

$$p_1 p_2 \cdots p_n - 1$$

can be represented as a perfect power of an integer with $n > 1$.
    Then we obtain

$$p_1 p_2 \cdots p_n - 1 = a^k, \text{ where } a, k \in \mathbb{N}. \tag{1}$$

• If $k \geq n$, then we would have

$$p_{n+1}^n > p_1 p_2 \cdots p_n > a^k \geq a^n. \tag{2}$$

We claim that if p is a prime factor of a, then

$$p \geq p_{n+1}.$$

This is true because if p was one of the prime numbers $p_1, p_2, \ldots, p_n$, then p should divide the integer

$$p_1 p_2 \cdots p_n - 1 \text{ (because of (1))},$$

which is impossible.

Thus, it follows that

$$a^n \geq p^n \geq p_{n+1}^n.$$ (3)

Therefore, from (2), (3) we obtain that

$$p_{n+1}^n > p_{n+1}^n,$$

which is not possible.

Thus, it follows that

$$k < n < p_n.$$

From this relation, it follows that if q is a prime factor of k, then

$$q = p_i, \text{ where } 1 \leq i < n.$$

Let $x = a^{k/q}$. Then from (1) it follows that

$$p_1 p_2 \cdots p_n - 1 = x^q, \text{ where } x \in \mathbb{N}.$$ (4)

We will prove that q must necessarily be an odd integer.

• If $q = 2$, then one has:

The integer

$$p_1 p_2 \cdots p_n - 1$$

is odd and thus

$$x^q = x^2$$

is also odd, that is, x is an odd integer.

Therefore, because of (4) we derive

$$p_1 p_2 \cdots p_n = x^q + 1 = (2\lambda + 1)^2 + 1$$

$$= 4\lambda^2 + 4\lambda + 2, \lambda \in \mathbb{N}.$$

It is true that $p_2 = 3$, that is, $3 \mid p_1 p_2 \cdots p_n$, and therefore $3 \mid (4\lambda^2 + 4\lambda + 2)$.

Hence,

$$3 \mid (2\lambda^2 + 2\lambda + 1), \lambda \in \mathbb{N}.$$

It is possible to express the positive integer λ in the form

$$\lambda = 3\mu + r,$$

where $\mu \in \mathbb{N}$ and $r = 0, 1, 2$.

Then

$$2\lambda^2 + 2\lambda + 1 = 2(3\mu + r)^2 + 2(3\mu + r) + 1$$

$$= 18\mu^2 + 12\mu r + 2r^2 + 6\mu + 2r + 1.$$

Therefore,

$$3 \mid (18\mu^2 + 12\mu r + 2r^2 + 6\mu + 2r + 1),$$

that is,

$$3 \mid (2r^2 + 2r + 1), \text{ where } r = 0 \text{ or } r = 1 \text{ or } r = 2.$$

If $r = 0$: $2r^2 + 2r + 1 = 1$, but $3 \nmid 1$, that is impossible.
If $r = 1$: $2r^2 + 2r + 1 = 5$, but $3 \nmid 5$, that is impossible.
If $r = 2$: $2r^2 + 2r + 1 = 13$, but $3 \nmid 13$, that is impossible.
Therefore, q must necessarily be an odd prime number.
Thus, from (4), it follows

$$x^q + 1 = p_1 p_2 \cdots p_n$$

and therefore

$$x^q - (-1)^q = p_1 p_2 \cdots p_n$$

(here we use the fact that q is an odd number).

But, generally it holds that if $a^p - b^p \equiv 0 \pmod{p}$, where $a, b \in \mathbb{N}$ and p is a prime number, then $a^p - b^p \equiv 0 \pmod{p^2}$. Indeed this is true.

From Fermat's Little Theorem it follows

$$a^p \equiv a \pmod{p} \text{ and } b^p \equiv b \pmod{p}.$$

Thus,

$$a^p - b^p \equiv a - b \pmod{p},$$

that is,

$$p \mid a^p - b^p - (a - b).$$

But $p \mid (a^p - b^p)$ and thus $p \mid (a - b)$.

Consequently,

$$a - b = kp, \text{ where } k \in \mathbb{Z}$$

and thus

$$a^p = (b + kp)^p$$

$$= b^p + \binom{p}{1} b^{p-1} kp + \binom{p}{2} b^{p-2} (kp)^2 + \cdots + (kp)^p$$

$$= b^p + p\, b^{p-1} kp + \binom{p}{2} b^{p-2} k^2 p^2 + \cdots + k^p p^2 p^{p-2}.$$

Therefore, we get

$$a^p - b^p = b^{p-1} kp^2 + \binom{p}{2} b^{p-2} k^2 p^2 + \cdots + k^p p^2 p^{p-2} = k_1 p^2,$$

where

$$k_1 = b^{p-1} k + \binom{p}{2} b^{p-2} k^2 + \cdots + k^p p^{p-2}.$$

Thus,

$$a^p - b^p \equiv 0 \,(\mathrm{mod}\, p^2).$$

But, it is true that

$$x^q - (-1)^q = p_1 p_2 \cdots p_n,$$

that is,

$$x^q - (-1)^q \equiv 0 \,(\mathrm{mod}\, q) \tag{5}$$

and according to the above ·

$$x^q - (-1)^q \equiv 0 \,(\mathrm{mod}\, q^2). \tag{6}$$

From (6), it follows that

$$q^2 \mid (x^q + 1),$$

that is,

$$q^2 \mid p_1 p_2 \cdots p_n,$$

which is impossible, since q should appear twice in the product $p_1 p_2 \cdots p_n$, which is a contradiction. $\square$

85) Let $f(x) = ax^2 + bx + c$ be a quadratic polynomial with integer coefficients such that $f(0)$ and $f(1)$ are odd integers. Prove that the equation $f(x) = 0$ does not accept an integer solution.

Proof. Assume that r is an integer solution of the equation $f(x) = 0$. Then $f(x)$ is divisible by $x - r$ and

$$f(x) = (x - r)g(x),$$

where $g(x)$ is a polynomial with integer coefficients.

It follows that

$$\begin{cases} f(0) = -rg(0) \\ f(1) = (1 - r)g(1). \end{cases} \tag{1}$$

Then

$$-rg(0) = 2k + 1 \quad \text{and} \quad (1 - r)g(1) = 2l + 1, \tag{2}$$

where k, l are integers, since by hypothesis $f(0)$ and $f(1)$ are odd integers. Thus,

$$-r, 1 - r, g(0) \quad \text{and} \quad g(1)$$

are integers.

From (2) it follows that

$$-r \quad \text{and} \quad 1 - r$$

are odd integers, which is impossible, because if r is odd, then the integer $1 - r$ is even.

Hence, r cannot be an integer solution of the equation $f(x) = 0$. $\square$

86) A function $f : \mathbb{N} \to \mathbb{N}$ **is defined as follows: writing a number** $x \in \mathbb{N}$ **in its decimal expansion and replacing each digit by its square we obtain the decimal expansion of the number** $f(x)$. **For example,** $f(2) = 4, f(35) = 925, f(708) = 49064$. **Solve the equation** $f(x) = 29x$.

(Vladimir Protasov, Moscow State University; Newsletter of the European Mathematical Society, Issue 77, 2010, Problem 67)

Solution. Let $x = d_1 \dots d_n$ be the decimal expansion of x. We shall find successively all the digits from d_n to d_1.

Let us suppose that $d_n = 0$, then if we divide x by 10, the number $f(x)$ will be divided by 10 also. Thus, the relation $f(x) = 29x$ remains true. Therefore, it suffices to consider the case $d_n \neq 0$, and then multiply all the solutions by $10^m, m \geq 0$.

Hence, $d_n \neq 0$. Since the last digit of $f(x)$ is the last digit of $29x$, we obtain

$$d_n^2 \equiv 9d_n \,(\mathrm{mod}\ 10).$$

Therefore, $d_n = 4, 5$ or 9.

• If $d_n = 4$, since $4^2 = 16$ and $4 \times 29 = 116$, we obtain the following equation on the digit d_{n-1}:

$$1 + 9d_{n-1} \equiv 1 \,(\mathrm{mod}\ 10).$$

This equation has a unique solution on the set of digits: $d_{n-1} = 0$. Then we obtain the equation on d_{n-2}:

$$1 + 9d_{n-1} \equiv 0 \,(\mathrm{mod}\ 10),$$

whose unique solution is $d_{n-2} = 1$.

At every step we get the next digit in a unique way:

$$d_{n-3} = 2, d_{n-4} = 2, d_{n-5} = 2, \dots.$$

We see that the sequence cycles, so this process will never terminate, which means that there is no solution.

• If $d_n = 5$, by computing the digits successively, we get $d_{n-1} = 2, d_{n-2} = 3$, and the process terminates. Thus, $x = 325$ is the only solution in this case.

• If $d_n = 9$, by computing successively the last four digits: $x = \dots 9189$, we see that three of these digits have two-digit squares, therefore the number $f(x)$ has, at least, three digits more than x, and hence $f(x) > 100x$. Thus, there is no solution in this case.

Therefore,

$$x = 325 \cdot 10^m.   \qquad \square$$

87) Prove that the only integer solution of the equation

$$y^2 = x^3 + x$$

is $x = 0, y = 0$.

Proof. Let $x = x_0$, $y = y_0$ be an integer solution of the equation

$$y^2 = x^3 + x.$$

Then

$$y_0^2 = x_0^3 + x_0 = x_0(x_0^2 + 1).$$

However, the integers x_0, $x_0^2 + 1$ are relatively prime. This is the case because if there was a prime number p such that

$$p \mid x_0 \text{ and } p \mid (x_0^2 + 1),$$

then

$$p \mid x_0^2 \text{ and } p \mid (x_0^2 + 1).$$

Therefore, $p \mid 1$, which is impossible.

Since $\gcd(x_0, x_0^2 + 1) = 1$ and the product $x_0(x_0^2 + 1)$ is a perfect power of an integer, then necessarily the integers $x_0, x_0^2 + 1$ are perfect powers of an integer.

Therefore, there exists an integer k such that

$$x_0^2 + 1 = k^2$$

and thus

$$(k - x_0)(k + x_0) = 1.$$

But, this can happen only when

$$k - x_0 = k + x_0 = 1 \text{ or } k - x_0 = k + x_0 = -1.$$

In both cases we obtain that

$$x_0 = 0.$$

Thus, it follows that $y_0 = 0$.   □

88) Prove that the equation $7x^3 - 13y = 5$ does not have any integer solutions.

(S. E. Louridas, Athens, Greece)

Proof. Let us assume that there exist integers x_0, y_0, such that

$$7x_0^3 - 13y_0 = 5. \tag{1}$$

Then,

$$7x_0^3 \equiv 5 \,(\mathrm{mod}\,13)$$

or

$$x_0^3 \equiv -3 \,(\mathrm{mod}\,13)$$

and thus
$$x_0^{12} \equiv 81 \,(\mathrm{mod}\ 13). \tag{2}$$

However, by Fermat's Little Theorem we obtain
$$x_0^{12} \equiv 1 \,(\mathrm{mod}\ 13).$$

Therefore, by (2) and the above relation, it follows that
$$0 \equiv 80 \,(\mathrm{mod}\ 13),$$

which is obviously impossible. □

89) Show that for any $n \in \mathbb{N}$, the equation $q = 2p^{2n} + 1$, where p and q are prime numbers, has at most one solution.

(Konstantinos Drakakis, University College Dublin, Ireland; Newsletter of the European Mathematical Society, Issue 67, 2008, Problem 23, p. 46)

Solution. It is evident that $p\ (\mathrm{mod}\ 3)$ can take three different values, namely, 0, 1, and -1. The former case $p \equiv 0\ (\mathrm{mod}\ 3)$ occurs if and only if $p = 3$, while the remaining two cases $p \equiv \pm 1\ (\mathrm{mod}\ 3)$ both lead to $p^2 \equiv 1\ (\mathrm{mod}\ 3)$, so that $2p^{2n} + 1 \equiv 2 \cdot 1 + 1 \equiv 0\ (\mathrm{mod}\ 3)$: in other words, whenever $p \equiv \pm 1\ (\mathrm{mod}\ 3)$, $3 \mid q > 3$, hence q cannot be a prime number. Therefore, q can be a prime number only if $p = 3$, so that the given equation has indeed at most one solution.

For example,
$$q = 2 \cdot 3^2 + 1 = 19, q = 2 \cdot 3^4 + 1 = 163 \text{ and } q = 2 \cdot 3^6 + 1 = 1,459$$

are all prime numbers, but
$$q = 2 \cdot 3^8 + 1 = 13,123 = 11 \cdot 1,193$$

is not a prime number. □

90) Find all positive integers x, y, z such that
$$x^3 + y^3 + z^3 - 3xyz = p,$$

where p is a prime number with $p > 3$.

(Titu Andreescu and Dorin Andrica, Problem 27, Newsletter of the European Mathematical Society, 69(2008), p. 24)

Proof. From Euler's identity it follows that
$$x^3 + y^3 + z^3 - 3xyz = p \Leftrightarrow (x + y + z)(x^2 + y^2 + z^2 - xy - yz - zx) = p.$$

Obviously one has

$$x + y + z > 1.$$

Since p is a prime number we obtain

$$x + y + z = p \text{ and } x^2 + y^2 + z^2 - xy - yz - zx = 1.$$

But

$$x^2 + y^2 + z^2 - xy - yz - zx = 1 \Leftrightarrow 2x^2 + 2y^2 + 2z^2 - 2xy - 2yz - 2zx = 2$$
$$\Leftrightarrow (x - y)^2 + (y - z)^2 + (z - x)^2 = 2. \quad (1)$$

Without loss of generality we consider $x \geq y \geq z$.

If $x > y > z$, we get

$$\left.\begin{matrix} x - y > 0 \\ y - z > 0 \end{matrix}\right\} \text{ or } \left.\begin{matrix} x - y \geq 1 \\ y - z \geq 1 \end{matrix}\right\}$$

from which it follows that

$$x - z \geq 2.$$

Thus,

$$(x - y)^2 + (y - z)^2 + (z - x)^2 \geq 6 > 2,$$

which is impossible, due to (1).

Thus, because of the fact that $x > y > z$ does not hold, it follows that

$$x = y = z \text{ or } x = y > z \text{ or } x > y = z.$$

However, $x = y = z$ cannot hold due to (1).

It follows that

$$x = y = z + 1 \text{ or } x - 1 = y = z,$$

because of the following reasoning.

We have

$$x = y > z \implies x = y = z + a, \text{ where } a \in \mathbb{N}$$

or

$$x > y = z \implies x - a = y = z, \text{ where } a \in \mathbb{N}.$$

However, in both cases, if $a \geq 2$, then (1) is not satisfied. Thus, necessarily we must have $a = 1$.

- If $x = y = z + 1$, then

$$x + y + z = p \Leftrightarrow 2x + (x - 1) = p \Leftrightarrow x = \frac{p + 1}{3} \implies y = \frac{p + 1}{3}$$

and therefore

$$z = \frac{p - 2}{3}.$$

Hence,

$$x = \frac{p + 1}{3}, y = \frac{p + 1}{3}, z = \frac{p - 2}{3},$$

such that $p = 3k + 2$, $k \in \mathbb{N}$ and p is a prime number.

- If $x - 1 = y = z$, then

$$x + y + z = p \Leftrightarrow x + 2(x - 1) = p \Leftrightarrow x = \frac{p + 2}{3}$$

$$\Rightarrow y = \frac{p - 1}{3} \Rightarrow z = \frac{p - 1}{3}.$$

Therefore,

$$x = \frac{p + 2}{3}, y = \frac{p - 1}{3}, z = \frac{p - 1}{3},$$

where $p = 3k + 1$, $k \in \mathbb{N}$ and p is a prime number.   □

91) Prove that there exists an integer n such that

$$p^{(p+3)/2} \left| \left[(p - 1)^{p-1} - p - 1 + \sum_{i=0}^{2a} (p^2 - a + i)^m \right] \right.$$

$$\times [(2 \cdot 4 \cdots (p - 1))^{p-1} - n(-1)^{(p+1)/2}],$$

where m is an odd positive integer, $a \in \mathbb{N}$ and p is an odd prime number.

Proof. Set

$$A = (p - 1)^{p-1} - p - 1 + \sum_{i=0}^{2a} (p^2 - a + i)^m$$

and

$$B = (2 \cdot 4 \cdots (p - 1))^{p-1} - n(-1)^{(p+1)/2}.$$

Firstly, we shall prove that p^2 divides A. We have

$$\sum_{i=0}^{2a} (p^2 - a + i)^m = (p^2 - a)^m + (p^2 - a + 1)^m + \cdots$$

$$+ (p^2 + a - 1)^m + (p^2 + a)^m$$

$$= [(p^2 - a)^m + (p^2 + a)^m]$$

$$+ [(p^2 + a + 1)^m + (p^2 + a - 1)^m] + \cdots + p^2. \quad (1)$$

However, since m is an odd integer, it is evident that

$$(p^2 - r) + (p^2 + r) \mid (p^2 - r)^m + (p^2 + r)^m$$

or

$$p^2 \mid (p^2 - r)^m + (p^2 + r)^m,$$

for any positive integer r. Therefore, by (1) it follows that

$$p^2 \left| \sum_{i=0}^{2a} (p^2 - a + i)^m. \right. \tag{2}$$

Furthermore, we have

$$(p-1)^{p-1} = p^{p-1} + \binom{p-1}{1} p^{p-2} + \cdots + \binom{p-1}{p-2} p(-1)^{p-2} + (-1)^{p-1}$$

$$\equiv (-1)^{p-1} + (p-1)p\,(-1)^{p-2} \,(\mathrm{mod}\,p^2)$$

or

$$(p-1)^{p-1} \equiv 1 - (p-1)p \equiv 1 + p \,(\mathrm{mod}\,p^2). \tag{3}$$

Thus, by (2) and (3) we obtain

$$p^2 \left| (p-1)^{p-1} - p - 1 + \sum_{i=0}^{2a} (p^2 - a + i)^m \right.$$

or

$$p^2 \mid A. \tag{4}$$

We shall now prove that there exists an integer n such that $p^{(p-1)/2} \mid B$.

By Wilson's theorem we know that

$$p \mid (p-1)! + 1.$$

Hence,

$$p \mid 1 \cdot 2 \cdot 3 \cdots (p-1) + 1 = (1 \cdot 3 \cdots (p-2))(2 \cdot 4 \cdots (p-1)) + 1$$

or equivalently

$$(1 \cdot 3 \cdots (p-2))(2 \cdot 4 \cdots (p-1)) \equiv (-1) \,(\mathrm{mod}\,p). \tag{5}$$

But

$$(p-1) \equiv -1 \,(\mathrm{mod}\,p)$$

$$(p-3) \equiv -3 \,(\mathrm{mod}\,p)$$

$$\vdots$$

$$p - (p-2) \equiv -(p-2) \,(\mathrm{mod}\,p).$$

Therefore, by multiplying the above relations, we get

$$2 \cdot 4 \cdots (p-3)(p-1) \equiv (-1)^{(p-1)/2}(1 \cdot 3 \cdots (p-2)) \,(\mathrm{mod}\,p). \tag{6}$$

If we now multiply (5) and (6), we obtain

$$(2^2 \cdot 4^2 \cdots (p-1)^2)(1 \cdot 3 \cdots (p-2)) \equiv (-1)^{(p+1)/2}(1 \cdot 3 \cdots (p-2)) \,(\mathrm{mod}\, p)$$

or

$$p \mid 2^2 \cdot 4^2 \cdots (p-1)^2 - (-1)^{(p+1)/2}.$$

Hence,

$$p^{(p-1)/2} \mid (2^2 \cdot 4^2 \cdots (p-1)^2 - (-1)^{(p+1)/2})^{(p-1)/2}.$$

Thus, there exists an integer n such that

$$p^{(p-1)/2} \mid (2^2 \cdot 4^2 \cdots (p-1)^2)^{(p-1)/2} - n(-1)^{(p+1)/2}$$

or

$$p^{(p-1)/2} \mid (2 \cdot 4 \cdots (p-1))^{p-1} - n(-1)^{(p+1)/2}$$

or

$$p^{(p-1)/2} \mid B. \tag{7}$$

By (4) and (7) it is clear that

$$p^2 \cdot p^{(p-1)/2} \mid AB$$

or

$$p^{(p+3)/2} \mid AB. \qquad \square$$

92) Find the minimum value of the product xyz over all triples of positive integers x, y, z for which 2010 divides $x^2 + y^2 + z^2 - xy - yz - zx$.

(Titu Andreescu, The University of Texas at Dallas, USA; Newsletter of the European Mathematical Society, Issue 77, 2010, Problem 70)

Solution. Without loss of generality assume that $x > y > z$ and write

$$x^2 + y^2 + z^2 - xy - yz - zx = \frac{1}{2}[(x-y)^2 + (y-z)^2 + (z-x)^2].$$

Let $x - y = a$ and $y - z = b$. Then

$$x^2 + y^2 + z^2 - xy - yz - zx = \frac{1}{2}[a^2 + b^2 + (a+b)^2] = a^2 + ab + b^2$$

and therefore the condition of the problem becomes $2010 \mid a^2 + ab + b^2$.

It is clear that if a prime number $p \equiv 2 \pmod 3$ divides $a^2 + ab + b^2$, then p divides a and p divides b. Hence, 2 and 5 divide both a and b and thus $a = 10u$ and $b = 10v$ for some positive integers u and v.

It follows that

$$u^2 + uv + v^2 = 201k,$$

for some positive integer k.

Since we are seeking the minimum of

$$xyz = (z + 10u + 10v)(z + 10v)z,$$

we must have $z = 1$ and $v \leq u$. For $k = 1$ the only pair (u, v) of positive integers with $v \leq u$ satisfying this equation is $(11, 5)$. Therefore, we obtain $xyz = 161 \cdot 51$.

If $k > 1$ and $v \geq 5$, then for each solution (u, v), we would have $u > 11$, implying $xyz > 161 \cdot 51$. Thus, we need to check what happens for $v \leq 4$.

If $2 \mid k$, then $2 \mid u$ and $2 \mid v$ and the equation $u^2 + uv + v^2 = 201k$ is either not solvable or it reduces to the equation

$$U^2 + UV + V^2 = 201 \cdot \frac{k}{4},$$

where $u = 2U$ and $v = 2V$. In the latter situation we obtain the solution $(2 \cdot 11, 2 \cdot 5)$, making xyz greater than $161 \cdot 51$.

The same argument works for 5 and all other prime numbers of the form $3k + 2 : 11, 17, 23, 29, \ldots$.

If $3 \mid k$, then since

$$(u - v)^2 + 3uv = 9 \cdot 67\frac{k}{3},$$

it is evident that 3 divides $u - v$ or 9 divides $(u - v)^2$ and thus 9 divides $3uv$.

Therefore, 3 divides u or v. However, since 3 also divides $u - v$, it follows that

$$u = 3s \text{ and } v = 3t,$$

for some positive integers s and t.

Hence, we obtain the equation

$$s^2 + st + t^2 = 67,$$

whose only solution (s, t) with $s \geq t$ is $(7, 2)$. Thus, $(u, v) = (21, 6)$ and $xyz = 271 \cdot 61 > 161 \cdot 51$.

For $k \leq 30$, the only values we have to worry about are $k = 7, 13, 19$. For $k > 30$, $201k \geq 6231$. If $v = 1$, $u(u + 1) \geq 6230$, implying $u \geq 79$ and $xyz \geq 801 \cdot 11 > 161 \cdot 51$. Because none of the equations $u(u + 1) = 201 \cdot 7 - 1$, $u(u + 1) = 201 \cdot 13 - 1$, $u(u + 1) = 201 \cdot 19 - 1$ has integer solution, we are done in the case $v = 1$. If $v = 2$ and $k \leq 12$, we only need to check that the equation $u(u + 1) = 201 \cdot 7 - 4$ has no solution in positive integers, because $k > 12$ implies $201k - 4 \geq 2609$, so $u(u + 1) \geq 2609$ and so, again, $u \geq 79$, implying $xyz \geq 801 \cdot 21 > 161 \cdot 51$. Cases $v = 3$ and $v = 4$ are now easy.

In conclusion, the minimum value of xyz is $161 \cdot 51$. $\square$

93) Find all pairs (x, y) of positive integers x, y for which it holds

$$\frac{1}{x} + \frac{1}{y} = \frac{1}{pq},$$

where p, q are prime numbers.

Solution. The given equation

$$\frac{1}{x} + \frac{1}{y} = \frac{1}{pq}$$

can be written in an equivalent form as follows:

$$\frac{1}{x} + \frac{1}{y} = \frac{1}{pq} \Leftrightarrow (x + y)pq - xy = 0$$

$$\Leftrightarrow p^2q^2 + xpq + ypq - xy = p^2q^2$$

$$\Leftrightarrow p^2q^2 = p^2q^2 - xpq - ypq + xy$$

$$\Leftrightarrow p^2q^2 = p^2q^2 + y(x - pq) - xpq$$

$$\Leftrightarrow p^2q^2 = pq(pq - x) + y(x - pq)$$

$$\Leftrightarrow (x - pq)(y - pq) = p^2q^2. \qquad (1)$$

From relation (1) one obtains all possible values of the expressions

$$x - pq, \quad y - pq.$$

Thus, we get

$$\left.\begin{array}{l} x - pq = p^2q^2 \\ y - pq = 1 \end{array}\right\} \text{ or } \left.\begin{array}{l} x - pq = 1 \\ y - pq = p^2q^2 \end{array}\right\} \text{ or } \left.\begin{array}{l} x - pq = p^2 \\ y - pq = q^2 \end{array}\right\}$$

$$\text{or } \left.\begin{array}{l} x - pq = q^2 \\ y - pq = p^2 \end{array}\right\} \text{ or } \left.\begin{array}{l} x - pq = p \\ y - pq = pq^2 \end{array}\right\} \text{ or } \left.\begin{array}{l} x - pq = pq^2 \\ y - pq = p \end{array}\right\}$$

$$\text{or } \left.\begin{array}{l} x - pq = q \\ y - pq = qp^2 \end{array}\right\} \text{ or } \left.\begin{array}{l} x - pq = qp^2 \\ y - pq = q \end{array}\right\} \text{ or } \left.\begin{array}{l} x - pq = pq \\ y - pq = pq. \end{array}\right\}$$

Hence, the solutions of the equation are the following:

$$(x_1, y_1) = (pq(pq + 1), pq + 1) \quad (x_5, y_5) = (p(1 + q), pq(q + 1))$$

$$(x_2, y_2) = (pq + 1, pq(pq + 1)) \quad (x_6, y_6) = (pq(q + 1), p(1 + q))$$

$$(x_3, y_3) = (p(p + q), q(q + p)) \quad (x_7, y_7) = (q(1 + p), pq(p + 1))$$

$$(x_4, y_4) = (q(q + p), p(p + q)) \quad (x_8, y_8) = (pq(p + 1), q(1 + p))$$

$$(x_9, y_9) = (2pq, 2pq).$$

$\square$

94) Let n be a positive integer. Prove that the equation

$$x + y + \frac{1}{x} + \frac{1}{y} = 3n$$

does not accept solutions in the set of positive rational numbers.

(66th Panhellenic Mathematical Competition, "ARCHIMEDES")

Proof. Assume that the equation

$$x + y + \frac{1}{x} + \frac{1}{y} = 3n$$

has one solution in the set of rational numbers and let

$$x = \frac{a}{b}, y = \frac{c}{d}, \text{ where } \gcd(a, b) = 1, \ \gcd(c, d) = 1,$$

with a, b, c, $d \in \mathbb{N}$ and $bd \neq 0$, be that solution. Thus, we obtain that

$$\frac{a}{b} + \frac{c}{d} + \frac{b}{a} + \frac{d}{c} = 3n \Leftrightarrow \frac{a^2 + b^2}{ab} + \frac{c^2 + d^2}{cd} = 3n$$

$$\Leftrightarrow cd(a^2 + b^2) + ab(c^2 + d^2) = 3nabcd. \tag{1}$$

Since

$$cd \mid 3nabcd, \text{ that is, } cd \mid [cd(a^2 + b^2) + ab(c^2 + d^2)],$$

it follows that

$$cd \mid ab(c^2 + d^2). \tag{2}$$

But $\gcd(cd, c^2 + d^2) = 1$ because if there were a prime number p such that $p \mid cd$ and $p \mid (c^2 + d^2)$, then we would have

- If $p \mid c$, since $p \mid (c^2 + d^2)$ it would imply that $p \mid d$, which is not possible since $\gcd(c, d) = 1$.
- If $p \mid d$, then similarly $p \mid c$ since $p \mid \gcd(c^2 + d^2)$, which is also not possible.

Thus, from (2) it follows that $cd \mid ab$.

Similarly, from (1) one obtains

$$ab \mid 3nabcd, \text{ that is, } ab \mid cd(a^2 + b^2)$$

and therefore $ab \mid cd$ since the $\gcd(ab, a^2 + b^2) = 1$ for exactly the same reasons for which the $\gcd(cd, c^2 + d^2) = 1$.

Consequently, since $ab \mid cd$ and $cd \mid ab$ it follows that $ab = cd$.
Thus,

$$(1) \Leftrightarrow 3n(ab)^2 = ab(a^2 + b^2) + ab(c^2 + d^2)$$

$$\Leftrightarrow 3nab = a^2 + b^2 + c^2 + d^2.$$

Therefore,

$$3 \mid (a^2 + b^2 + c^2 + d^2). \tag{3}$$

From (3) one obtains the following cases:

(α) None of the numbers a, b, c, d is divisible by 3.
(β) Only one of the numbers a, b, c, d is divisible by 3.
(γ) Two of the numbers a, b, c, d are divisible by 3.
(δ) Three of the numbers a, b, c, d are divisible by 3.
(ϵ) All numbers a, b, c, d are divisible by 3.

Cases (δ), (ϵ) are not considered because otherwise the conditions

$$\gcd(a, b) = 1, \ \gcd(c, d) = 1$$

would not be satisfied.

Case (β) can be included in case (γ), because from (β) if for example $3 \mid a$ then $3 \mid c$ or $3 \mid d$, since $ab = cd$.

- If case (α) is satisfied, then

$$a^2 + b^2 + c^2 + d^2 \equiv 1 \text{ or } 2 \,(\mathrm{mod}\,3),$$

because

$$a, b, c, d = 3k + 1 \text{ or } 3k + 2.$$

But

$$(3k + 1)^2 = 9k^2 + 6k + 1$$

and

$$(3k + 2)^2 = 9k^2 + 12k + 4,$$

therefore there does not exist a linear combination of $(3k + 1)^2$, $(3k + 2)^2$ equal to a multiple of 3, since there does not exist a linear combination of 1 and 4 equal to a multiple of 3.
- If case (γ) is satisfied, then two of the integers a, b, c, d have the form $3k_1 + 1$, $3k_2 + 2$ for $k_1, k_2 \in \mathbb{N}$. However, none of the summations

$$(3k_1 + 1)^2 + (3k_2 + 1)^2, (3k_1 + 1)^2 + (3k_2 + 2)^2, (3k_1 + 2)^2 + (3k_2 + 2)^2$$

is a multiple of 3.

Therefore, it holds

$$a^2 + b^2 + c^2 + d^2 \equiv 1 \text{ or } 2 \,(\mathrm{mod}\,3).$$

Therefore, the equation $x + y + 1/x + 1/y = 3n$ does not accept positive rational solutions. □

95) Find all integers n, $n \geq 2$, for which it holds

$$1^n + 2^n + \cdots + (n - 1)^n \equiv 0 \,(\mathrm{mod}\,n).$$

Solution. We will consider two cases:
- If n is an odd integer, then $1, 2, \ldots, \frac{n-1}{2}$ are all integers.

  Consider now the number

$$A_\lambda = \lambda^n + (n - \lambda)^n, \lambda \in \mathbb{N}.$$

Then

$$A_1 + A_2 + \cdots + A_{\frac{n-1}{2}} = 1^n + (n-1)^n$$
$$+ 2^n + (n-2)^n$$
$$+ 3^n + (n-3)^n + \cdots$$
$$+ \left(\frac{n-3}{2}\right)^n + \left(\frac{n+3}{2}\right)^n$$
$$+ \left(\frac{n-1}{2}\right)^n + \left(\frac{n+1}{2}\right)^n,$$

that is,

$$A_1 + A_2 + \cdots + A_{\frac{n-1}{2}} = 1^n + 2^n + \cdots + (n-1)^n. \tag{1}$$

However,

$$A_\lambda = \lambda^n + n^n + \binom{n}{1}n^{n-1}(-\lambda) + \cdots + (-\lambda)^n \equiv 0 \,(\mathrm{mod}\,n),$$

since $(-\lambda)^n = -\lambda^n$, because n is an odd number.

Therefore, n divides each term of the set of numbers

$$A_1, A_2, \ldots, A_{\frac{n-1}{2}}.$$

Thus, from (1) it follows that

$$n \mid (1^n + 2^n + \cdots + (n-1)^n)$$

if n is an odd integer.
- If n is an even integer, then it can be expressed in the form $n = 2^r k$, where $r, k \in \mathbb{N}$.[5]

  Let r_0 be the greatest possible value of r, such that $n = 2^{r_0}k$. Then the integer k must necessarily be an odd number, because if k was an even number, then $n = 2^{r_0+\lambda} \cdot m$ for $\lambda, m \in \mathbb{N}$, which contradicts the assumption that r_0 is the greatest possible value.

  We will prove that

$$1^n + 2^n + \cdots + (n-1)^n \equiv \frac{n}{2} \,(\mathrm{mod}\,2^{r_0}).$$

---

[5] Since $0 \notin \mathbb{N}$, the number n is always even.

If λ is an odd number, then according to Euler's theorem it holds

$$\lambda^{\phi(2^{r_0})} \equiv 1 \,(\mathrm{mod}\, 2^{r_0}) \text{ or } \lambda^{2^{r_0} - 2^{r_0-1}} \equiv 1 \,(\mathrm{mod}\, 2^{r_0})$$

or

$$\lambda^{2^{r_0-1}} \equiv 1 \,(\mathrm{mod}\, 2^{r_0}).$$

Therefore, we obtain

$$\lambda^n = \lambda^{2^{r_0}k} = (\lambda^{2^{r_0-1}})^{2k} \equiv 1^{2k} \,(\mathrm{mod}\, 2^{r_0})$$

that is,

$$\lambda^n \equiv 1 \,(\mathrm{mod}\, 2^{r_0}).$$

Therefore,

$$\left.\begin{array}{r} 1^n \equiv 1 \,(\mathrm{mod}\, 2^{r_0}) \\ 3^n \equiv 1 \,(\mathrm{mod}\, 2^{r_0}) \\ \vdots \\ (n-3)^n \equiv 1 \,(\mathrm{mod}\, 2^{r_0}) \\ (n-1)^n \equiv 1 \,(\mathrm{mod}\, 2^{r_0}). \end{array}\right\}$$

Adding by parts we get

$$1^n + 3^n + \cdots + (n-3)^n + (n-1)^n \equiv 0 \,(\mathrm{mod}\, 2^{r_0}). \tag{2}$$

If λ is an even number (that is, $\lambda = 2\nu$, $\nu \in \mathbb{N}$), then we have

$$\lambda^n = (2\nu)^{2^{r_0}k} = (2\nu)^{(r_0+q)k}, q \in \mathbb{N},$$

because

$$2^{r_0} > r_0$$

and thus

$$2^{r_0} = r_0 + q$$

(this can be easily proved by mathematical induction).

Hence,

$$\lambda^n = (2\nu)^{r_0k+qk} = (2\nu)^{r_0k} \cdot (2\nu)^{qk} = (2^{r_0})^k \nu^{r_0k}(2\nu)^{qk}.$$

Therefore, it follows

$$2^{r_0} \mid \lambda^n \text{ or } \lambda^n \equiv 0 \,(\mathrm{mod}\, 2^{r_0}).$$

Thus,

$$\left.\begin{array}{r} 2^n \equiv 0 \,(\mathrm{mod}\, 2^{r_0}) \\ 4^n \equiv 0 \,(\mathrm{mod}\, 2^{r_0}) \\ \vdots \\ (n-2)^n \equiv 0 \,(\mathrm{mod}\, 2^{r_0}). \end{array}\right\}$$

Adding by parts the above relations it follows

$$2^n + 4^n + \cdots + (n-2)^n \equiv 0 \,(\mathrm{mod}\, 2^{r_0}). \tag{3}$$

Adding by parts (2) and (3) we obtain

$$1^n + 2^n + \cdots + (n-2)^n + (n-1)^n \equiv \frac{n}{2} \,(\mathrm{mod}\, 2^{r_0}). \tag{4}$$

Thus,

$$n \nmid (1^n + 2^n + \cdots + (n-1)^n),$$

because otherwise

$$2^{r_0} \mid (1^n + 2^n + \cdots + (n-1)^n), \text{ since } 2^{r_0} \mid n$$

and thus by (4), it should hold $2^{r_0} \mid \frac{n}{2}$, that is, $2^{r_0+1} \mid n$, which leads to a contradiction, due to the property of r_0, which we defined above.

Hence,

$$1^n + 2^n + \cdots + (n-1)^n \equiv 0 \,(\mathrm{mod}\, n)$$

holds true only for all odd numbers n. $\square$

96) Prove that for every positive integer k, the equation

$$x_1^3 + x_2^3 + \cdots + x_k^3 + x_{k+1}^2 = x_{k+2}^4$$

has infinitely many solutions in positive integers, such that $x_1 < x_2 < \cdots < x_{k+1}$.

(Dorin Andrica, "Babeş-Bolyai" University, Cluj-Napoca, Romania; Newsletter of the European Mathematical Society, Issue 77, 2010, Problem 71)

Proof. It is a standard fact that for every positive integer n, it holds

$$1^3 + 2^3 + \cdots + n^3 + (n+1)^3 + \cdots + (n+k)^3 = \left(\frac{(n+k)(n+k+1)}{2}\right)^2,$$

that is,

$$\left(\frac{n(n+1)}{2}\right)^2 + (n+1)^3 + \cdots + (n+k)^3 = \left(\frac{(n+k)(n+k+1)}{2}\right)^2.$$

Consider the positive integers n, such that the triangular number

$$t_{n+k} = \frac{(n+k)(n+k+1)}{2}$$

is a perfect square. There are infinitely many such integers since the relation $t_{n+k} = u^2$ is equivalent to Pell's equation

$$(2n + 2k + 1)^2 - 2w^2 = 1,$$

where $w = 2u$.

The fundamental solution to this equation is $(3, 2)$, i.e., $2n + 2k + 1 = 3$ and $w = 2$. Hence, all these integers are given by the sequence (n_s), where

$$2n_s + 2k + 1 + w_s\sqrt{2} = (3 + 2\sqrt{2})^s,$$

for sufficiently large values of s, such that $n_s \geq 1$. We can take

$$x_1 = n_s + 1, \ldots, x_k = n_s + k, x_{k+1} = \frac{n_s(n_s + 1)}{2}, x_{k+2} = w_s.$$

It is clear that for s large enough we have $n_s \geq 1$ and $n(n + 1)/2 > n + k$. Therefore, we get an infinite family of solutions.   □

97) Prove that for every prime number p, the equation

$$2^p + 3^p = a^n$$

does not have integer solutions for all a, n with $a, n \in \mathbb{N} - \{1\}$.

Proof. Let us assume that the equation has integer solutions for a, n with $a, n \in \mathbb{N} - \{1\}$.

- If $p = 2$, the equation

$$2^p + 3^p = a^n$$

  can be written in the form

$$a^n = 13,$$

  which does not accept integer solutions with respect to a, n with

$$a, n > 1.$$

- If $p \neq 2$, then p is an odd integer and thus

$$5 \mid (2^p + 3^p).$$

Therefore,

$$5 \mid a^n \Rightarrow 5 \mid a \Rightarrow 25 \mid a^2.$$

But $n \geq 2$, therefore

$$25 \mid a^n \Rightarrow 25 \mid (2^p + 3^p).$$

However,

$$2^p + 3^p = 2^p + (5 - 2)^p$$

$$= 2^p + 5^p + \binom{p}{1}5^{p-1}(-2) + \cdots + \binom{p}{p-2}5^2(-2)^{p-2}$$

$$+ \binom{p}{p-1}5(-2)^{p-1}(-2)^p$$

$$= 2^p + \text{mult.}25 + \binom{p}{p-1} 5 \cdot 2^{p-1} - 2^p, \text{ since } p \geq 3$$

$$= 5p\, 2^{p-1} + \text{mult.}25$$

(by mult.k we denote an integer which is a multiple of k.)
    Thus,

$$2^p + 3^p = 5p\, 2^{p-1} + \text{mult.}25.$$

Therefore,

$$25 \mid (5p\, 2^{p-1} + \text{mult.}25) \Rightarrow 25 \mid 5p\, 2^{p-1}$$

$$\Rightarrow 5 \mid p\, 2^{p-1}.$$

This implies that $5 \mid p$ or $5 \mid 2^{p-1}$. Thus, $5 \mid p$.
    Therefore, $p = 5$. Hence, the equation $2^p + 3^p = a^n$ reduces to the form

$$a^n = 2^5 + 3^5 = 275 = 5^2 \cdot 11,$$

which does not accept integer solutions with respect to a, n with $a, n > 1$.
    Hence, the equation

$$2^p + 3^p = a^n$$

does not accept integer solutions for all a, n with $a, n \in \mathbb{N} - \{1\}$.   □

98) Let p_1, p_2 be two odd prime numbers and a, n integers such that $a > 1$ and $n > 1$. Prove that the equation

$$\left(\frac{p_2 - 1}{2}\right)^{p_1} + \left(\frac{p_2 + 1}{2}\right)^{p_1} = a^n$$

accepts integer solutions for a, n only in the case $p_1 = p_2$.

(M. Th. Rassias, Proposed problem W. 5, Octogon Mathematical Magazine, 17(1) (2009), p. 307)

Proof. By the hypothesis it follows that it is not possible to have $p_1 = 2$ and $p_2 = 2$. Thus, we consider $p_2 = 2x + 1$, where $x \in \mathbb{N}$.
    Suppose that the equation accepts at least one integer solution for a, n with $a > 1$ and $n > 1$. Then the given equation

$$\left(\frac{p_2 - 1}{2}\right)^{p_1} + \left(\frac{p_2 + 1}{2}\right)^{p_1} = a^n$$

can be written as follows

$$x^{p_1} + (x + 1)^{p_1} = a^n. \tag{1}$$

Because of the fact that p_1 is odd, one has

$$a^n = x^{p_1} + (x+1)^{p_1} = (x+x+1)A = (2x+1)A, \text{ where } A \in \mathbb{N}.$$

Thus,

$$(2x+1) \mid a^n.$$

Since the integer $2x+1$ is a prime number, it follows that

$$(2x+1) \mid a,$$

and therefore

$$(2x+1)^2 \mid a^2.$$

However, $n > 1$, thus

$$(2x+1)^2 \mid a^n \text{ or } (2x+1)^2 \mid x^{p_1} + (x+1)^{p_1}. \tag{2}$$

Hence,

$$x^{p_1} + (x+1)^{p_1} = x^{p_1} + [(2x+1)-x]^{p_1}$$

$$= x^{p_1} + (2x+1)^{p_1} + \binom{p_1}{1}(2x+1)^{p_1-1}(-x) + \dots$$

$$+ \binom{p_1}{p_1-1}(2x+1)(-x)^{p_1-1} + (-x)^{p_1}$$

$$= x^{p_1} + (2x+1)^2 B + \binom{p_1}{p_1-1}(2x+1)x^{p_1-1} - x^{p_1}$$

$$= \binom{p_1}{p_1-1}(2x+1)x^{p_1-1} + (2x+1)^2 B, \text{ where } B \in \mathbb{Z}.$$

Therefore,

$$x^{p_1} + (x+1)^{p_1} = \binom{p_1}{p_1-1}(2x+1)x^{p_1-1} + (2x+1)^2 B.$$

From (2) it follows

$$(2x+1)^2 \left| \binom{p_1}{p_1-1}(2x+1)x^{p_1-1} \right.$$

or

$$(2x+1) \left| \binom{p_1}{p_1-1}x^{p_1-1} \right.$$

or

$$(2x+1) | p_1 x^{p_1-1}.$$

Because of the fact that the number $2x + 1$ is prime, it yields

$$(2x + 1) \mid p_1 \text{ or } (2x + 1) \mid x.$$

The second case though is not possible since

$$2x + 1 > x \text{ and } \gcd(2x + 1, x) = 1.$$

Finally, $(2x + 1) \mid p_1$, that is,

$$p_1 = 2x + 1, \text{ since } p_1 \text{ is a prime number.}$$

Thus,

$$p_1 = p_2. \qquad \qquad \square$$

99) Find all integer solutions of the equation

$$\frac{a^7 - 1}{a - 1} = b^5 - 1.$$

(Shortlisted, 47th IMO, Slovenia, 2006)

Solution. We shall first investigate whether the equation

$$\frac{a^7 - 1}{a - 1} = b^5 - 1$$

has integer solutions. If it does, then we will determine them. Before we begin our investigation, we will prove a useful lemma.

Lemma. *If $a \in \mathbb{N}$ and p is a prime number for which p divides $(a^7 - 1)/(a - 1)$, then it holds*

$$p \equiv 1 \pmod 7 \text{ or } p = 7.$$

Proof. Let us suppose that $p \equiv 1 \pmod 7$ does not hold. In this case, the integers $p - 1$ and 7 are co-prime. Therefore, by Bezout's lemma we obtain that there exist integers x, y for which

$$7x + (p - 1)y = 1.$$

Hence,

$$a = a^{7x + (p-1)y} = a^{7x} \cdot a^{(p-1)y}. \qquad (1)$$

But, by the hypothesis we know that $p \mid a^7 - 1$, which means that

$$a^{7x} \equiv 1 \pmod p.$$

Furthermore, by Fermat's Little Theorem we also obtain that

$$a^{(p-1)y} \equiv 1 \pmod p.$$

Therefore, by (1) we get

$$a \equiv 1 \,(\mathrm{mod}\, p). \qquad (2)$$

However,

$$\frac{a^7 - 1}{a - 1} = 1 + a + a^2 + \cdots + a^6$$

and by (2) we derive that

$$\frac{a^7 - 1}{a - 1} \equiv 7 \,(\mathrm{mod}\, p).$$

Since p also divides $(a^7 - 1)/(a - 1)$, it is evident that $p = 7$. So, from the above, we can conclude that either

$$p \equiv 1 \,(\mathrm{mod}\, 7) \text{ or } p = 7.$$

We shall now proceed to the solution of the initial problem.
If d is a positive divisor of

$$(a^7 - 1)/(a - 1)$$

and

$$d = p_1^{q_1} p_2^{q_2} \cdots p_k^{q_k}$$

is its standard form, then either

$$d = p_1^{q_1} p_2^{q_2} \cdots p_k^{q_k} \equiv 1 \,(\mathrm{mod}\, 7)$$

or

$$d = p_1^{q_1} p_2^{q_2} \cdots p_k^{q_k} = 7^{q_1} 7^{q_2} \cdots 7^{q_k} \equiv 0 \,(\mathrm{mod}\, 7).$$

In addition, if (a_0, b_0) is a solution of the equation, then

$$\frac{a_0^7 - 1}{a_0 - 1} = b_0^5 - 1 = (b_0 - 1)(1 + b_0 + b_0^2 + b_0^3 + b_0^4).$$

Hence, $b_0 - 1$ is a positive divisor of $(a_0^7 - 1)/(a_0 - 1)$. So we get

$$b_0 \equiv 2 \,(\mathrm{mod}\, 7)$$

or

$$b_0 \equiv 1 \,(\mathrm{mod}\, 7).$$

By the above, we obtain also that

$$1 + b_0 + b_0^2 + b_0^3 + b_0^4 \equiv 1 + 2 + 4 + 1 + 2 \equiv 3 \,(\mathrm{mod}\, 7)$$

or

$$1 + b_0 + b_0^2 + b_0^3 + b_0^4 \equiv 5 \,(\mathrm{mod}\, 7).$$

Both cases contradict the fact that $1 + b_0 + b_0^2 + b_0^3 + b_0^4$ is a positive divisor of

$$(a_0^7 - 1)/(a_0 - 1).$$

Therefore, the equation $(a^7 - 1)/(a - 1) = b^5 - 1$ does not have any integer solutions. $\qquad \square$

100) Find all integer solutions of the system

$$x + 4y + 24z + 120w = 782 \tag{1}$$

$$0 \le x \le 4 \tag{2}$$

$$0 \le y \le 6 \tag{3}$$

$$0 \le z \le 5 \tag{4}$$

Solution. From equation (1), we obtain

$$782 - x = 4y + 24z + 120w,$$

that is,

$$782 - x = 4(y + 6z + 30w), \tag{5}$$

which means that

$$782 - x = \text{mult. } 4. \tag{6}$$

From (2) and because of the fact that x is an integer it follows that

$$x = 0, 1, 2, 3, 4. \tag{7}$$

From (6) and (7) by trial, we get only $x = 2$.
    If we substitute $x = 2$ in (5), we have

$$4(y + 6z + 30w) = 780,$$

which implies

$$6(z + 5w) = 195 - y \tag{8}$$

and therefore

$$195 - y = \text{mult.}6. \tag{9}$$

From (3) and because of the fact y is an integer it follows that

$$y = 0, 1, 2, 3, 4, 5, 6. \tag{10}$$

From (9) and (10) by trial, we only obtain $y = 3$.
    For $y = 3$, equation (8) becomes

$$6(z + 5w) = 192,$$

that is,

$$5w = 32 - z. \tag{11}$$

From (4) and the fact z is an integer it follows that

$$z = 0, 1, 2, 3, 4, 5. \tag{12}$$

From (11) and (12) by trial, we only get $z = 2$.

For $z = 2$, equation (11) implies $w = 6$.

Hence, the only integer solutions for x, y, z, w of the given system of (1), (2), (3) and (4) are

$$x = 2, y = 3, z = 2, w = 6,$$

that is,

$$(x, y, z, w) = (2, 3, 2, 6). \qquad \square$$

101) Find all integer solutions of the system

$$35x + 63y + 45z = 1 \tag{1}$$

$$|x| < 9 \tag{2}$$

$$|y| < 5 \tag{3}$$

$$|z| < 7. \tag{4}$$

Solution. From equation (1), we obtain

$$1 - 63y = 35x + 45z,$$

that is,

$$1 - 63y = 5(7x + 9z), \tag{5}$$

which implies that

$$1 - 63y = \text{mult.}5. \tag{6}$$

Inequalities (2), (3) and (4) can be also written in the form

$$-9 < x < 9 \tag{7}$$

$$-5 < y < 5 \tag{8}$$

$$-7 < z < 7. \tag{9}$$

From (8) and the fact that y is an integer it follows that

$$y = -4, -3, -2, -1, 0, 1, 2, 3, 4. \tag{10}$$

From (6) and (10) by trial, we get only

$$y = -3 \text{ and } y = 2.$$

For these integer values of y, equation (5) yields the equations

$$7x + 9z = 38 \tag{11}$$

and

$$7x + 9z = -25, \tag{12}$$

respectively.

- Equation (11) is a diophantine equation with integer solutions

$$x = -1 - 9k, z = 5 + 7k, \text{ where } k \in \mathbb{Z}. \tag{13}$$

From $x = -1 - 9k$ and (7), we derive

$$-9 < -1 - 9k < 9,$$

that is,

$$-\frac{10}{9} < k < \frac{8}{9}, \text{ where } k \in \mathbb{Z}.$$

Thus, $k = -1, k = 0$.
From $z = 5 + 7k$ and (9), it follows that

$$-7 < 5 + 7k < 7,$$

that is,

$$-\frac{12}{7} < k < \frac{2}{7}, \text{ where } k \in \mathbb{Z}.$$

Therefore, $k = -1, k = 0$.
Hence, equations (13) for $k = -1$ imply

$$x = 8, z = -2$$

and for $k = 0$ imply

$$x = -1, z = 5.$$

Therefore, the integer solutions of (1) are the following:

$$x = 8, y = -3, z = -2$$

and

$$x = -1, y = -3, z = 5.$$

- Equation (12) is a diophantine equation with integer solutions

$$x = -10 - 9\lambda, z = 5 + 7\lambda, \text{ where } \lambda \in \mathbb{Z}. \tag{14}$$

From (14) and (7), (9) it follows that

$$-\frac{19}{9} < \lambda < -\frac{1}{9}$$

as well as

$$-\frac{12}{7} < \lambda < \frac{2}{7}$$

for $\lambda \in \mathbb{Z}$, which are satisfied for $\lambda = -1$.
    For $\lambda = -1$, equations (14) imply

$$x = -1, z = -2$$

and from (2), we get $y = 2$.
    Thus, an integer solution of (1) is the following:

$$x = -1, y = 2, z = -2.$$

Hence, the integer solutions of the system of (1), (2), (3) and (4) are the following:

$$(x, y, z) = (8, -3, -2)$$

$$(x, y, z) = (-1, -3, 5)$$

$$(x, y, z) = (-1, 2, -2).$$ □

102) Find the integer solutions of the system

$$x^2 + 2yz < 36 \tag{1}$$

$$y^2 + 2zx = -16 \tag{2}$$

$$z^2 + 2xy = -16. \tag{3}$$

(National Technical University of Athens, Entrance Examinations, 1946)

Solution. It is evident that if $xyz = 0$, equations (2) and (3) are not satisfied for integer values of x, y, z. Thus, for the system of (1), (2) and (3) to accept integer solutions, it must hold that

$$xyz \neq 0.$$

From (2) it follows that

$$2zx = -16 - y^2$$

and therefore

$$zx < 0. \tag{4}$$

Similarly from (3) it follows that

$$2xy = -16 - z^2$$

and thus

$$xy < 0. \tag{5}$$

From (4) and (5) it yields $x^2 yz > 0$ and thus

$$yz > 0. \tag{6}$$

From (4), (5) and (6) we get either

$$x > 0, y < 0, z < 0$$

or

$$x < 0, y > 0, z > 0.$$

Therefore, the numbers

$$-x, y, z$$

are of the same sign.

The system of (1), (2) and (3) can be written in the following equivalent forms:

$$\begin{cases} x^2 + y^2 + z^2 + 2xy + 2yz + 2zx < 4 \\ y^2 + 2zx = -16 \\ z^2 + 2xy = -16 \end{cases}$$

or

$$\begin{cases} (x + y + z)^2 < 4 \\ y^2 + 2zx = -16 \\ z^2 + 2xy = -16 \end{cases}$$

or

$$\begin{cases} |x + y + z| < 2 \\ y^2 + 2zx = -16 \\ z^2 + 2xy = -16 \end{cases}$$

or

$$\begin{cases} -2 < x + y + z < 2 \\ y^2 + 2zx = -16 \\ z^2 + 2xy = -16. \end{cases} \tag{7}$$

If we subtract (3) from (2) by parts we derive

$$(y - z)(y + z - 2x) = 0. \tag{8}$$

However,

$$-2x, y, z$$

are of the same sign, therefore

$$y + z - 2x \neq 0.$$

Thus, from (8), it follows that

$$y - z = 0.$$

Hence, the system (7) can be written in the form

$$\begin{cases} -2 < x + y + z < 2 \\ y - z = 0 \\ z^2 + 2xy = -16. \end{cases} \tag{9}$$

Because of the fact that $x, y, z \in \mathbb{Z}$, it yields $x + y + z \in \mathbb{Z}$ and since

$$-2 < x + y + z < 2$$

one has
$$x + y + z = -1 \text{ or } x + y + z = 0 \text{ or } x + y + z = 1. \tag{10}$$

From (9) and (10), we obtain the systems

$$S_1 : \begin{cases} x + y + z = -1 \\ y - z = 0 \\ z^2 + 2xy = -16 \end{cases} \qquad S_2 : \begin{cases} x + y + z = 0 \\ y - z = 0 \\ z^2 + 2xy = -16 \end{cases} \qquad S_3 : \begin{cases} x + y + z = 1 \\ y - z = 0 \\ z^2 + 2xy = -16. \end{cases}$$

For S_1: From $x + y + z = -1$ and $y - z = 0$, we get $x = -1 - 2y$. However, $z^2 + 2xy = -16$, therefore

$$3y^2 + 2y - 16 = 0,$$

whose roots are the numbers $y = 2$, $y = -8/3$.
  The only integer solution of the equation is $y = 2$.
  The integer solution of S_1 is

$$(x, y, z) = (-5, 2, 2).$$

For S_2: From $x + y + z = 0$ and $y - z = 0$, we obtain $x = -2y$. However, $z^2 + 2xy = -16$, thus

$$3y^2 = 16,$$

from which we obtain

$$y = \pm\sqrt{\frac{16}{3}}.$$

These values are not integers, and thus not acceptable.
  For S_3: Following the same method as for S_1, the integer solution of S_3 is

$$(x, y, z) = (5, -2, -2).$$

Hence, the only solutions of the given system are

$$(x, y, z) = (-5, 2, 2), (x, y, z) = (5, -2, -2). \qquad \square$$

103) Let a, b, c be real numbers which are not all equal. Prove that positive integer solutions of the system

$$(b - a)x - (c - b)z = 3b \tag{1}$$

$$(c - b)y - (a - c)x = 3c \tag{2}$$

$$(a - c)z - (b - a)y = 3a \tag{3}$$

do not exist, except the trivial solution

$$(x, y, z) = (1, 1, 1),$$

which occurs only when $a + b + c = 0$.

Proof.

Case I. Suppose that $b \neq c$.

From (1) and (2) we obtain

$$y = \frac{(a-c)x + 3c}{c-b}, \quad z = \frac{(b-a)x - 3b}{c-b}. \tag{4}$$

But these values of y, z satisfy (3). Therefore, the given system of (1), (2) and (3) is indefinite and accepts infinite many solutions.

From the two equalities in (4), adding by parts we get

$$y + z = \frac{(a-c+b-a)x + 3(c-b)}{c-b}$$

$$= \frac{-(c-b)x + 3(c-b)}{c-b}.$$

That is,

$$y + z = -x + 3,$$

or

$$x + y + z = 3. \tag{5}$$

The only positive integer solution of (5) is

$$(x, y, z) = (1, 1, 1).$$

that is, $x = y = z = 1$.

Substituting these values to the initial system and summing the equations by parts, we obtain that

$$a + b + c = 0.$$

Case II. If $b = c \neq a$, from (1) we derive

$$x = -\frac{3b}{a-b} \tag{6}$$

and from (3) we get

$$y + z = \frac{3a}{a-b}. \tag{7}$$

If we add (6) and (7) by parts, we have

$$x + y + z = \frac{3a - 3b}{a-b} = 3,$$

that is,

$$x + y + z = 3$$

and therefore $x = y = z = 1$. Then (6) implies

$$\frac{-3b}{a - b} = 1,$$

that is,

$$a - b = -3b$$

or

$$a + 2b = 0$$

and hence

$$a + b + c = 0. \qquad \qquad \square$$

104) Show that, for any $n \in \mathbb{N}$, any $k \in \mathbb{N}$ which is not equal to a power of 10, and any sequence of (decimal) digits $x_0, x_1, \ldots, x_{n-1}$ in $\{0, 1, \ldots, 9\}$, there exists an $m \in \mathbb{N} \cup \{0\}$ such that the first n decimal digits of the power k^m are, from left to right, $x_{n-1}x_{n-2} \ldots x_1 x_0$. As an example, a power of 2 beginning with the digits 409 is $2^{12} = 4096$.

(Konstantinos Drakakis, University College Dublin, Ireland; Newsletter of the European Mathematical Society, Issue 69, 2008, Problem 37, p. 23)

Solution. Because k is not a power of 10, the number $z = \log_{10} k$ is irrational: for $z = r/s$, $r, s \in \mathbb{Z}$, implies that $k^s = 10^r$, which is impossible, as the two sides of the equation do not have the same prime factors. This, in turn, implies that the sequence $S = \{iz \, (\text{mod } 1), \ i \in \mathbb{N} \cup \{0\}\}$ has infinitely many points in $[0, 1]$, or else there would be two distinct values of u, say N_1 and N_2, for which $(N_2 - N_1)z = N \in \mathbb{N}$, implying that $z \in \mathbb{Q}$, a contradiction.

Even more, the sequence S is actually dense in $[0, 1]$. To see this, choose $l \in \mathbb{N}$, consider the $l+1$ terms of the sequence corresponding to $i = 1, \ldots, l+1$, and place them into the l intervals $(j/l, (j+1)/l)$, $j = 0, \ldots, l-1$. Necessarily at least two terms fall into the same interval, say those for $i = i_1$ and $i = i_2 > i_1$, hence, for $u = i_2 - i_1$ and some $v \in \mathbb{N} \cup \{0\}$,

$$|ux - v| < \frac{1}{l},$$

and, as u is the difference of two integers between 1 and $l+1$, it further follows that $u \leq l$, so that, finally,

$$\left| z - \frac{v}{u} \right| < \frac{1}{lu} \leq \frac{1}{u^2}.$$

Now, for any $\epsilon > 0$, consider l such that $1/l < \epsilon$, and find the rational v/u as above: it follows that

$$\left| iz - i\frac{v}{u} \right| = \left| iz \, (\text{mod } 1) - \left(i\frac{v}{u} \right) (\text{mod } 1) \right| < \frac{i}{lu} < \frac{1}{l} < \epsilon,$$

for $i < u$. At this point, note that, if

$$\left| z - \frac{v}{u} \right| > \frac{1}{l'},$$

repeating the process with l' instead of l yields a new fraction v'/u', as

$$\left| z - \frac{v'}{u'} \right| < \frac{1}{l'u'} < \frac{1}{l'} < \left| z - \frac{v}{u} \right|,$$

hence infinitely many rational approximations of z are obtainable this way, whose denominators must therefore grow arbitrarily large. Assuming, without loss of generality, that u and v are relatively prime, and that $1/u < \epsilon$, for any number $w \in (0, 1)$ it is true that

$$\left| w - i\frac{v}{u} \,(\mathrm{mod}\ 1) \right| < \epsilon, \tag{8}$$

for some $i < u$, and, therefore,

$$|w - iz \,(\mathrm{mod}\ 1)| \le \left| w - i\frac{v}{u} \,(\mathrm{mod}\ 1) \right| + \left| iz - i\frac{v}{u} \right| < 2\epsilon.$$

This implies that, for every interval I of $[0, 1]$, there exist infinitely many pairs $(u, v) \in \mathbb{N} \times \mathbb{N}$ such that $uz - v$ lies in I. Choose then $n \in \mathbb{N}$, choose a sequence of decimal digits $x_0, x_1, \ldots, x_{n-1}$, form the number

$$x = \sum_{k=0}^{n-1} x_k 10^k,$$

and set $x' = \log_{10} x \,(\mathrm{mod}\ 1)$. For a small $\epsilon' > 0$, to be further specified later, it follows that a number of the form $uz - v$ lies in $I = [x', x' + \epsilon']$ (in fact, infinitely many such numbers do so):

$$x' \le u \log_{10} k - v \le x' + \epsilon' \Leftrightarrow x \le \frac{k^u}{10^v} \le x 10^{\epsilon'}.$$

Setting $\epsilon' = \log_{10}(1 + \epsilon)$:

$$x \le \frac{k^u}{10^v} \le x(1 + \epsilon) < x + 1 \Leftrightarrow 0 \le k^u - 10^v x < 10^v, \text{ for } \epsilon < \frac{1}{x}.$$

This, however, implies that k^u has at least $n + v$ decimal digits, and that at most the v least significant ones differ from the decimal digits of $10^v x$; therefore, at least the n most significant digits of k^u and $10^v x$ are in agreement. $\qquad\square$

105) In order to file a collection of n books, each book needs a number label from 1 to n. To form this number, digit stickers are used: for example, the number **123** will be formed by the three stickers 1, 2, and 3 side by side (unnecessary zeros in the beginning, such as 00123, are not added, as this would be a terrible waste).

These stickers are sold in sets of 10, and each decimal digit $\{0, 1, 2, \ldots, 9\}$ appears exactly once in the set. How many sets of stickers are needed? As an example, for $n = 21$ books, digit 1 appears 13 times (in numbers 1, 10–19, and 21—note that it appears twice in 11!), 2 appears 4 times (2, 12, 20, and 21), and every other digit from 3 to 9 appears exactly twice, so overall 13 sets are needed.

(Konstantinos Drakakis, University College Dublin, Ireland; Newsletter of the European Mathematical Society, Issue 73, 2009, Problem 45, p. 52)

Solution. Assuming n has m digits, it can be written in the form

$$n = \sum_{k=0}^{m-1} n_k 10^k, n_k \in \{0, 1, \ldots, 9\}, k \in \{0, 1, \ldots, m-1\}, n_{m-1} \neq 0.$$

Since unnecessary initial 0s are not used, the first time a particular digit needs to be considered is when it is equal to 1, and therefore, for any n, no digit appears more times than 1. As each set contains exactly one sticker of each decimal digit value, the number of sets needed will equal the number of 1s needed, and, to determine this, it is enough to count the number of times each digit becomes equal to 1, and then sum. Consider below the general kth digit:

$n_k = 0$: The kth digit goes through a complete cycle $\sum_{l=k+1}^{m-1} n_l 10^l$ times, and each cycle contains 1 once, so that the number of 1s needed is

$$\sum_{l=k+1}^{m-1} n_l 10^{l-1}.$$

Note that $k = m - 1$ is not allowed in this case.

$n_k = 1$: In addition to the number of 1s counted in the previous case, there are some extra due to the numbers whose kth digit is actually 1. These are precisely

$$1 + \sum_{l=0}^{k-1} n_l 10^l.$$

The added 1 accounts for the number whose kth digit is 1 and all less significant digits are 0.

$n_k > 1$: Here, in addition to the number counted in the case $n_k = 0$, all of the numbers whose jth digit is n_j, $j > k$, and whose kth digit is 1 appear as well. These are clearly 10^k.

To summarize, using the boolean function

$$[P] = \begin{cases} 1, & P \text{ true} \\ 0, & P \text{ false,} \end{cases}$$

the number of times the kth digit equals 1 can be written as

$$\sum_{l=k+1}^{m-1} n_l 10^{l-1} + [n_k = 1]\left(1 + \sum_{l=0}^{k-1} n_l 10^l\right) + [n_k > 1]10^k,$$

and therefore the total number of times digit 1 appears is written as

$$S(n) = \sum_{k=0}^{m-1}\left(\sum_{l=k+1}^{m-1} n_l 10^{l-1} + [n_k = 1]\left(1 + \sum_{l=0}^{k-1} n_l 10^l\right) + [n_k > 1]10^k\right). \quad \square$$

12

Appendix

The shortest path between two truths in the real domain
passes through the complex domain.
Jacques Hadamard (1865–1963)

12.1 Prime number theorem

A step-by-step analysis of Newman's proof of the Prime Number Theorem

We shall present a step-by-step analysis of D. J. Newman's proof [44] of the Prime Number Theorem, which we mentioned in Chapter 6. Within the proof, some other theorems are going to be introduced, some of which may seem elementary to the reader. However, we present them for the sake of completeness.

Theorem.

$$\pi(x) \sim \frac{x}{\log x}, \ \text{as } x \to +\infty.$$

Proof. We are going to show that for the proof of the Prime Number Theorem, it is sufficient to prove that $\vartheta(x) \sim x$, where

$$\vartheta(x) = \sum_{p \leq x} \log p$$

is the Chebyshev function (where p stands for the prime numbers).

One has

$$\vartheta(x) = \sum_{p \leq x} \log p \geq \sum_{x^{1-k} < p \leq x} \log p \,, \ \text{for every } k, \text{ with } 0 < k < 1.$$

M.Th. Rassias, *Problem-Solving and Selected Topics in Number Theory: In the Spirit of the Mathematical Olympiads*, DOI 10.1007/978-1-4419-0495-9_12,
© Springer Science+Business Media, LLC 2011

From $x^{1-k} \leq p \leq x$ we obtain that

$$(1 - k) \log x \leq \log p \leq \log x.$$

Hence,

$$\vartheta(x) \geq \sum_{x^{1-k} < p \leq x} (1 - k) \log x = [\pi(x) - \pi(x^{1-k})](1 - k) \log x. \qquad (1)$$

It is obvious that $\pi(x) = O(x)$, and hence $\pi(x^{1-k}) = O(x^{1-k})$.
   Therefore, from (1), we obtain that

$$\vartheta(x) \geq [\pi(x) + O(x^{1-k})](1 - k) \log x. \qquad (2)$$

Similarly,

$$\vartheta(x) = \sum_{p \leq x} \log p \leq \sum_{p \leq x} \log x = \pi(x) \log x. \qquad (3)$$

From (2) and (3) we have

$$[\pi(x) + O(x^{1-k})](1 - k) \log x \leq \vartheta(x) \leq \pi(x) \log x.$$

We consider here x to be a positive real number.
   Therefore,

$$\frac{O(x^{1-k})(1 - k) \log x}{x} + \frac{\pi(x)(1 - k) \log x}{x} \leq \frac{\vartheta(x)}{x} \leq \frac{\pi(x) \log x}{x},$$

that is,

$$\frac{O(x^{1-k})}{x^{1-k/2}} \frac{(1 - k) \log x}{x^{k/2}} + \frac{\pi(x)(1 - k) \log x}{x} \leq \frac{\vartheta(x)}{x} \leq \frac{\pi(x) \log x}{x}. \qquad (4)$$

For $x \to +\infty$ it follows that $\frac{O(x^{1-k})}{x^{1-k/2}} \to 0$ and $\frac{(1-k) \log x}{x^{k/2}} \to 0$. Thus, for $x \to +\infty$ if $\vartheta(x) \sim x$, then from (4), one has

$$o(1) + \frac{\pi(x)(1 - k) \log x}{x} \leq 1 \leq \frac{\pi(x) \log x}{x} + o(1).$$

Therefore,

$$(1 + o(1)) \frac{x}{\log x} \leq \pi(x) \leq \frac{x}{(1 - k) \log x}(1 + o(1)), \quad \text{for every } k, \text{ with } k > 0,$$

which means that

$$\pi(x) \sim \frac{x}{\log x}, \quad \text{as } x \to +\infty.$$

Therefore, as we have just shown, in order to prove the Prime Number Theorem it is sufficient to prove that $\vartheta(x) \sim x$. To do so, we are going to use the following lemma, which is due to D. J. Newman.

Lemma. *Let $f(t)$ be a bounded and locally integrable function for $t \geq 0$. Let*

$$F(z) = \int_0^{+\infty} f(t)e^{-zt}dt$$

be a holomorphic function for $Re\{z\} > 0$ that can be extended to a function which is holomorphic for $Re\{z\} \geq 0$. In other words, we assume that there is a function $w(z)$ that equals $F(z)$ whenever $Re\{z\} > 0$ and the function $w(z)$ is holomorphic in the bigger region $Re\{z\} \geq 0$. Then, the integral $\int_0^{+\infty} f(t)dt$ exists and, in fact,

$$w(0) = \int_0^{+\infty} f(t)dt.$$

(Note: $Re\{z\}$ denotes the real part of the complex number z.)

Proof of the Lemma. With some abuse of notation, we will refer to $F(z)$ as the holomorphic extension of

$$\int_0^{+\infty} f(t)e^{-zt}dt \text{ to } Re\{z\} \geq 0.$$

We have to show that $\int_0^{+\infty} f(t)dt$ exists and is equal to $F(0)$. In other words, $\int_0^{+\infty} f(t)dt$ is equal to the value of the holomorphic extension of $F(z)$ evaluated at $z = 0$.

It suffices to prove that

$$\lim_{L\to\infty} F_L(0) = F(0), \text{ where } F_L(z) = \int_0^L f(t)e^{-zt}dt.$$

Set

$$h(z) = [F(z) - F_L(z)]e^{zL}\left(1 + \frac{z^2}{R^2}\right),$$

in the region $|z| \leq R$, subject to the condition $Re\{z\} > -\varepsilon$, where $\varepsilon > 0$ so that the function $F(z)$ is holomorphic for a specific value of R.

If C is the boundary of that region, then from the Cauchy integral formula we have

$$\begin{aligned}
h(0) &= \frac{1}{2\pi i}\int_C \frac{1}{z}[F(z) - F_L(z)]e^{zL}\left(1 + \frac{z^2}{R^2}\right)dz \\
&= \frac{1}{2\pi i}\left\{\int_{C_1} \frac{1}{z}[F(z) - F_L(z)]e^{zL}\left(1 + \frac{z^2}{R^2}\right)dz \right. \\
&\quad \left. + \int_{C_2} \frac{1}{z}[F(z) - F_L(z)]e^{zL}\left(1 + \frac{z^2}{R^2}\right)dz\right\},
\end{aligned} \tag{L1}$$

where $C_1 = C \cap \{z | Re\{z\} > 0\}$ and $C_2 = C \cap \{z | Re\{z\} \leq 0\}$.

On the curve C_1 we have

$$\left| \int_{C_1} \frac{1}{z}[F(z) - F_L(z)]e^{zL}\left(1 + \frac{z^2}{R^2}\right)dz \right|$$

$$\leq \int_{C_1} |F(z) - F_L(z)|\left|\frac{e^{zL}}{z}\left(1 + \frac{z^2}{R^2}\right)\right||dz|$$

$$\leq \int_{C_1} \left|\int_L^{+\infty} f(t)e^{-zt}dt\right|\left|1 + \frac{z^2}{R^2}\right|\frac{e^{Re\{z\}L}}{R}|dz|. \qquad (L2)$$

But

$$\left|\int_L^{+\infty} f(t)e^{-zt}dt\right| \leq \int_L^{+\infty} |f(t)|e^{-Re\{z\}t}dt$$

$$\leq \int_L^{+\infty} Me^{-Re\{z\}t}dt$$

$$= \frac{Me^{-Re\{z\}L}}{Re\{z\}},$$

for a certain constant M, since $f(t)$ is a bounded function.

Furthermore, if we express z in the form $z = Re^{i\theta}$, where $-\pi/2 \leq \theta \leq \pi/2$ since $z \in C_1$, we obtain

$$\left|1 + \frac{z^2}{R^2}\right| = |1 + e^{2i\theta}| = \sqrt{(1 + \cos 2\theta)^2 + \sin^2 2\theta}$$

$$= \sqrt{2(1 + \cos 2\theta)} = 2\cos\theta = \frac{2Re\{z\}}{R}.$$

Hence, (L2) can be written in the form

$$\left|\int_{C_1} \frac{1}{z}[F(z) - F_L(z)]e^{zL}\left(1 + \frac{z^2}{R^2}\right)dz\right| \leq \frac{2M}{R^2}\pi R = \frac{2\pi M}{R}. \qquad (L3)$$

Similarly, on the curve C_2 we have

$$\left|\int_{C_2} \frac{1}{z}[F(z) - F_L(z)]e^{zL}\left(1 + \frac{z^2}{R^2}\right)dz\right|$$

$$\leq \left|\int_{C_2} \frac{1}{z}F(z)e^{zL}\left(1 + \frac{z^2}{R^2}\right)dz\right| + \left|\int_{C_2} \frac{1}{z}F_L(z)e^{zL}\left(1 + \frac{z^2}{R^2}\right)dz\right|. \qquad (L4)$$

Note that with L fixed, the function $F_L(z)$ is entire. By an application of Cauchy's integral formula, we can deduce that

$$\int_{C_2} \frac{1}{z}F_L(z)e^{zL}\left(1 + \frac{z^2}{R^2}\right)dz = \int_{C_3} \frac{1}{z}F_L(z)e^{zL}\left(1 + \frac{z^2}{R^2}\right)dz,$$

where $C_3 = \{z : |z| = R$ and $Re\{z\} \leq 0\}$. Here, one considers both C_2 and C_3 as being oriented counterclockwise. To see the above, we can subtract the left-hand side from both sides above. The resulting integral on the right will be an integral over a closed curve bounding a region in which the integrand has no poles (since the only pole of the integrand is at 0, and 0 is not in the region).

For $z \in C_3$, we can deduce that

$$\left|\frac{e^{zL}}{z}\right| = \frac{e^{Re\{z\}L}}{R} \quad \text{and} \quad \left|1 + \frac{z^2}{R^2}\right| \leq \frac{-2Re\{z\}}{R},$$

where the negative sign is needed as $Re\{z\} < 0$.

But

$$|F_L(z)| \leq \left|\int_0^L f(t)e^{-zt}dt\right| \leq \int_0^L |f(t)||e^{-zt}|dt$$

$$\leq M \int_0^L |e^{-zt}|dt = M \int_0^L e^{-Re\{z\}t}dt$$

$$\leq M \int_{-\infty}^L e^{-Re\{z\}t}dt = M\frac{e^{-Re\{z\}L}}{-Re\{z\}}.$$

The first integral on the right-hand side of (L4) is over C_2. Now, we break up C_2 into three pieces C_4, C_5 and C_6 as follows:

$$C_4 = \{z \in C_2 : |z| = R, \quad Re\{z\} \geq -1/\sqrt{L}, \; Im(z) > 0\},$$

$$C_6 = \{z \in C_2 : |z| = R, \quad Re\{z\} \geq -1/\sqrt{L}, \; Im(z) < 0\},$$

$$C_5 = \{z \in C_2 : z \notin C_4 \cup C_6\}.$$

We can observe that each of C_4 and C_6 has length $O(1/\sqrt{L})$. Also, as $Re\{z\} < 0$ on C_4 and C_6, we easily have

$$|e^{zL}| = e^{Re\{z\}L} \leq 1, \quad \text{for } z \in C_4 \cup C_6.$$

On the other hand, for $z \in C_6$ and L large enough (that is, $L > 1/\varepsilon^2$), we have $Re\{z\} < -1/\sqrt{L}$ so that

$$|e^{zL}| = e^{Re\{z\}L} \leq e^{-\sqrt{L}}, \quad \text{for } z \in C_5.$$

Note that

$$\left|\int_{C_2} \frac{1}{z}F(z)e^{zL}\left(1 + \frac{z^2}{R^2}\right)dz\right| \leq \int_{C_2} \left|\frac{1}{z}F(z)\left(1 + \frac{z^2}{R^2}\right)\right||e^{zL}||dz|$$

$$= I_4 + I_5 + I_6,$$

where

$$I_j = \int_{C_j} \left| \frac{1}{z} F(z) \left(1 + \frac{z^2}{R^2} \right) \right| |e^{zL}| |dz|.$$

We deduce from the above that

$$I_5 \leq \left(\int_{C_5} \left| \frac{1}{z} F(z) \left(1 + \frac{z^2}{R^2} \right) \right| |dz| \right) e^{-z\sqrt{L}}.$$

The function $F(z)(1+(z^2/R^2))/z$ is holomorphic in a closed region containing C_2. If M' (which depends on R) is a bound for this function in this region, we obtain for $j \in \{4, 6\}$ that

$$I_j = \int_{C_j} \left| \frac{1}{z} F(z) \left(1 + \frac{z^2}{R^2} \right) \right| |e^{zL}| |dz| \leq M' \int_{C_j} |dz| = O_R(1/\sqrt{L}),$$

where the subscript R indicates that the implied constant depends on R. It follows that, for R fixed, $I_4 + I_5 + I_6$ tends to 0 as L tends to infinity.

Hence,

$$\left| \int_{C_2} \frac{1}{z} F(z) e^{zL} \left(1 + \frac{z^2}{R^2} \right) dz \right| \to 0, \text{ as } L \to +\infty. \qquad (L5)$$

From the relations (L1), (L2), (L4) and (L5) one obtains

$$\limsup_{L \to +\infty} |h(0)| \leq \left| \frac{1}{2\pi i} \left(\frac{2\pi M}{R} + \frac{2\pi M}{R} \right) \right| = \frac{2M}{R},$$

that is,

$$\limsup_{L \to +\infty} |F(0) - F_L(0)| \leq \frac{2M}{R} \Rightarrow \lim_{L \to +\infty} F_L(0) = F(0),$$

because R can be arbitrarily large. This proves the lemma. $\square$

Let

$$f(t) = \frac{\vartheta(e^t)}{e^t} - 1, \quad t \geq 0$$

and

$$F(z) = \int_0^{+\infty} \left[\frac{\vartheta(e^t)}{e^t} - 1 \right] e^{-zt} dt, \quad z \in \mathbb{C}.$$

If we prove that $\vartheta(x) = O(x)$, then

$$f(t) \leq \frac{ce^t}{e^t} - 1 = c - 1,$$

where c is a constant, which means that $f(t)$ is a bounded function. It is also locally integrable. Thus, it satisfies the hypotheses of the lemma. We also want to prove that $F(z)$ is holomorphic for $Re\{z\} \geq 0$, so that all hypotheses of the lemma are satisfied.

We shall begin by proving that $\vartheta(x) = O(x)$.
For every $n \in \mathbb{N}$, we have

$$e^{2n \log 2} = 2^{2n} = \binom{2n}{0} + \binom{2n}{1} + \cdots + \binom{2n}{2n}$$

$$\geq \binom{2n}{n} = \frac{(n+1)(n+2)\cdots 2n}{n!}.$$

The number $n!$ divides factors of the product $(n+1)(n+2)\cdots 2n$, but none of the prime numbers $p_k \in [n+1, 2n)$ is affected and along with them, some composite integers remain unaffected, too. Hence,

$$\binom{2n}{n} \geq \prod_{n < p \leq 2n} p.$$

Therefore,

$$e^{2n \log 2} \geq \prod_{n < p \leq 2n} p = \exp\left(\sum_{p \leq 2n} \log p - \sum_{p < n} \log p\right) = e^{\vartheta(2n) - \vartheta(n)},$$

which means that

$$\vartheta(2n) - \vartheta(n) \leq 2n \log 2.$$

It follows that for every $\varepsilon > 0$ there exists $x(\varepsilon)$ such that for every $x \geq x(\varepsilon)$

$$\vartheta(x) - \vartheta(x/2) \leq x(\varepsilon + \log 2) \leq Bx,$$

for some constant $B > \log 2$.

Next, this last inequality implies

$$\vartheta(x) - \vartheta(x/2) \leq Bx,$$

for all positive values of x. Thus, we have

$$\left.\begin{array}{rl} \vartheta(x) - \vartheta(x/2) \leq & Bx \\ \vartheta(x/2) - \vartheta(x/2^2) \leq & Bx/2 \\ \vdots \\ \vartheta(x/2^r) - \vartheta(x/2^{r+1}) \leq & Bx/2^r. \end{array}\right\}$$

If we add up the above relations, we obtain

$$\vartheta(x) \leq Bx \cdot \left(1 + \frac{1}{2} + \cdots + \frac{1}{2^r}\right) + \vartheta(x/2^{r+1}).$$

Hence, for $r \to +\infty$ it follows

$$\vartheta(x) \leq 2Bx \Rightarrow \vartheta(x) = O(x).$$

Now, we are going to prove the fact that $F(z)$ is a holomorphic function, for $Re\{z\} \geq 0$.

We have

$$F(z) = \int_0^{+\infty} f(t)e^{-zt}dt = \int_0^{+\infty} \left[\frac{\vartheta(e^t)}{e^t} - 1\right]e^{-zt}dt$$

$$= \int_0^{+\infty} \vartheta(e^t)e^{-t(1+z)}dt - \int_0^{+\infty} e^{-zt}dt$$

$$= \int_0^{+\infty} \vartheta(e^t)e^{-t(1+z)}dt - \frac{1}{z}, \tag{5}$$

for $Re\{z\} > 0$. Set

$$A(z) = \sum_p \frac{\log p}{p^z}.$$

Then, we can express $A(z)$ by a Stieltjes integral as follows:

$$A(z) = \sum_p \frac{\log p}{p^z} = \int_1^{+\infty} \frac{d\vartheta(x)}{x^z}$$

$$= \left[\frac{1}{x^z}\vartheta(x)\right]_1^{+\infty} + \int_1^{+\infty} \frac{z}{x^{z+1}}\vartheta(x)dx. \tag{6}$$

For $x = 1$, we know that $\vartheta(1) = 0$ and for $x \to +\infty$ we have

$$\lim_{x \to \infty} \left|\frac{\vartheta(x)}{x^z}\right| \leq \lim_{x \to \infty} \left|\frac{Bx}{x^z}\right| = 0, \text{ for } Re\{z\} > 1,$$

since $\vartheta(x) = O(x)$.

Therefore, (6) takes the form

$$A(z) = z\int_1^{+\infty} \frac{\vartheta(x)}{x^{z+1}}dx$$

and for $x = e^t$, we can write

$$A(z) = z\int_0^{+\infty} \frac{\vartheta(e^t)}{e^{zt}}dt, \text{ for } Re\{z\} > 1.$$

Therefore, (5) can be written in the form

$$F(z) = \frac{A(z+1)}{z+1} - \frac{1}{z}, \text{ for } Re\{z\} > 0.$$

Thus, it suffices to prove that this function is holomorphic for $Re\{z\} \geq 0$. This is equivalent to prove that the function

$$\frac{A(z)}{z} - \frac{1}{z-1}$$

is holomorphic for $Re\{z\} \geq 1$.

By Euler's formula we know that

$$\zeta(z) = \prod_p \frac{1}{1 - \frac{1}{p^z}}, \quad \text{for } Re\{z\} > 1.$$

Hence, we can write

$$-\frac{d}{dz}[\log \zeta(z)] = -\frac{d}{dz}\left[\sum_p \log \frac{1}{1 - 1/p^z}\right] = \sum_p \frac{\log p}{p^z - 1} = \sum_p \frac{p^z \log p}{p^z(p^z - 1)}$$

$$= \sum_p \frac{(p^z - 1)\log p + \log p}{p^z(p^z - 1)} = \sum_p \frac{\log p}{p^z} + \sum_p \frac{\log p}{p^z(p^z - 1)}$$

$$= A(z) + \sum_p \frac{\log p}{p^z(p^z - 1)}. \tag{7}$$

It is obvious that

$$\left|\sum_p \frac{\log p}{p^z(p^z - 1)}\right| \leq \sum_p \left|\frac{\log p}{p^z(p^z - 1)}\right|.$$

Set $z = \alpha + i\beta$. Then

$$\left|\sum_p \frac{\log p}{p^z(p^z - 1)}\right| \leq \sum_p \frac{\log p}{p^\alpha(p^\alpha - 1)} < \sum_{n=2}^{+\infty} \frac{\log n}{n(n - 1)}.$$

The last series is convergent, as it can easily be proved by the Cauchy criterion. Hence, clearly the series

$$\sum_p \frac{\log p}{p^z(p^z - 1)}$$

is holomorphic for $Re\{z\} \geq 1$.

At this point, we are going to prove that the Riemann zeta function has no zeros in the half plane $Re\{z\} \geq 1$.

By expanding the function

$$\log \frac{1}{(1 - p^{-z})}$$

in a Taylor series, we get

$$\log \zeta(z) = \sum_{p} \log \frac{1}{1-p^{-z}} = \sum_{p} \sum_{n=1}^{+\infty} \frac{1}{np^{nz}}.$$

Thus,

$$\zeta(z) = \exp\left\{\sum_{p} \sum_{n=1}^{+\infty} \frac{1}{np^{nz}}\right\}, \quad \text{for } Re\{z\} > 1.$$

But

$$\sum_{n=1}^{+\infty} \frac{1}{np^{nz}} = \sum_{n=1}^{+\infty} \frac{p^{-ni\beta}}{np^{n\alpha}} = \sum_{n=1}^{+\infty} \frac{e^{-ni\beta \log p}}{np^{n\alpha}}$$

$$= \sum_{n=1}^{+\infty} \frac{\cos(n\beta \log p) - i\sin(n\beta \log p)}{np^{n\alpha}}$$

and hence

$$|\zeta(z)| = \left|\exp\left\{\sum_{p} \sum_{n=1}^{+\infty} \frac{\cos(n\beta \log p)}{np^{n\alpha}}\right\} \cdot \exp\left\{-i\sum_{p} \sum_{n=1}^{+\infty} \frac{\sin(n\beta \log p)}{np^{n\alpha}}\right\}\right|$$

$$= \exp\left\{\sum_{p} \sum_{n=1}^{+\infty} \frac{\cos(n\beta \log p)}{np^{n\alpha}}\right\}.$$

Therefore, according to the above relation, we get

$$|\zeta^3(\alpha)\zeta^4(\alpha + i\beta)\zeta(\alpha + 2i\beta)|$$

$$= \exp\left\{\sum_{p} \sum_{n=1}^{+\infty} \frac{3 + 4\cos(n\beta \log p) + \cos(2n\beta \log p)}{np^{n\alpha}}\right\}.$$

We know that

$$2(\cos\theta + 1)^2 = 2\cos^2\theta + 4\cos\theta + 2 = 3 + 4\cos\theta + (2\cos^2\theta - 1)$$

$$= 3 + 4\cos\theta + \cos 2\theta.$$

Therefore,

$$3 + 4\cos(n\beta \log p) + \cos(2n\beta \log p) \geq 0$$

and hence

$$|\zeta^3(\alpha)\zeta^4(\alpha + i\beta)\zeta(\alpha + 2i\beta)| \geq 1, \quad \text{for } \alpha > 1.$$

If we could find even one point z_0 of the half plane $Re\{z\} > 1$ for which $\zeta(z_0) = 0$, then the above inequality would not hold for that point, which is a contradiction.

Now, we are also going to prove that the Riemann zeta function has no zeros on the line $Re\{z\} = 1$. In order to do so, we shall begin by proving that the function

$$\zeta(z) - \frac{1}{z-1}$$

can be extended to a holomorphic function in $Re\{z\} > 0$.

For $Re\{z\} > 1$, the series $\zeta(z)$ converges and therefore

$$\zeta(z) - \frac{1}{z-1} = \sum_{n=1}^{+\infty} \frac{1}{n^z} - \int_1^{+\infty} \frac{1}{x^z} dx$$

$$= \sum_{n=1}^{+\infty} \int_n^{n+1} \frac{1}{n^z} dx - \sum_{n=1}^{+\infty} \int_n^{n+1} \frac{1}{x^z} dx$$

$$= \sum_{n=1}^{+\infty} \int_n^{n+1} \left(\frac{1}{n^z} - \frac{1}{x^z} \right) dx.$$

Now

$$\left| \sum_{n=1}^{+\infty} \int_n^{n+1} \left(\frac{1}{n^z} - \frac{1}{x^z} \right) dx \right| \leq \sum_{n=1}^{+\infty} \left| \int_n^{n+1} \left(\frac{1}{n^z} - \frac{1}{x^z} \right) dx \right|$$

$$= \sum_{n=1}^{+\infty} \left| z \int_n^{n+1} \int_n^x \frac{1}{y^{z+1}} dy dx \right|$$

$$\leq \sum_{n=1}^{+\infty} |z| \int_n^{n+1} \int_n^x \left| \frac{1}{y^{z+1}} \right| dy dx. \tag{8}$$

But, from the limits of the integrals we know that $n \leq y \leq n+1$ and therefore the maximum value of the integral

$$\int_n^{n+1} \int_n^x \left| \frac{1}{y^{z+1}} \right| dy dx$$

is bounded by $|1/n^{z+1}| = 1/n^{\alpha+1}$, where $\alpha = Re\{z\}$.

Hence, by (8), we obtain

$$\left| \zeta(z) - \frac{1}{z-1} \right| \leq \sum_{n=1}^{+\infty} \frac{|z|}{n^{\alpha+1}} = \sum_{n=1}^{+\infty} \frac{\sqrt{\alpha^2 + \beta^2}}{n^{\alpha+1}}, \quad \text{where } \beta = Im\{z\}.$$

But, the latter series is convergent for $\alpha > 0$. Thus, by analytic continuation, we can say that the function

$$\zeta(z) - \frac{1}{z-1}$$

can be extended holomorphically for $a > 0$. Hence, we can write

$$\zeta(z) - \frac{1}{z - 1} = g(z),$$

where $g(z)$ is a holomorphic function.

Therefore,

$$(\alpha - 1)\zeta(a) = 1 + (\alpha - 1)g(a)$$

from which it follows that

$$\lim_{\alpha \to 1^+} (\alpha - 1)\zeta(\alpha) = 1.$$

Similarly,

$$|\zeta(1 + 2ik)| = \left| \frac{1}{2ik} + g(1 + 2ik) \right| , \quad k \neq 0.$$

Furthermore,

$$\frac{d}{dz}\zeta(z) = -\frac{1}{(z - 1)^2} + \frac{d}{dz}g(z),$$

that is,

$$\left| \frac{d}{dz}\zeta(1 + ik) \right| = \left| \frac{1}{k^2} + \frac{d}{dz}g(1 + ik) \right|.$$

Let us suppose that $1 + i\beta$, $\beta \neq 0$, is a root of $\zeta(z)$ in the complex plane.

We have already proved that

$$|\zeta^3(\alpha)\zeta^4(\alpha + i\beta)\zeta(\alpha + 2i\beta)| \geq 1, \quad \text{for } \alpha > 1.$$

Therefore, we get

$$L = [(\alpha - 1)\zeta(\alpha)]^3 \left| \frac{\zeta(\alpha + i\beta)}{\alpha - 1} \right|^4 |\zeta(\alpha + 2\beta i)| \geq \frac{1}{\alpha - 1}, \quad \text{for } \alpha > 1. \qquad (9)$$

Hence,

$$\lim_{\alpha \to 1^+} [(\alpha - 1)\zeta(\alpha)]^3 \left| \frac{\zeta(\alpha + i\beta)}{\alpha - 1} \right|^4 |\zeta(\alpha + 2\beta i)|$$

$$= 1 \cdot \left(\lim_{\alpha \to 1^+} \left| \frac{\zeta(\alpha + i\beta)}{\alpha - 1} \right|^4 \right) \cdot \left| \frac{1}{2i\beta} + g(1 + 2i\beta) \right|.$$

However, we have assumed that $1 + i\beta$ is a zero of $\zeta(s)$. Therefore, we can write

$$\lim_{\alpha \to 1^+} [(\alpha - 1)\zeta(\alpha)]^3 \left| \frac{\zeta(\alpha + i\beta)}{\alpha - 1} \right|^4 |\zeta(\alpha + 2\beta i)|$$

$$= \left| \frac{1}{2i\beta} + g(1 + 2i\beta) \right| \cdot \lim_{\alpha \to 1^+} \left| \frac{\zeta(\alpha + i\beta) - \zeta(1 + i\beta)}{\alpha - 1} \right|^4$$

$$= \left| \frac{1}{2i\beta} + g(1 + 2i\beta) \right| \cdot \left| \frac{d}{dz} \zeta(1 + i\beta) \right|^4$$

$$= \left| \frac{1}{2i\beta} + g(1 + 2i\beta) \right| \cdot \left| \frac{1}{\beta^2} + \frac{d}{dz} g(1 + i\beta) \right|^4.$$

That is a contradiction, because by (9) we can easily see that $L \to +\infty$ when $\alpha \to 1^+$. Therefore, in general the Riemann zeta function has no zeros in the half plane $Re\{z\} \geq 1$.

Moreover, by the formulae

$$\zeta(z) = g(z) + \frac{1}{z - 1}, \quad \text{for } Re\{z\} > 0$$

and

$$\frac{d}{dz} \zeta(z) = \frac{d}{dz} g(z) - \frac{1}{(z-1)^2}, \quad \text{for } Re\{z\} > 0$$

where $g(z)$ is a holomorphic function in the positive complex plane, we get

$$-\frac{d}{dz}[\log \zeta(z)] = -\frac{\frac{d}{dz}\zeta(z)}{\zeta(z)} = -\frac{[\frac{d}{dz}g(z)](z-1)^2 - 1}{(z-1)[1 + g(z)(z-1)]}$$

$$= \frac{-1}{z-1} \cdot \frac{[\frac{d}{dz}g(z)](z-1)^2 - 1}{1 + g(z)(z-1)}, \quad \text{for } Re\{z\} \geq 1.$$

If we expand the function defined by

$$G(z) = \frac{[\frac{d}{dz}g(z)](z-1)^2 - 1}{1 + g(z)(z-1)}$$

in a Taylor series around the point $1 + 0i$ we get

$$-\frac{d}{dz}[\log \zeta(z)] = \frac{-1}{z-1} \cdot [c_0 + c_1(z-1) + \cdots] = \frac{-c_0}{z-1} + E(z)$$

$$= \frac{-G(1)}{z-1} + E(z) = \frac{1}{z-1} + E(z),$$

where $E(z)$ is a holomorphic function for $Re\{z\} \geq 1$.

Hence,

$$-\frac{d}{dz}[\log \zeta(z)]$$

is a holomorphic function in the half plane $Re\{z\} \geq 1$, except for the point $z = 1$ which is a simple pole.

Therefore, from (7) it follows that $A(z)$ is a holomorphic function in the half plane $Re\{z\} \geq 1$, except for $z = 1$, which is a simple pole. Clearly, the function $A(z)/z$ is also holomorphic in the half plane $Re\{z\} \geq 1$, except for $z = 1$. More precisely,

$$A(z) = \frac{1}{z-1} + B(z),$$

for some function $B(z)$ which is holomorphic for $Re\{z\} \geq 1$.

This implies that

$$\frac{A(z)}{z} = \frac{1}{z(z-1)} + \frac{B(z)}{z} = \frac{1}{z-1} - \frac{1}{z} + \frac{B(z)}{z}.$$

Therefore,

$$\frac{A(z)}{z} - \frac{1}{z-1}$$

is a holomorphic function in the half plane $Re\{z\} \geq 1$.

But,

$$F(z) = \frac{A(z+1)}{z+1} - \frac{1}{z},$$

for $Re\{z\} > 0$.

Since $A(z+1)/(z+1) - 1/z$ is a holomorphic function for $Re\{z\} \geq 0$ and this function agrees with $F(z)$ for $Re\{z\} > 0$, the function $A(z+1)/(z+1)-1/z$ extends $F(z)$ to a holomorphic function for $Re\{z\} \geq 0$.

Thus, the second hypothesis of the lemma is also satisfied.

Hence, the integral

$$\int_0^{+\infty} f(t)dt = \int_0^{+\infty} \left(\frac{\vartheta(e^t)}{e^t} - 1 \right) dt \tag{10}$$

exists.

If we set $x = e^t$ in (10), we obtain that the integral

$$\int_1^{+\infty} \frac{\vartheta(x) - x}{x^2} dx$$

exists.

Let us now suppose that the assertion $\vartheta(x) \sim x$ is false. Then, for some arbitrarily large values of x we can distinguish two cases.

First case. There exists $\nu > 1$ and arbitrarily large values of x for which

$$\vartheta(x) \geq \nu x.$$

In this case, we get

$$\int_x^{\nu x} \frac{\vartheta(y) - y}{y^2} dy \geq \int_x^{\nu x} \frac{\nu x - y}{y^2} dy = \nu - 1 - \log \nu > 0$$

for every possible positive value of x.

That is a contradiction, because

$$\int_x^{\nu x} \frac{\vartheta(y) - y}{y^2} dy \to 0, \quad \text{when } x \to +\infty$$

since the integral

$$\int_0^{+\infty} (\vartheta(x) - x)/x^2 dy$$

exists.

Second case. There exists $\nu < 1$ and arbitrarily large values of x for which

$$\vartheta(x) \leq \nu x.$$

In this case

$$\int_{\nu x}^x \frac{\vartheta(y) - y}{y^2} dy \leq \int_{\nu x}^x \frac{\nu x - y}{y^2} dy = 1 - \nu + \log \nu < 0$$

for every possible positive value of x.

That is a contradiction, because

$$\int_{\nu x}^x (\vartheta(y) - y)/y^2 dy \to 0, \quad \text{when } x \to +\infty$$

since the integral

$$\int_0^{+\infty} (\vartheta(x) - x)/x^2 dx$$

exists. Therefore, the statement $\vartheta(x) \sim x$ must be true. Hence,

$$\pi(x) \sim \frac{x}{\log x}, \quad \text{as } x \to +\infty.$$

This completes the proof of the Prime Number Theorem. $\qquad \square$

12.2 A brief history of Fermat's last theorem

> *I hope that seeing the excitement of solving this problem*
> *will make young mathematicians realize that*
> *there are lots and lots of other problems in mathematics*
> *which are going to be just as challenging in the future.*
>
> Andrew Wiles (1953–)

Pierre de Fermat was born in Beaumont de Lomagne, France, in August 1601. He was a professional lawyer and started working on mathematics as an amateur after the age of 30. He contributed both in pure and applied mathematics.

Independently of Descartes (also known as Cartesius, 1596–1650), Fermat discovered analytic geometry. However, in contrast to Descartes who examined two-dimensional problems, Fermat was the first to apply methods of analytic geometry to three-dimensional problems. In addition, Fermat along with Blaise Pascal (1623–1662) developed the theory of probability. We shall talk about his contribution to number theory later on.[1]

His most important contribution in applied mathematics was in optics. His work in this area is considered to be the foundation of wave mechanics. Fermat is also considered as the founder of modern number theory. An important achievement of his is the discovery of the *method of infinite descent*, which he used in order to prove several theorems. One of the most well-known theorems he proved by the use of that method is that every prime number of the form $4n + 1$ can be expressed as the sum of two squares of integers in a unique way. Fermat mainly dealt with problems related to prime numbers, divisibility and diophantine equations.

After his death on the 12th of January, 1665, his son Samuel collected his archives of books and manuscripts. In the second book of Diophantus' *Arithmetica*, next to the 8th problem, which is related to the rational solutions of the diophantine equation

$$x^2 + y^2 = z^2,$$

---

[1] For further historical remarks the reader is referred to D. J. Struik, *A Concise History of Mathematics*, Dover Publications, Fourth revised edition, London, 1987, pp. 99–103; B. V. Gnedenko (translated from the Russian by B. D. Seckler), *The Theory of Probability*, Chelsea Publishing Company, New York, 1968, p. 15; and D. Abbott (ed.), *The Biographical Dictionary of Scientists: Mathematicians*, Blond Educational, London, 1985, pp. 50–51.

Fermat had made a note written in Latin. The note said the following:

> It is impossible to separate a cube into two cubes, a fourth power into two fourth powers, or in general, any power higher than the second into two like powers. I have discovered a marvelous proof of this fact, but this margin is too narrow to contain it.

That note, dated approximately in 1637, known today as Fermat's Last Theorem, would remain unproved for approximately 350 years. The problem can be stated more rigorously as follows:

$$x^n + y^n \neq z^n,$$

for every x, y, $z \in \mathbb{Z}^+$, where n is an integer greater than 2.

Several great mathematicians investigated special cases of the problem. Among them were Euler (1707–1783), Legendre (1752–1833), Dirichlet (1805–1859) and Gauss (1777–1855).

During the 18th century, Euler presented a proof of the theorem for the case when $n = 3$. However, due to a small gap in his proof, in the beginning of the 19th century, Gauss presented a complete proof for the same case. A proof of the theorem for the case when $n = 5$ was presented independently by Dirichlet and Legendre.

Some years later, in 1839, Lamé presented a proof for the case when $n = 7$. But, the most important progress during the 19th century was made by Kummer, who proved the theorem for all prime numbers less than or equal to 31. However, the research which contributed the most toward the proof of Fermat's Last Theorem was conducted during the 20th century.

In 1983, the German mathematician Gerhard Faltings presented a very important result. He proved that the diophantine equation

$$x^n + y^n = z^n$$

has at most finite rational solutions for every positive integer n, with $n > 2$. In order to obtain this result, Faltings proved another very important conjecture known as Mordell's conjecture. According to this conjecture, every nonsingular projective curve with genus greater than 1 has at most finitely many rational points.

In 1984, in a conference held in Oberwolfach, Germany, Gerhard Frey presented an important observation. Assuming the conjecture of Shimura–Taniyama–Weil to hold, then Fermat's Last Theorem follows. According to the Shimura–Taniyama–Weil conjecture, every elliptic curve over the rational numbers is modular, that is, there is a nonconstant morphism defined over the rationals from a modular curve to the elliptic curve.

According to Frey's argument, if Fermat's Last Theorem was not true, then there would exist a triple of integers x_0, y_0, z_0, such that

$$x_0^n + y_0^n = z_0^n,$$

for $n > 2$, where n can be considered to be a prime number.[2] However, Frey observed that by the triples of solutions, the elliptic curve

$$y^2 = x(x - a^n)(x + b^n) \tag{1}$$

occurs. He claimed that this elliptic curve was not modular. But, if the Shimura–Taniyama–Weil conjecture was true, then the elliptic curve (1) should be modular. Therefore, if someone managed to prove Frey's argument,[3] then (1) should not exist. Hence, the solutions x_0, y_0, z_0 should not exist and thus Fermat's Last Theorem would hold true.

After the presentation of Frey's argument, Jean-Pierre Serre in a letter to J. F. Mestre stated some other relevant conjectures. When that letter went public, Ken Ribet started investigating Serre's conjectures and in 1990 he proved that the Frey curve, if it exists, cannot be modular. Therefore, the only thing remaining in order to prove Fermat's Last Theorem was to prove the Shimura–Taniyama–Weil conjecture.

When Andrew Wiles was informed about Ribet's result, he decided to fulfill his childhood dream and prove Fermat's Last Theorem.

While Wiles was still at the beginning of his research for the proof, he realized that it was not necessary to prove the complete Shimura–Taniyama–Weil conjecture. It would be enough to prove the conjecture for semistable elliptic curves. That was the type of the elliptic curve that Frey had described. Wiles, based on the results of several mathematicians, such as Mazur, Serre, Hida, Flach, Kolyvagin, Langlands, Tunnell and Ribet, managed to prove the Shimura–Taniyama–Weil conjecture for the cases he needed. Hence, Fermat's Last Theorem followed as a result.

On Wednesday, 23 June, 1993, Andrew Wiles presented his proof at the conference of Number Theory with title *L-Functions and Arithmetic*, which was held in Newton's Institute at Cambridge University. The news that Fermat's Last Theorem was finally proved staggered the whole mathematical community and many researchers started examining his long proof (of about 200 typed pages) in order to verify its validity.

One of the experts who examined the proof was a friend of Wiles named Nick Katz. In July and August of 1993, Katz went through the proof step by step and communicated with Wiles whenever he reached a point he did not understand. Everything seemed to be fine for most of the proof, until Katz reached an argument he could not understand and Wiles was not able

---

[2] If n is a composite number and $n = p_1^{a_1} p_2^{a_2} \cdots p_k^{a_k}$ is its canonical representation with at least two of $a_1, a_2, \ldots, a_k$ being nonzero, then the diophantine equation $x^{p_i} + y^{p_i} = z^{p_i}$ accepts the solution $(x_0^\lambda, y_0^\lambda, z_0^\lambda)$, where $\lambda = p_1^{a_1} p_2^{a_2} \cdots p_{i-1}^{a_{i-1}} p_i^{a_i - 1} p_{i+1}^{a_{i+1}} \cdots p_k^{a_k}$.

[3] More specifically, Frey observed that the curves of the form (1) are very unlikely to exist. This observation breaks down into two arguments. The first was that (1) cannot be modular and the second was the proof of the Shimura–Taniyama–Weil conjecture for semisimple elliptic curves.

to explain. Quite simply, Wiles had assumed at that point of the proof that an Euler system existed, which was not true. Thus, the proof had a gap.

Wiles after hard work and in collaboration with Richard Taylor, managed to complete the proof of Fermat's Last Theorem on the morning of the 19th of September, 1994. He then sent it for publication to the journal *Annals of Mathematics* of Princeton University. The whole issue was dedicated to the proof. After more than 350 years, Fermat's Last Theorem was finally proved.

12.3 Catalan's conjecture

> *A mathematician, like a painter or a poet, is a maker of patterns.*
> *If his patterns are more permanent than theirs,*
> *it is because they are made with ideas.*
> Godfrey Harold Hardy (1877–1947)

Whoever has frequented numbers in a playful manner, has certainly encountered more than once the particular property of the successive integers 8 and 9 of being a cube and a square, respectively: they satisfy the intriguing equality

$$3^2 - 2^3 = 1. \tag{1}$$

Like every simple equality, it can be perceived as a special case of various *patterns*. Since the roles of 2 and 3 interchange in the two terms of the left-hand side, between bases and exponents, one may ask how often can one encounter the general pattern

$$x^y - y^x = 1$$

with $x, y \in \mathbb{Z}$? This is an exercise that can be solved. One may then preserve the bases and ask if

$$3^m - 2^n = 1, \quad m, n > 1$$

has other solutions with integer m, n, other than the above pair $(m, n) = (2, 3)$. This question was solved by the 13th century Jewish philosopher and astronomer Ben Gershon. The complementary approach consists in fixing the exponents and letting the bases vary, thus obtaining the equation

$$y^2 = x^3 + 1,$$

which was considered by Euler in the 18th century. Here too, it turned out that the only integer solutions were the ones in (1). We have thus already three different Diophantine equations which generalize the property (1) in various ways, and they all have this identity as their unique solution. In view of this, the Franco–Belgian mathematician Eugène Charles Catalan (1814–1894) made in 1844 the step of allowing all parameters to vary, thus asking whether the equation

$$x^m - y^n = 1 \tag{2}$$

has any other nontrivial integer solutions except (1). An immediate observation shows that we may restrict our attention to prime values of m, n—at least if we expect that there will be no other solutions than the known one. Indeed, if $(x, y; m, n)$ is a solution and $p \mid m, q \mid n$ are primes, then $(x^{m/p}, y^{n/q}; p, q)$ is another nontrivial solution, with prime exponents. The first prime to look at might be the oddest prime of all, the even $p = 2$. In fact, for $n = 2$, Victor Lebesgue could prove, less than 10 years after Catalan's statement of his question, that there are no other nontrivial solutions except the ones in (1); he used the factorization of numbers in the Gaussian integers. It took, however, more than 100 years until Chao Ko, a Chinese mathematician who studied at Cambridge University with Mordell, could prove in the early 1960s that $x^2 = y^q + 1$ has no other solutions than the ones in (1). The proof used recent results on continued fractions and Pell's equation $x^2 + d = y^2$.

The first general result for odd exponents in this equation was obtained by J. Cassels. He proved in 1961 that if (2) has an integer solution with prime exponents m, n, then $m \mid y$ and $n \mid x$ and ruled out the first case $(p, x-1) = 1$. In fact, Cassels' fundamental result can be stated as follows:

Suppose that (x, y) are two integers and p, q two odd primes such that

$$x^p - y^q = 1.^4$$

Then

$$x - 1 = p^{q-1}a^q \quad \text{and} \quad \frac{x^p - 1}{x - 1} = pv^q, \quad y = pav \qquad (3)$$

$$y + 1 = q^{p-1}b^p \quad \text{and} \quad \frac{y^q + 1}{y + 1} = qu^p, \quad x = qbu,$$

where a, b and u, v are integers for which $(pa, u) = (qb, v) = 1$. In particular, the solutions are particularly large, since $|x| > p^{q-1}$ and $|y| > q^{p-1}$. In view of the approximate equality

$$\log(x^p) = p\log(x) = \log(y^q + 1) \sim \log(y^q) = q\log(y),$$

we see that the solutions verify $p\log(x) \sim q\log(y)$ with a *high degree of accuracy*. The relations of Cassels allow one to give an explicit lower bound in this approximation. This opens the door for the application of a field of Diophantine approximation which went through a massive renewal in the 1960s, when Alan Baker proved his famous theorem on *linear forms in logarithms*, for which he was awarded the Fields medal in 1967. His result essentially states that the linear form

$$F(\boldsymbol{\alpha}, \boldsymbol{\beta}) = \sum_i \alpha_i \log(\beta_i),$$

in which both α_i, β_i are *algebraic numbers*, i.e., zeros of polynomials with integer coefficients, only vanishes in trivial cases.

Some years later, in 1973, Baker sharpened his result by stating some explicit lower bounds for the absolute values of $F(\boldsymbol{\alpha}, \boldsymbol{\beta})$. This was used by

---

[4] Note that Cassels allows also negative values for x, y, which brings a nice symmetry in equation (2).

R. Tijdeman, who proved herewith that (2) accepts *at most finitely many integer* solutions. From a qualitative point of view, shared by some mathematicians, the equations were *solved*, since knowing that it has finitely many solutions was theoretically sufficient for finding these solutions in some time.[5] However, Tijdeman had not even given an upper bound for these solutions in his initial proof. The first such bound which was developed soon by Langevin was on the order of $|x| < 10^{10^{700}}$—quite a large number indeed. For those who wanted to know more about the *at most finitely many solutions of Catalan's equation*, the work had to go on. The method of linear forms in logarithms was successfully improved, thus dramatically lowering the size of the upper bound. Many authors worked at this problem in the period since Tijdeman's breakthrough. Among all, Maurice Mignotte from Strasbourg is most noteworthy for his continuous strive and number of improvements and partial results which contributed to keeping the interest in Catalan's conjecture alive. The most recent result of Mignotte used linear forms in three logarithms for proving the following reciprocal bound between p and q:

If

$$3000 < q < p,$$

then

$$p < 2.77 \cdot q(\log(p/\log(q)) + 2.333)^2 \cdot \log(q).$$

As a consequence,

$$q < p \text{ and } q < 7.15 \times 10^{11} \text{ and } p < 7.78 \times 10^{16}. \tag{4}$$

Although these are more *tangible* numbers, if one imagines that for each pair of exponents (p, q) in the above range, one should prove that there either are no nontrivial solutions to (2), or find all existing ones, then one sees that these important improvements were still in themselves insufficient for a successful completion of the answer to Catalan's question.

Without additional algebraic methods, however, the investigation of Catalan's equation would stand only on one leg. While linear forms in logarithms helped reduce the upper bound on the solutions, the algebraic conditions increased lower bounds. The algebraic ideas used could draw back on the long experience and bag of tricks which had been (with only partial results) applied to Fermat's equation by myriads of mathematicians, since the seminal works of Kummer in the 1850s.[6]

In order to understand the favor of the results that one may obtain herewith, we have to make some remarks on the arithmetic of these fields. Let

---

[5] The situation was comparable to the one for Fermat's equation after Falting's proof. Mordell's conjecture implies for Catalan's conjecture that there should exist finitely many solutions for fixed p and q.

[6] Methods from the field of cyclotomy, i.e., from the study of the algebraic properties of the fields obtained by adjoining to the rationals $\mathbb{Q}$ a complex pth root of unity.

this $\zeta \in \mathbb{C}$ be a pth root of unity—one may envision this, for instance, as $\zeta = \exp(2\pi i/p)$, but the algebraist prefers to consider it as a solution of the equation

$$\Phi_p(X) = \frac{X^p - 1}{X - 1} = 0.$$

One way or the other, the nice improvement brought about by this extension is the fact that in the field $\mathbb{K} = \mathbb{Q}[\zeta]$, we have the following factorization, induced from Cassels' relations in (3):

$$v^q = \frac{x^p - 1}{p(x - 1)} = \prod_{c=1}^{p-1} \frac{x - \zeta^c}{1 - \zeta^c}. \tag{5}$$

If one observes that the factors in the product of the right-hand side of this equation are all mutually coprime, then a tempting conclusion arises: all of them must be qth powers. This is almost true, but not really the whole truth. The reason is that in the *integers* of $\mathbb{K}$ (which are the zeros of polynomials over $\mathbb{Z}$ with leading coefficient 1), we do not have unique factorization any more. This was observed already by Kummer, who encountered a similar factorization in his work on Fermat's equation, but was careful enough to recognize in the example of $21 = (1 + 2\sqrt{-5})(1 - 2\sqrt{-5})$ a suggestive apprehension of the loss of unique factorization of integers. This loss is replaced by the unique factorization of ideals (which were first called *ideal factors*, by Kummer). In Kummer's sight, an ideal factor in the above example would be a factor which divides both 3 and $(1 + 2\sqrt{-5})$. Such numbers do not exist in the field $\mathbb{Q}[\sqrt{-5}]$, thus he called them ideal. In some sense, the ideal is the greatest common divisor of the two.

The notion of ideal is now general and simple: the numbers $a, b \in \mathbb{K}$ generate the ideal (a, b) which consists of all linear combinations of the two over $\mathbb{Z}_{\mathbb{K}}$, which are the integers of this field. Since we have unique factorization of ideals, the above equation shows at least that each

$$\alpha_c = \frac{x - \zeta^c}{1 - \zeta^c}$$

is the qth power of an ideal, say $\alpha_c = \mathfrak{A}_c^q$. This is not completely correct, since it is not the number which is a power, but the ideal that it generates, namely, the ideal of all its multiples, which is also written as $(\alpha_c) = \mathbb{Z}_{\mathbb{K}} \alpha_c = \mathfrak{A}_c^q$. The ideals that are generated by one single integer, like (α_c), are particularly interesting and simple. These ideals are called *principal ideals*. We have seen for instance the ideal $(3, 1 + 2\sqrt{-5})$, which cannot be principal. It is releaving to know that if not all ideals are principal, at least a finite, fixed power $h(\mathbb{K})$ of theirs is always a principal ideal. The constant $h(\mathbb{K})$ which only depends on the field $\mathbb{K}$ is called the class number of the field, and it measures in some sense the deviation from unique factorization in the given field. Since (α_c) is the qth power of an ideal, but also $\mathfrak{A}_c^{h(\mathbb{K})}$ is principal, while the constant $h(\mathbb{K})$ only depends on $\mathbb{K}$—and thus hardly on q—there is an obvious conclusion: for most primes q, namely, for all the ones that do not divide $h(\mathbb{Q}[\zeta])$, the ideal $\mathfrak{A}_c$ itself must be principal, and thus $(\alpha_c) = (\beta_c^q)$. This type of observation nurtured results of the type: If (2) has a nontrivial solution, then either $p^{q-1} \equiv 1(\mathrm{mod} q^2)$ or $q \mid h(\mathbb{K})$. Such results, and refinements thereof, which would require more details in order to be explained properly, were derived first by the Finnish number theorist Kustaa Inkeri and his school, and then by various followers. Note that the two conditions occurring above have the potential of being computable: while it is impossible to search for a fixed pair of primes p, q, among all integer pairs (x, y) in order to ascertain that none solves (2), it is conceivable to verify only for the pair of primes (p, q) that the two conditions $p^{q-1} \equiv 1(\mathrm{mod} q^2)$ and $q \mid h(\mathbb{K})$ are not fulfilled. For the remaining few counterexamples, one then needs some additional criteria.

This way, between lowering the upper bounds obtained with forms in logarithms, and improving the algebraic criteria—both of general kind, like the one of Inkeri quoted above, and special ones, designed to rule out particular cases—the domain of possible exceptions to Catalan's conjecture continued to be restricted until in 1999 two new results allowed, on the algebraic side, to *separate* the conditions $q \mid h(\mathbb{K})$ and $p^{q-1} \equiv 1(\mathrm{mod} q^2)$. In that period, Bugeaud and Hanrot first proved that for $p > q$, the condition $q \mid h(\mathbb{K})$ had to hold necessarily, for any solution to (2). Inspired by their work, P. Mihăilescu proved several months later that also

$$p^{q-1} \equiv 1(\mathrm{mod} q^2) \quad \text{and} \quad q^{p-1} \equiv 1(\mathrm{mod} p^2) \tag{6}$$

had to hold necessarily. It was this second condition which was particularly easy to verify on a computer: this triggered a massive effort of computations. The most successful were Mignotte and Grantham, who had succeeded by the year 2002 to give the lower bound $p, q > 2 \cdot 10^8$ for possible solutions. The road until lower and upper bounds would cross was, however, still long, if one compares to the best upper bounds known by then, which are in (4).

When the Catalan conjecture was eventually solved in 2002 by P. Mihăilescu, his new algebraic insights allowed him to reduce the analytic apparatus involved in his proof. The main ideas improved upon the methods from the algebraic track used before, by including deeper, recent insights in the field of cyclotomic fields, in particular the celebrated Theorem of Francisco Thaine, which had, in 1988, marked a major cross road in the development of the Iwasawa theory, by simplifying the proofs of important results of this field. Basically, the improved algebraic apparatus allowed one to eliminate the use of linear forms in logarithms, which were too general in order to provide optimal results for the specific equation under consideration. Instead, the analytic methods used were simpler, but tightly connected to the algebraic results. Without entering into the details of the proof, one may give the following overview of the ideas involved: starting from equation (5) and using the *Galois actions* of $\mathbb{K}$ which manifest by $\sigma_c : \zeta \mapsto \zeta^c$ and which, being automorphisms of the field $\mathbb{K}$, preserve the algebraic operations, one asks the following question.

Are there, apart from the class number $h(\mathbb{K})$, other expressions which make all ideal into principal ideals of $\mathbb{K}$?

One allows this time also expressions of the type $\theta = \sum_{c=1}^{p-1} n_c \sigma_c$, where $n_c \in \mathbb{Z}$. The world in which these expressions live is called the *group ring* $\mathbb{Z}[\text{Gal }(\mathbb{K}/\mathbb{Q})]$ and their action is given by

$$\alpha^\theta = \prod_{c=1}^{p-1} \alpha_c^{n_c}.$$

If θ has the desired property of making all ideals principal, then we have an equation of the type $\alpha^\theta = \varepsilon \nu^q$, were the incomodating factor ε here is called a unit. These are the integers of $\mathbb{K}$ that are invertible, such as for instance $\frac{1-\zeta^k}{1-\zeta}$. Unlike $\mathbb{Z}$ which has only the units ± 1, the units of $\mathbb{K}$ are *numerous*. However, by combining the Theorem of Thaine, with some consequences of Cassels' result and his own theorem on double Wieferich primes in (6), Mihăilescu proved that if (2) has a nontrivial solution, then there is $\theta \in \mathbb{Z}^+[\text{Gal }(\mathbb{K}/\mathbb{Q})]$ such that

$$\alpha^\theta = \nu^q. \tag{7}$$

The exponent $+$ stands for the fact that this time the expression α^θ is a real number. Plainly, the previous equation implies that $\nu = \alpha^{\theta/q}$ is a real, algebraic integer. The fact that it is a real number has the major consequence, that the *power series* development for $\alpha^{\theta/q}$ will necessarily yield the correct answer ν. Thus, instead of using linear forms in logarithms, the analytic apparatus is reduced to an accurate investigation of the power series related to the qth root—this series was deduced by Abel and is sometimes named the binomial series, or Abel series. Following the simple principle that if a certain equation is meant not to have solutions, then the assumption of some nontrivial solution should raise a sequence of consequences which eventually should break up into a contradiction, Mihăilescu pursued the arithmetic investigation of the Abel series related to (7) and, using the lower bounds on $|x|, |y|$ mentioned above, showed that the algebraic integer ν would need to satisfy the following contradictory properties: it is not vanishing, yet its norm $|\mathbf{N}_{\mathbb{K}/\mathbb{Q}}(\alpha)| < 1$. Since this norm must be an integer, the relations contradict $\nu \neq 0$. This eventually shows that the assumption that (2) has nontrivial solutions must be wrong.

While the proof of Fermat's Last Theorem required the use of a highly technical apparatus involving modular forms and, mainly, the proof of the Shimura–Taniyama–Weil conjecture related to the L-series of these forms, the proof of Catalan's conjecture appears to be almost elementary. However, it is interesting that, similar as the two equations are in appearance, the methods used for the proof of Fermat's Last Theorem fail when applied to Catalan's conjecture—so the elementary methods were, somehow, necessary.

Open problems, generalizing the above two major solved conjectures, concern the following Diophantine equations:

$$x^m + y^m = z^n, \quad \gcd(x, y, z) = 1, \quad m, n > 2$$

and

$$x^p + y^q = z^r, \quad \gcd(x, y, z) = 1 \quad \text{and} \quad p, q, r \geq 2, \ 1/p + 1/q + 1/r < 1.$$

A more recent, deep conjecture which implies in particular the fact that all the above-mentioned equations have no solutions is the ABC conjecture. This was proposed in 1985 by D. Masser and J. Oesterlé based on an analogy to a similar fact which holds in function fields. It claims that if an equality $A + B = C$ holds between three positive integers A, B, C, then for every $\varepsilon > 0$ there is some constant k_ε such that

$$|A \cdot B \cdot C| < k_\varepsilon \cdot \mathrm{rad}(ABC)^\varepsilon,$$

where the radical of an integer n, which we denoted by $\mathrm{rad}(n)$, is the product of all the prime numbers dividing n.

References

1. A. Adler and J. E. Coury, *The Theory of Numbers*, Jones and Bartlett Publishers, Boston, 1995.
2. M. Aigner and G. M. Ziegler, *Proofs from the BOOK*, Springer-Verlag, New York, 1999.
3. G. L. Alexanderson, L. F. Klosinski and L. C. Larson, *The William Lowell Putnam Mathematical Competition, Problems and Solutions, 1965–1984*. The Mathematical Association of America, Washington, DC, 1985.
4. T. Andreescu and D. Andrica, *An Introduction to Diophantine Equations*, Gil Publishing House, Romania, 2002.
5. T. Andreescu and D. Andrica, *Number Theory: Structures, Examples and Problems*, Birkhäuser, Boston, 2009.
6. R. Apéry, *Irrationaliti de $\zeta(2)$ et $\zeta(3)$*, Astérisque, 61(1979), 11–13.
7. T. Apostol, *Introduction to Analytic Number Theory*, Springer-Verlag, New York, 1984.
8. E. J. Barbeau, *Polynomials*, Springer-Verlag, New York, 1989.
9. E. J. Barbeau, *Power Play*, The Mathematical Association of America, Washington, DC, 1997.
10. C. Bays and R. H. Hudson, *A new bound for the smallest x with $\pi(x) > li(x)$*, Math. Comp., 69(1999), 1285–1296.
11. A. Bremner, R. J. Stroeker and N. Tzanakis, *On sums of consecutive squares*, J. Number Theory, 62(1997), 39–70.
12. T. T. Bell, *Men of Mathematics*, Simon & Schuster, New York, 1986.
13. B. Bollobás, *The Art of Mathematics–Coffee Time in Memphis*, Cambridge University Press, Cambridge, 2006.
14. R. P. Brent, G. L. Cohen and H. J. J. te Riele, *Improved techniques for lower bounds for odd perfect numbers*, Math. Comp., 61(1993), 857–868.
15. D. M. Burton, *Elementary Number Theory*, Allyn and Bacon, Boston, 1980.
16. H. Cohn, *Advanced Number Theory*, Dover Publications, New York, 1962.
17. R. Crandall and C. Pomerance, *Prime Numbers–A Computational Perspective*, Springer-Verlag, New York, 2005.
18. J. B. Dence and T. P. Dence, *Elements of the Theory of Numbers*, Harcourt Academic Press, London, 1999.
19. D. Djukic, V. Jankovic, I. Matic and N. Petrovic, *The IMO Compendium*, Springer-Verlag, New York, 2006.

20. H. M. Edwards, *Riemann's Zeta Function*, Dover Publications, New York, 1974.
21. P. Erdős and J. Surányi, *Topics in the Theory of Numbers*, Springer-Verlag, New York, 2003.
22. L. Euler, *Institutiones Calculi Differentialis*, Pt. 2, Chapters 5 and 6. Acad. Imp. Sci. Petropolitanae, St. Petersburg, *Opera* (1), Vol. 10, 1755.
23. G. Everst and T. Ward, *An Introduction to Number Theory*, Springer-Verlag, New York, 2005.
24. B. Fine and G. Rosenberger, *Number Theory: An Introduction via the Distribution of Primes*, Birkhäuser, Boston, 2007.
25. C. F. Gauss, *Disquisitiones Arithmeticae*, Leipzig (English translation: A. F. Clarke, Yale University Press, New Haven, 1966).
26. A. M. Gleason, R. E. Greenwood and L. M. Kelly, *The William Lowell Putnam Mathematical Competition Problems and Solutions, 1938–1964*. The Mathematical Association of America, Washington, DC, 1980.
27. J. R. Goldman, *The Queen of Mathematics, A Historically Motivated Guide to Number Theory*, A. K. Peters, Natick, Massachusetts, 2004.
28. S. L. Greitzer, *International Mathematical Olympiads 1959–1977*, The Mathematical Association of America, Washington, DC, 1978.
29. R. K. Guy, *Unsolved Problems in Number Theory*, 2nd edition, Springer-Verlag, New York, 1994.
30. K. Hardy and K. S. Williams, *The Green Book of Mathematical Problems*, Dover Publications, New York, 1985.
31. G. H. Hardy and E. W. Wright, *An Introduction to the Theory of Numbers*, 5th edition, Clarendon Press, Oxford, 1979.
32. R. Honsberger, *From Erdös to Kiev, Problems of Olympiad Caliber*, The Mathematical Association of America, Washington, DC, 1966.
33. K. S. Kedlaya, B. Poonen and R. Vakil, *The William Lowell Putnam Mathematical Competition, 1985–2000*. The Mathematical Association of America, Washington, DC, 2002.
34. N. Koblitz, *A Course in Number Theory and Cryptography*, Springer-Verlag, New York, 1994.
35. M. Křížek, F. Luca and L. Somer, *17 Lectures on Fermat Numbers: From Number Theory to Geometry*, Springer-Verlag, New York, 2001.
36. E. Landau, *Elementary Number Theory*, 2nd edition, Chelsea, New York, 1966.
37. L. C. Larson, *Problem-Solving Through Problems*, Springer-Verlag, New York, 1983.
38. L. Lovász, *Combinatorial Problems and Exercises*, North-Holland, Amsterdam, 1979.
39. E. Lozansky and C. Rousseau, *Winning Solutions*, Springer-Verlag, New York, 1996.
40. D. S. Mitrinović, *Analytic Inequalities*, Springer-Verlag, New York, 1968.
41. D. S. Mitrinović, J. Sándor and B. Crstici, *Handbook of Number Theory*, Kluwer Academic, Dordrecht, 1996.
42. C. J. Moreno and S. S. Wagstaff, *Sums of Squares of Integers*, Chapman & Hall/CRC, London, 2006.
43. M. R. Murty, *Problems in Analytic Number Theory*, Springer-Verlag, New York, 2001.
44. D. J. Newman, *Simple analytic proof of the prime number theorem*, Amer. Math. Monthly, 87(1980), 693–696.

45. I. Niven, H. S. Zuckerman and H. L. Montgomery, *An Introduction to the Theory of Numbers*, John Wiley & Sons, Toronto, 1991.

46. A. Papaioannou and M. Th. Rassias, *An Introduction to Number Theory* (in Greek), Symeon, Athens, 2010.

47. G. Pòlya and G. Szegö, *Problems and Theorems in Analysis II*, Springer-Verlag, New York, 1976.

48. P. Ribenboim, *The Little Book of Big Primes*, Springer-Verlag, New York, 1991.

49. J. B. Rosser and L. Schoenfeld, *Approximate formulas for some functions of prime numbers*, Illinois J. Math., 6(1962), 64–94.

50. K. H. Rosen, *Elementary Number Theory and its Applications*, 3rd edition, Addison-Wesley, Reading, Massachusetts, 1993.

51. R. Schoof, *Catalan's Conjecture*, Springer-Verlag, New York, 2008.

52. D. Shanks, *Solved and Unsolved Problems in Number Theory*, AMS, Chelsea, Rhode Island, 2001.

53. D. O. Shklarsky, N. N. Chentzov and I. M. Yaglom, *The USSR Olympiad Problem Book, Selected Problems and Theorems of Elementary Mathematics*, Dover Publications, New York, 1962.

54. W. Sierpiński, *250 Problèmes de Théorie Élémentaire des Nombres*, Panstwoew Wydawnictwo, Warsaw, 1970.

55. J. H. Silverman, *A Friendly Introduction to Number Theory*, 3rd edition, Pearson Prentice Hall, Upper Saddle River, New Jersey, 2006.

56. S. Skewes, *On the difference $\pi(x) - Li(x)$*, J. London Math. Soc., 8(1933), 277–283.

57. S. Skewes, *On the difference $\pi(x) - Li(x)$ (II)*, Proc. London Math. Soc., 5(1955), 48–70.

58. D. R. Stinson, *Cryptography Theory and Practice*, Chapman & Hall/CRC, London, 2006.

59. D. J. Struik, *A Concise History of Mathematics*, Dover Publications, New York, 1987.

60. G. J. Székely (ed.), *Contests in Higher Mathematics, Miklós Schweitzer Competitions 1962–1991*, Springer-Verlag, New York, 1996.

61. D. Wells, *The Penguin Book of Curious and Interesting Puzzles*, Penguin Books, New York, 1992.

62. K. S. Williams and K. Hardy, *The Red Book of Mathematical Problems*, Dover Publications, New York, 1988.

63. H. N. Wright, *First Course in Theory of Numbers*, John Wiley & Sons, London, 1939.

64. I. M. Vinogradov, *Elements of Number Theory*, Dover Publications, New York, 1954.

Index of Symbols

$\mathbb{N}$: The set of natural numbers $1, 2, 3, \ldots, n, \ldots$

$\mathbb{Z}$: The set of integers

$\mathbb{Z}^+$: The set of nonnegative integers

$\mathbb{Z}^-$: The set of nonpositive integers

$\mathbb{Z}^*$: The set of nonzero integers

$\mathbb{Q}$: The set of rational numbers

$\mathbb{Q}^+$: The set of nonnegative rational numbers

$\mathbb{Q}^-$: The set of nonpositive rational numbers

$\mathbb{R}$: The set of real numbers

$\mathbb{R}^+$: The set of nonnegative real numbers

$\mathbb{R}^-$: The set of nonpositive real numbers

$\mathbb{C}$: The set of complex numbers

$D(f, s)$: Dirichlet series with coefficients $f(n)$

$\mu(n)$: Möbius function

$\sigma_a(n)$: The sum of the ath powers of the positive divisors of n

$\tau(n)$: The number of positive divisors of n

$\phi(n)$: Euler phi function

$\zeta(s)$: Riemann zeta function

$\pi(x)$: The number of primes not exceeding x

π: Ratio of the circumference of circle to diameter, $\pi \cong 3.14159265358\ldots$

e: Base of natural logarithm, $e \cong 2.718281828459\ldots$

F_n: Fermat numbers, $F_n = 2^{2^n} + 1$

M_n: Mersenne numbers, $M_n = 2^n - 1$

$f(x) \sim g(x)$: $\lim_{x \to +\infty} f(x)/g(x) = 1$, where $f, g > 0$

$f(x) = o(g(x))$: $\lim_{x \to +\infty} f(x)/g(x) > 0$, where $g > 0$

$f(x) = O(g(x))$: There exists a constant c, such that $|f(x)| < c\,g(x)$ for sufficiently large values of x

$a \equiv b \pmod{m}$: $a - b$ is divisible by m

$\gcd(a, b)$: The greatest common divisor of a and b

$\left(\frac{a}{p}\right)$: Legendre symbol

$\left(\frac{a}{P}\right)$: Jacobi symbol

$a \in A$: a is an element of the set A

$a \notin A$: a is not an element of the set A

$A \cup B$: Union of two sets A, B

$A \cap B$: Intersection of two sets A, B

$A \times B$: Direct product of two sets A, B

$a \Rightarrow b$: if a then b

$a \Leftrightarrow b$: a if and only if b

$\emptyset$: Empty set

$A \subseteq B$: A is a subset of B

$n! = 1 \cdot 2 \cdot 3 \cdots n$, where $n \in \mathbb{N}$

$d \mid n$: d divides n

$d \nmid n$: d does not divide n

$p^k \parallel n$: p^k divides n, but p^{k+1} does not divide n

$\lfloor x \rfloor$: The greatest integer not exceeding x

$\lceil x \rceil$: The least integer not less than x

$\square$: End of the solution or the proof

Index